ADVANCES IN
CELL AND MOLECULAR
BIOLOGY OF MEMBRANES

Volume 2 • 1993 (Part B)

MEMBRANE TRAFFIC
IN PROTOZOA

ADVANCES IN CELL AND MOLECULAR BIOLOGY OF MEMBRANES

MEMBRANE TRAFFIC IN PROTOZOA

Series Editor: ALAN M. TARTAKOFF
Institute of Pathology
Case Western Reserve University

Editor: HELMUT PLATTNER
Faculty of Biology
University of Konstanz

VOLUME 2 • 1993 (Part B)

JAI PRESS INC.

Greenwich, Connecticut *London, England*

55 Old Post Road No. 2
Greenwich, Connecticut 06830

JAI PRESS LTD.
The Courtyard
28 High Street
Hampton Hill
Middlesex TW12 1PD
England

ISBN: 1-55938-628-2 (Set: Part A and Part B)

Transferred to digital printing 2006

CONTENTS

LIST OF CONTRIBUTORS

André Adoutte	Laboratory of Cellular Biology University of Paris—South Orsay, France
Richard D. Allen	Department of Microbiology University of Hawaii Honolulu
Claude Antony	Jacques Monod Institute CNRS, University of Paris
Yvonne Capdeville	Center for Molecular Genetics, CNRS Gif-sur-Yvette, France
Renée Charret	Laboratory of Cellular Biology University of Paris—South Orsay, France
Melpo Christofidou-Solomidou	Department of Physiology and Cell Biology Albany Medical College
Jean Cohen	Center for Molecular Genetics, CNRS Gif-sur-Yvette, France
Julienne Delorme	Laboratory of Cellular Biology University of Paris—South Orsay, France
Jean François Dubremetz	INSERM Villeneuve D'Ascq, France
Michael Duszenko	Biochemical Institute University of Tübingen Germany
Rolf Entzeroth	Zoological Institute University of Bonn Germany

Agnes K. Fok	Department of Microbiology University of Hawaii Honolulu
Nicole Garreau de Loubresse	Center for Molecular Genetics, CNRS Gif-sur-Yvette, France
Klaus Hausmann	Faculty of Biology German Institute for Zoology Free University of Berlin
Kwang W. Jeon	Department of Zoology University of Tennessee Knoxville
Daniel Kerboeuf	Center for Molecular Genetics, CNRS Gif-sur-Yvette, France
Thomas Kiy	Institute for General Zoology and Genetics University of Münster Germany
Gerd Knoll	Faculty of Biology University of Konstanz Germany
Pierre Nahon	University of Paris—South Orsay, France
Yoshinori Nozawa	Department of Biochemistry Gifu University School of Medicine Japan
Regina Pape	Faculty of Biology University of Konstanz Germany
Robert K. Peck	Department of Animal Biology University of Geneva Switzerland
Helmut Plattner	Faculty of Biology University of Konstanz Germany

Robert D. Prusch
Biology Department
Gonzaga University
Spokane, Washington

Renate Radek
Faculty of Biology
Institute of Zoology
Free University of Berlin

Leif Rasmussen
Institute of Medical Biology
Odense University
Odense, Denmark

Werner Reisser
Institute of Plant Physiology
University of Göttingen
Germany

Andreas Seyfang
Biochemical Institute
University of Tübingen
Germany

Yoshinobu Shigenaka
Laboratory of Cell Physiology
Faculty of Integrated Arts & Sciences
Hiroshima University
Japan

Christian Slomianny
INSERM
Villeneue D'Ascq
France

Wilhelm Stockem
Institute of Cell Biology
University of Bonn
Germany

Toshinobu Suzaki
Laboratory of Cell Physiology
Faculty of Integrated Arts & Sciences
Hiroshima University
Japan

Barbara Swiderski
Department of Animal Biology
University of Geneva
Switzerland

Arno Tiedtke
Institute for General Zoology and Genetics
University of Münster
Germany

Anne-Marie Tourmel — Department of Animal Biology
University of Geneva
Switzerland

Shigenobu Umeki — Department of Biochemistry
Gifu University School of Medicine
Gifu, Japan

Christian Vosskühler — Institute for General Zoology and Genetics
University of Münster
Germany

Jason Wolfe — Biology Department
Wesleyan University
Middletown, Connecticut

INTRODUCTION TO THE SERIES

The remarkable vigor and central importance of cell biology result from the realization that emphasis on structure/function relations at the cellular and subcellular levels is essential for a rigorous and satisfactorily complete understanding.

Unlike many subdivisions of biomedical science, cell biology is not linked to any one methodology. It often emphasizes topological or topographic questions, and it is concerned with the structure, biogenesis, and turnover of macromolecular structures; however, there is no limit to the techniques and conceptual approaches which it brings to bear on these issues. Indeed, this has even been true since the term *cell biology* was first used. Certain scientists trace its origins back to ultrastructure and histology; others consider E.B. Wilson's *The Cell in Development and Heredity* to epitomize the foundations of cell biology; while others consider cell biology as an outgrowth of somatic cell genetics or the extension of biophysics to objects of increasingly large size. This varied and often changing identity makes it easier to specify what cell biology is not, rather than to describe it in a positive sense. It also suggests the rich intellectual mix which underlies today's successes in cell biology research.

This series aims to match the continuing evolution of cell biology—with particular emphasis on cell membranes—by treating coherent areas in multiauthored volumes. This approach allows a multifaceted coverage of topics which benefits from the unity of vision of the volume editor(s), but does not rely on one individual to synthesize an entire subject. Hopefully, the result will be more readable and intrinsically richer than lengthy review chapters which attempt to be all encompassing.

Alan M. Tartakoff
Series Editor

PREFACE

It is trivial to say that protozoa, solely as members of the living world, deserve the interest of cell biologists. Far beyond this, some protozoa offer unique advantages as model systems to analyze some specific aspects of cell function. Indeed, many fundamental aspects have been unraveled first in protozoa.

Just a few examples may be given. Consider ciliates—what admirable cells they are. About 600 million years of age, they now demonstrate to us their own solution to the soma-Keimbahn (germ track) problem by "inventing" their macro-micronucleus system. From protozoa the fist telomere has been isolated, sequenced, and then used in heterologous systems. Amplification of genes for rRNA can be induced in some ciliates. In the ciliate *Tetrahymena*, catalytic RNA (ribozyme) has been discovered, which is now of considerable interest in such different fields as evolutionary biology and AIDS research. *Tetrahymena* has also provided a model for synchronized stages of meiosis and for chromosome pairing mechanisms. Recently, in *Paramecium*, the readiness for integration of injected DNA in the macronuclear chromosome fragments, possibly with important positioning effects, has been shown. This organism is also unique in providing mutants in regulated secretion. First evidence for the involvement of proteins in the regulation of membrane fusion, for example, during exocytosis, was also found in *Paramecium* and several aspects could be extended to metazoan cells. Recent mutagenesis studies have indicated that the repertoire of proteins involved might be larger than expected.

This has already led us to the topic of the present book. Membranes are highly dynamic and operate not only as inert boundaries, but the packages they carry around in a cell are well addressed for appropriate delivery. This holds for a variety of endomembrane systems engaged in exo- and endocytosis, for organelles along the biosynthetic pathway, phagosomes, and lysosomes. It also holds for the establishment of functional surface properties. Cell pairing (conjugation) phenomena are a good model for the problem of how a cell can discriminate between "self" and "non-self." On the other hand parasitic sporozoa developed to experts in masking their molecular sur-"face" by frequent shedding of their variant antigens.

The discovery of their glycosyl-phosphatidylinositol (GPI) anchor has led to the discovery, in polar epithelial cells, of a specific targeting mechanism for GPI-anchored surface glycoproteins. Over one hundred such proteins have been detected in metazoans since then.

Many of these basic aspects are dealt with in this book using quite different methodical approaches. Of course, there are many more aspects previously known from metazoan systems and such aspects then had to be verified also for protozoa. In this context, it is fascinating to see how basic cellular functions are maintained—with variations of the basic theme—throughout evolution.

However, sometimes cell biologists dedicated to work with protozoa have to live with a regrettable phenomenon. Colleagues working with "higher" eukaryotic cells are frequently unaware of the fact that the primary input may have come from work with protozoa. Some phenomena may even be rediscovered inadvertently. In this sense, *our* book should address not only colleagues working with protozoa but also many of our fellow cell biologists working with metazoan cells.

I say *our book*, since we owe this book to all the authors of the individual chapters. I thank them all for having followed, without any hesitation and even with enthusiasm (and *almost* within the deadline), my invitation to contribute a competent chapter. We all feel that our book may fill a real gap in the literature.

Finally I should like to anticipate my thanks to Mr. Fred Verhoeven from JAI Press for all his efforts to have this volume published without delay, thus guaranteeing its actuality.

Helmut Plattner
Editor

ENDOCYTOSIS AND INTRACELLULAR TRANSPORT OF VARIANT SURFACE GLYCOPROTEINS IN TRYPANOSOMES

Michael Duszenko and Andreas Seyfang

Advances in Cell and Molecular Biology of Membranes, Volume 2B
Membrane Traffic in Protozoa, pages 227–258

ISBN: 1-55938-628-2

I. INTRODUCTION

To cope with the threatening attacks of the sophisticated and well-elaborated immune defense system in mammals, parasites have evolved many effective strategies for being able to maintain a host–parasite relationship of unbalanced benefit. Those strategies include hiding by living within the protective environment of host cells, camouflage by insertion of host antigens into parasite membranes, preemptive strikes by immunosuppression, or mock attacks by antigenic variation. In any case, the plasma membrane of protozoan parasites is of special interest because, as the primary parasite–host interface, it plays essential roles in protecting the invader from host defense mechanisms, in maintaining the intracellular milieu, and in carrying out requisite functions such as nutrient uptake, endocytosis, or secretion. Membrane constituents and membrane-associated functions are thus increasingly being viewed as possible targets for chemotherapy or vaccination. See also chapter on plasmodia, by Slomianny in this volume.

Among mammalian trypanosomiases African trypanosomes are best studied for their economical importance. These protozoan parasites afflict a wide belt of equatorial Africa, causing human sleeping sickness (*Trypanosoma brucei*-group) and livestock diseases of vast economical importance (*T. vivax, T. congolense, T. brucei*-group). Other important pathogens of this group include species with cosmopolitan distribution in tropical Africa, Asia, and South America (*T. evansi, T. vivax*), and the South American *T. equinum*, which all share the phenomenon of antigenic variation. These parasites inhabit the blood circulation and lymphatic system of their mammalian hosts, and (except for *T. equiperdum*) are cyclically transmitted by the tsetse fly or other arthropod vectors. Characteristic for a human or livestock infection with *T. brucei* is a relapsing parasitemia, which, if untreated, usually leads to death. However, game or other natural hosts often appear to be much less affected by trypanosome infections.

All bloodstream forms as well as the metacyclic insect form, which is also infective to the mammalian host, are covered by a densely packed surface coat consisting of about 10^7 identical copies of a single glycoprotein, the variant surface glycoprotein (VSG). This coat obviously is a protective barrier, saving the parasite from lytic attacks of different serum components like the complement system or macrophages. VSG, on the other hand, is highly immunogenic and leads to the production of specific antibodies, which finally will kill all trypanosomes expressing the respective antigen. Nevertheless, a sustained infection is yet possible because trypanosomes are able to express a repertoire of antigenicly distinct VSGs and periodically switch from the production of one VSG type to another. They are thus able to steadily keep one step ahead of the formation of specific antibodies. This phenomenon, called antigenic variation, was first recognized by Franke performing his doctoral thesis in Paul Ehrlich's laboratory. He observed that blood, collected from an infected monkey (*Cercopithecus callitrichus*), led to the lysis of the trypanosome population used for the initial inoculation, whereas parasites appearing after 17 days were not killed, although the protective effect could still

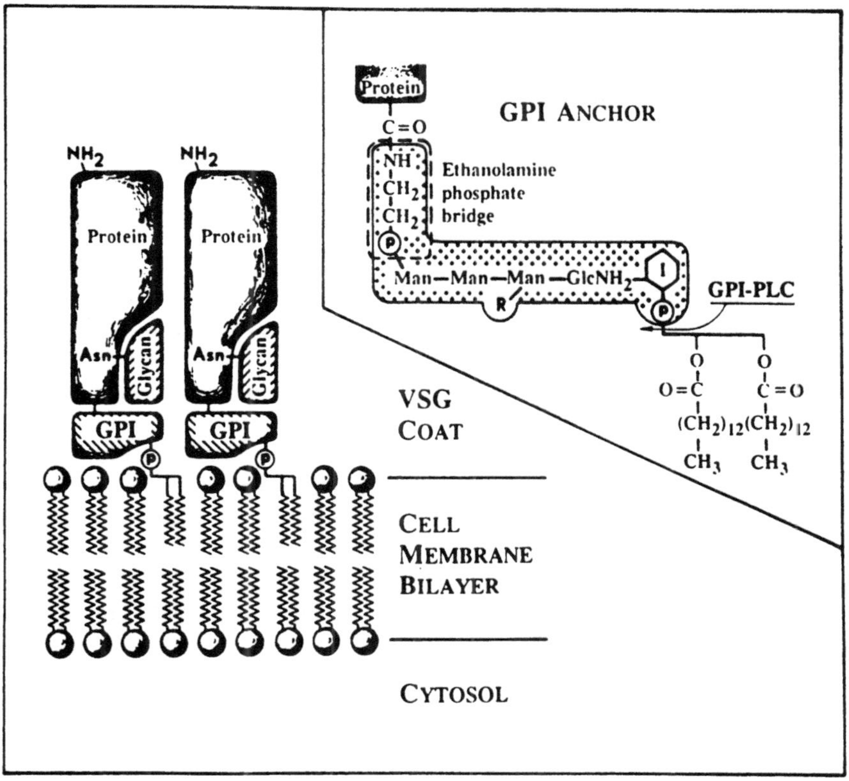

Figure 1. Structural features of variant surface glycoprotein (VSG) and the glycosyl phosphatidylinositol (GPI) membrane anchor (inset). Membrane form VSG (mfVSG) of *T. brucei* bloodstream stages, covalently anchored to the GPI anchor at the carboxy-terminal amino acid, forms a dense 10–15 nm thick surface coat. *N*-glycosylation of the protein via asparagine is found predominately at the inner C-terminal part of the protein and may function as structural "spacers" to fill internal areas of the VSG coat. They are highly immunogenic but not exposed to antibody recognition on living parasites. **Inset:** The GPI membrane anchor expresses a glycan structure which is different from conventional *N*- or *O*-glycosylations of proteins. The anchoring diacylglycerol moiety exclusively consists of two myristic acid molecules ($C_{14:0}$ fatty acid) in trypanosomes. mfVSG can be cleaved by the GPI-specific phospholipase C (GPI-PLC) into soluble VSG (sVSG) and the diacylglycerol moiety. The consensus structure of the GPI anchor is stippled. P = phosphate, Man = mannose, $GlcNH_2$ = glucosamine, I = inositol, R = additional galactosylation site in most VSGs (2–8 galactose residues).

be demonstrated in the parasite-free medium (Franke, 1903, 1905). Today it has been well established that many parasites, ranging from viruses to eukaryotic cells, perform antigenic variation in a more or less elaborate way (Cruse and Lewis, 1987). In African trypanosomes, however, antigenic variation has been evolved to perfection, unrivaled elsewhere in nature (Borst, 1986). *Trypanosoma brucei*, for example, seems able to sequentially express up to several 1000 serologically different antigen variants, encoded within the genome (Van der Ploeg et al., 1982; Borst, 1991a). Thus, VSG diversity parallels the antibody diversity in mammalian cells, although the genetic mechanism for diversion is clearly different in either case since VSGs are single gene products and not assembled from several genes (Boothroyd, 1985).

Interestingly, despite nonhomologous sequences, all VSG types seem structurally equipped to build up a uniform and densely packed surface coat (Fig. 1). Although the complex functionality of this coat has yet to be defined, two major functions seem obvious: first, as stated above, it protects the cell from unspecific immune attacks, and, second, it probably masks trypanosomal plasma membrane molecules from being recognized by their antigenic epitopes (Ziegelbauer et al., 1992). In live cells not even internal VSG epitopes are recognized by monoclonal antibodies raised against purified VSG molecules (reviewed by Cross, 1990a). Forming and maintaining the surface coat is a highly dynamic process, depending on VSG biosynthesis and transport to account for the growing surface during cell division, and on VSG removal to allow an exchange of the surface coat during antigenic switching. Intracellular VSG traffic is immense and includes export to the cell surface as well as internalization (Duszenko et al., 1988; Frevert and Reinwald, 1988b; Webster, 1989; Seyfang et al., 1990). Indeed, VSG transport might well serve as a paradigm for membrane traffic in trypanosomes, which otherwise is not well documented in this parasite. In this review, we will thus focus on VSG transport, although other known examples for trypanosomal surface proteins (like the transferrin receptor) will also briefly be summarized.

Variant surface glycoprotein is the most dominant protein in bloodstream-form trypanosomes, accounting for about 10% of total protein, and is constitutively synthesized at a high rate. The surface coat consists of about 10^7 identical VSG molecules per cell, which, following antigenic switching, are replaced by another variant. Interestingly, VSG is linked to the membrane via a glycosyl-phosphatidylinositol (GPI) membrane anchor, which basically is a modified phospholipid. (See also chapter by Capdeville et al.) Since this membrane anchor may alter physical properties, especially the lateral diffusion rate, special emphasis will be given to GPI biosynthesis and the membrane anchor exchange.

II. THE EXOCYTIC PATHWAY

A. VSG Biosynthesis and Posttranslational Modifications

VSGs are glycoproteins with molecular weights of about 60 kDa. They are variably glycosylated, containing one, two, or several *N*-linked glycans of the high mannose or complex type within the polypeptide portion of the molecule (Holder, 1985; Zamze et al., 1990; 1991). As judged from cDNA sequences, precursor VSG (pVSG) is synthesized with an N-terminal leader sequence and a C-terminal peptide tail, both of which are missing from mature VSG (Allen et al., 1982; Boothroyd et al., 1982). The former represents a conventional signal peptide, necessary for the correct translocation of the nascent polypeptide across the rough endoplasmic reticulum (rER) membrane and is cleaved off cotranslationally by a signal peptidase (McConnell et al., 1981; McConnell et al., 1982). The latter is remarkably conserved between different VSGs and, owing to a marked hydrophobicity, seems to represent a transmembrane domain (Cross, 1984). However, the C-terminal peptide serves only as a transient membrane anchor for the nascent VSG molecule and is replaced during maturation by a glycosyl-phosphatidylinositol (GPI) residue. This unusual glycolipid, which is linked via ethanolamine phosphate bridge to the C-terminal amino acid of processed VSG (Fig. 1, inset), was an enigma for several years. It replaced a possible transmembrane segment of the polypeptide, but isolated VSG did not show any hydrophobic domain, and thus membrane insertion of mature VSG seemed obscure. The mystery was solved by the observation that membrane-form VSG (mfVSG) contains a diacylglycerol residue, which is lost during hypotonic or detergent-mediated lysis of the parasite but remains intact upon lysis in boiling detergent (Cardoso de Almeida and Turner, 1983). In trypanosome VSG diacylglycerol contains exclusively myristic acid residues ($C_{14:0}$ fatty acid) (Ferguson and Cross, 1984) and was shown to be part of a phosphatidylinositol residue (Ferguson et al., 1985). These experiments revealed the first conclusive evidence for a GPI membrane anchor covalently linked to a membrane protein. In the meantime, more than 100 examples of GPI-anchored proteins have been characterized in a broad variety of different cell types, ranging from protozoa to humans (reviewed by Ferguson and Williamson, 1988; Cross, 1990b; Ferguson, 1991). The GPI anchor has thus been recognized as an alternative anchoring mechanism for membrane proteins. The complete sequence of a GPI membrane anchor from a *T. brucei* clone is published (Ferguson et al., 1988b; Fig. 1, inset).

Since the GPI anchor is linked in a peptide-analogous bond to the α-carboxylic residue of the C-terminal amino acid (Holder, 1983), anchor exchange must be a posttranslational event and cannot occur before cleavage of the C-terminal peptide tail. The kinetic of this reaction was subject to pulse-chase experiments, showing that the GPI anchor could be demonstrated within the first minute after translation (Bangs et al., 1985; Ferguson et al., 1986). These results already suggested that the membrane anchor exchange might be a single step reaction occurring on the luminal site of the rER. It was thus promising to look for a putative GPI anchor precursor in trypanosomal membranes, which could be transferred en bloc to the VSG molecule. Appropriate experiments have been performed and clearly revealed the existence of a preformed GPI molecule (Krakow et al., 1986; Menon et al., 1988; Doering et al., 1990). Recently, assembling of the GPI membrane anchor precursor was also achieved in isolated microsomal membranes of *T. brucei in vitro*, showing that this reaction is energy (ATP) dependent and needs coenzyme A, probably to transfer activated myristic acid across the membrane (Menon et al., 1990b). The membrane anchor exchange reaction is not well understood so far. It is tempting to speculate, however, that it is analogous to a transamidase (factor XIIIa) or transpeptidase reaction, occurring in bloodclotting or in bacteria during cell wall assembling. In any case, both pVSG and GPI precursor will probably react in a nonactivated state and will remain membrane bound throughout the reaction. Although the necessary enzyme(s) has not been isolated or characterized to date, cleavage of the C-terminal peptide was shown *in vitro* using a simple protein substrate, prepromini-PALP (a shortened and modified human placental alkaline phosphatase) and rough microsomal membranes (Kodukula et al., 1991). Additionally, transfer of the GPI anchor to polypeptide acceptors has been shown in cell-free membranes (Mayor et al., 1991), and we have recently been able to demonstrate production of mfVSG in a cell-free trypanosome lysate, using a heterologous system including a nuclease-treated lysate and isolated mRNA from different trypanosome clones, thus allowing immunoprecipitation of newly formed VSG (Bühling et al., 1993). All these results are consistent with the hypothesis of a single enzyme which recognizes both pVSG and precursor GPI, cleaves the C-terminal peptide from pVSG, and concomitantly transfers the processed VSG to the GPI membrane anchor, resulting in functional mfVSG. Considering the rapidity of the necessary reactions as well as the results from *in vitro* experiments, it seems likely that this process occurs immediately after termination of translation when the protein is still within the rER membranes on its way to the Golgi apparatus. Interestingly, the GPI precursor is formed initially with diacylglycerol containing unspecific fatty acid residues. Obviously, remodeling reactions occur during maturation of the GPI precursor, leading to a specific replacement of any fatty acid residue by myristic acid (Masterson et al., 1990; Doering et al., 1990); this remodeling reaction may also occur on mature mfVSG (Bühling et al., 1993). Although no specific function has been found for the exclusive use of myristic acid in mfVSG so far, the chain length and the physical property of this fatty acid may possess certain advantages, for example, for lateral diffusion of the whole molecule or the dense packing of the surface coat.

B. General Aspects of Protein Targeting

A conventional intracellular transport route has been well established for plasma membrane proteins and lysosomal proteins, which are translocated cotranslationally into the lumen of the rER. This route includes transfer of proteins to the transient zone, facing the cis-side of the Golgi apparatus, and from then on transport by a process of vesicle budding and fusion to individual Golgi stacks, the trans-Golgi network (TGN), and eventually to the plasma membrane or certain intracellular compartments (Palade, 1975; Farquhar, 1985). At least some of these steps seem to be G-protein dependent (Goud and McCaffrey, 1991; Waters et al., 1991). Using an acylated tripeptide (*N*-acyl-Asn-Tyr-Thr-NH_2) containing an acceptor sequence for *N*-linked glycosylation (-Asn-x-Ser/Thr-), the bulk flow rate for secretion of the glycosylated peptide after initial glycosylation within rER membranes was estimated to be about 5 to 10 min (Wieland et al., 1987). Although it

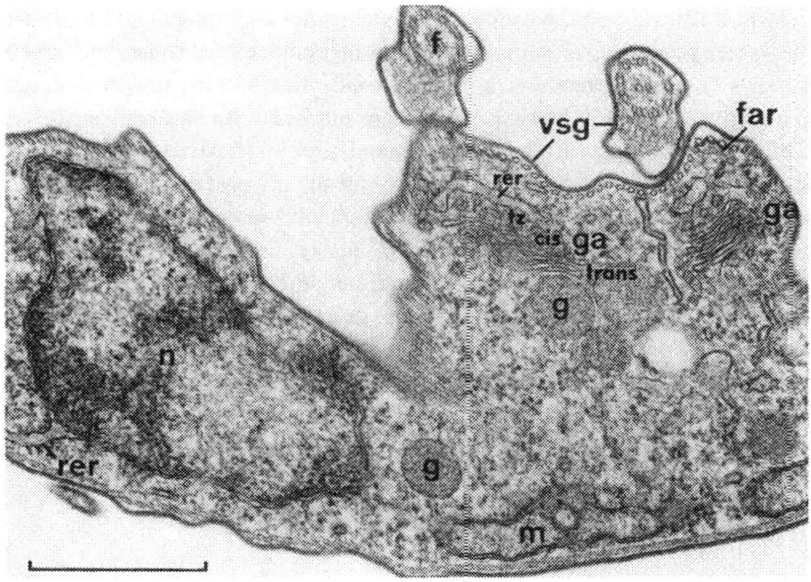

Figure 2. Electron micrograph of Epon-embedded *T. b. brucei* bloodstream form. Representative section through a dividing cell. The pellicle membrane and flagellum (f) are both coated by a 10–15 nm thick electron-dense VSG coat (vsg). Subcellular structures include nucleus (n), rough endoplasmic reticulum (rer), the transient zone (tz) between ER and Golgi, Golgi apparatus (ga) oriented by cis- and trans-stacks, flagellum-associated rER (far), mitochondrion (m), and glycosomes (g). The entire surface membrane, except the flagellar pocket (not shown), is stabilized by a subpellicular corset of microtubules. Unpublished micrograph by M. Duszenko. Bar: 0.5 µm.

has been questioned whether glycosylated peptides may act as a true bulk flow marker (Rose and Doms, 1988; Pelham, 1989), available evidence also supports a bulk flow rate in the range of transport halftimes for the most rapidly secreted proteins, indicating that no special sorting signal is needed to assure an efficient transport of glycoproteins from rER to the cell surface (Pelham, 1989). Instead, retaining and sorting signals seem to be important for organelle resident proteins, like the KDEL tetrapeptide signal for luminal ER proteins (Pelham, 1991), or the mannose-6-phosphate receptor for lysosomal proteins (von Figura, 1991). Additionally, epithelial cells express a plasma membrane divided into apical and basolateral domains to serve vectorial functions. Although default pathways to one or both surfaces may exist, signal mediated sorting of proteins to either surface area has been described: a 17-residue cytoplasmic peptide sequence for basolateral (Bomsel and Mostov, 1991) and, interestingly, the GPI membrane anchor for apical targeting (Lisanti et al., 1988, 1989).

Protein trafficking on the exocytic pathway is not well understood in trypanosomes, and it remains to be demonstrated whether intracellular routes and targeting sequences of parasitic proteins are similar or identical to their mammalian counterparts (Fig. 2). There is, however, at least one marked difference between delivery of cell surface proteins to the plasma membrane of trypanosomes and higher eukaryotes. Whereas vesicle fusion may occur on any part of the cell surface in mammalian cells, fusion of exocytic (and endocytic) vesicles with the cell membrane seems to be restricted to the flagellar pocket in trypanosomes. This is the only surface area where vesicle fusion and budding have been observed so far. Nevertheless, it has to be shown, if this is due to a simple spatial hindrance of vesicles by the rigid trypanosomal cytoskeleton (see Section III), or if vesicles are actively directed either by membrane asymmetry or by a mechanical guidance due to microtubules or filaments, as discussed for several steps of intracellular trafficking in mammalian cells (Breitfeld et al., 1990; Gilbert et al., 1991).

C. Membrane Traffic of Newly Formed VSG

Structural integrity of the surface coat seems of vital importance for bloodstream form trypanosomes, since coatless vector forms or trypsinized vital bloodstream forms are noninfectious to their vertebrate hosts. To account for the growing cell surface during cell division and for VSG loss by shedding or intracellular degradation (see Section IV), VSG is constitutively synthesized at a high rate (about 8% of total protein biosynthesis; M. Duszenko, unpublished observation) and transported to the cell surface with a half-time of about 15 min (Ferguson et al., 1986; Duszenko et al., 1988). This transport rate seems rather high as compared with respective half-times for other glycoproteins, which are in the range of about 15 min for some viral glycoproteins to ≥ 2 hr for several serum proteins (Pelham, 1989). Interestingly, the half-time for VSG transfer is quite similar to the bulk flow rate of about 10 min for glycosylated peptides (Wieland et al., 1987) and membrane lipids (Van Meer, 1989).

A physiological function of the GPI anchor is still unknown. Since the anchor contains a phospholipase C cleavage site (Fig. 1, inset), and a corresponding GPI-specific phospholipase C (GPI-PLC) was isolated from trypanosomes (Bülow and Overath, 1986; Fox et al., 1986; Hereld et al., 1986), it has been proposed that anchor cleavage, which converts mfVSG into soluble VSG (sVSG), might be effective during transformation from bloodstream to procyclic insect forms or during antigenic switching. However, available evidence does not suggest GPI-PLC activation in the course of transformation and it seems that coat release is mediated by other means (Bülow et al., 1989). On the other hand, complete coat release during antigenic switching seems also unlikely, because the "naked" trypanosomes which would occur until formation of a new coat are easily recognized by unspecific immune responses (Mosser and Roberts, 1982; Ferrante and Allison, 1983).

Instead of providing a cleavage site for an endogenous GPI-PLC, the GPI anchor may serve as a sorting signal, as described for glycoproteins in epithelial cells (Bomsel and Mostov, 1991). Therefore, intracellular transport of newly formed VSG was investigated using kinetic analyses and immunolocalization of VSG in ultrathin cryosections (Duszenko et al., 1988). These experiments clearly revealed that VSG can be localized by immunogold labeling within the rER, all Golgi stacks, post-trans-Golgi flattened cisternae (presumably equivalent to TGN in higher eukaryotes), vesicles of different sizes, and tubulovesicular elements (see Fig. 3). Labeling of the luminal face of the rER is consistent with the appearance of a conventional signal sequence in pVSG, which directs the nascent polypeptide to rER membranes. Labeling of all stacks of the Golgi apparatus and the transient zone (Fig. 3 a–c) clearly indicates that VSG is transported along the classical route for integral plasma membrane proteins. This is further supported by a recent structural analysis of *N*-linked complex type oligosaccharides (Zamze et al., 1991), which revealed terminal processing that probably occurs as VSG traverses the Golgi apparatus. Additionally, the GPI anchor is galactosylated in most VSG types, containing unusual α-glycosidic linkages (Ferguson et al., 1988; Pingel et al., 1992). Although in contrast to earlier findings (Bangs et al., 1988) galactosylated GPI molecules are found in trypanosome membranes (Mayor et al., 1992), further processing steps may occur on fully assembled mfVSG molecules within the Golgi apparatus (Doering et al., 1990). Further transport of VSG from Golgi to the plasma membrane is speculative at the moment. Immunogold labeling studies show flattened cisternae and tubulovesicular elements adjacent to the trans-Golgi stacks, both of which contain the gold label. Since a sorting compartment like the TGN should be necessary in all eukaryotic cells to facilitate directed delivery of lysosomal, secretory, and plasma membrane proteins, it seems likely that a corresponding organelle also exists in trypanosomes. As judged from electron microscopical analysis, the flattened cisternae seen in the post-trans-Golgi area show striking similarities in morphology and may resemble the TGN of higher eukaryotes (Duszenko et al., 1988). Interestingly, extensions of the flattened cisternae are often found in the vicinity of the flagellar pocket (Fig. 3 d and g). Newly synthesized VSG could thus be directly transported to the flagellar pocket, where incorporation into the existing coat would probably occur by budding and fusion of small vesicles. In contrast, large vesicles frequently found around the flagellar pocket contain a surface coat on their luminal face and represent endocytic vesicles (Section IV). Although further studies are necessary to define the precise route for VSG on the exocytic pathway, the results clearly demonstrate that the GPI anchor is not a sorting signal for VSG transport, which obviously follows the classical pathway with a half-time very close to the bulk flow.

Owing to spatial hindrance, an integral membrane protein may express a reduced lateral mobility if packed in a dense coat or may be retarded in intracellular membranes on its way to the cell surface. The GPI anchor, which inserts into only one leaflet of the lipid bilayer, could thus facilitate a rapid lateral diffusion rate. Measurements of fluorescence recovery after photobleaching was used to deter-

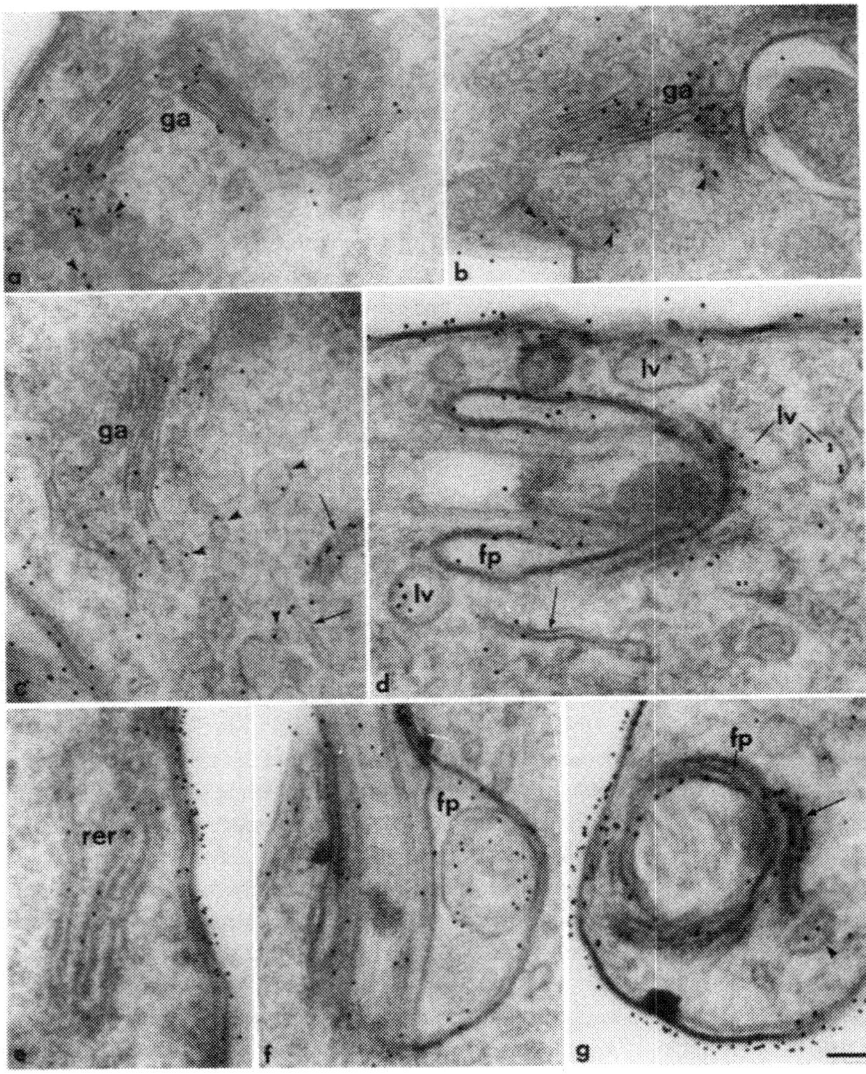

Figure 3. Immunolocalization of VSG in *T. b. brucei* bloodstream form. Thin frozen sections of cells fixed in 2% glutaraldehyde were labeled using anti-VSG antibody and protein A-gold conjugates (10 nm). The cell surface, including the flagellar pocket (**d–g**), was homogeneously labeled. Intracellularly, the membranes of the rough ER (rer) cisternae showed a low but significant staining of the luminal face of the membrane (**e**). A prominent amount of the marker was found over all stacks of the Golgi apparatus (ga) (**a–c**). Tubulovesicular elements (arrowheads) and flattened cisternae (arrows) adjacent to the trans-Golgi cisternae and the flagellar pocket (fp) were also labeled (**a–d**, **g**). Large vesicles (lv), about 100–150 nm in diameter and the same size and localization as coated vesicles in resin-embedded specimen (compare Figs. 4b, 5a), were found close to the flagellar pocket and were systematically labeled on the luminal side of their membranes (**d**). Micrographs reproduced from Duszenko et al. (1988) by permission of the Rockefeller University Press. Bar (**a–g**): 0.1 μm.

mine lateral diffusion coefficients for several GPI-anchored proteins, including VSG. Compared to respective coefficients for lipids (1×10^{-9}–5×10^{-8} cm^2/sec) and plasma membrane glycoproteins (1×10^{-11}–5×10^{-9} cm^2/sec) in biomembranes (McCloskey and Poo, 1986), lateral mobilities of the GPI-anchored proteins Thy-1 antigen, decay accelerating factor (DAF), and alkaline phosphatase were close to those usually observed for lipid probes (Ishihara et al., 1987; Noda et al., 1987; Thomas et al., 1987). MfVSG mobility in immobilized trypanosomes or, after reconstitution of isolated mfVSG in BHK cells, was between 1×10^{-10} and 7×10^{-9} cm^2/sec and thus within the overlapping range of integral membrane proteins and slower lipids (Bülow et al., 1988). The authors argue that these results do not indicate different mobility due to the anchor exchange, no matter if VSG is densely packed in the surface coat or artificially inserted into the BHK plasma membrane. However, lateral diffusion of PH-20, a GPI-anchored sperm protein, varies within a wide range during sperm development (Phelps et al., 1988), and two closely related antigens from a hepatoma cell line, which are conventionally (H-2D) or GPI-anchored (Qa-2), showed a restriction to microdomains for the transmembrane protein, whereas Qa-2 was randomly distributed (Edidin and Stroynowski, 1991). The GPI anchor may thus confer an intrinsic capability to move with a high diffusion rate (Cross, 1990b) and to pass boundaries which may exist for integral membrane proteins. This is further supported by kinetic studies performed in the presence of monensin or at low temperature (20°C). Both treatments usually result in a complete block for intracellular transport of membrane proteins (Farquhar, 1985). In trypanosomes, however, delivery of VSG to the cell surface was not inhibited by monensin and only moderately retarded at low temperature, although dramatic morphological alterations of the trans- and post-trans-Golgi area were caused by the ionophore, and a considerable increase of post-trans-Golgi membrane elements occurred due to a 20°C incubation (Duszenko et al., 1988). The main advantage of VSG GPI-anchorage may thus be related to intracellular transport, linking VSG to the bulk flow in order to assure a fast and unhampered transfer to the cell surface.

III. FLAGELLAR POCKET AND THE CYTOSKELETON

The trypanosome flagellar pocket is the checkpoint for membrane traffic at the border extracellular environment and cytosol, since endocytosis or exocytosis is found exclusively along the flagellar pocket membrane, a functionally discrete surface domain of trypanosomatids (reviewed by Balber, 1990). The flagellar pocket is coated by a VSG coat, as the pellicle membrane, and no differences in the optical density of the VSG coat are visible in electron micrographs (Fig. 4 a) or immunocytochemical labeling studies. However, membrane turnover and secretion events that could expose invariant surface proteins are limited to the pocket because binding antibodies and complement that can enter the flagellar pocket will not be able to mobilize host effector cells like lymphocytes or macrophages, which

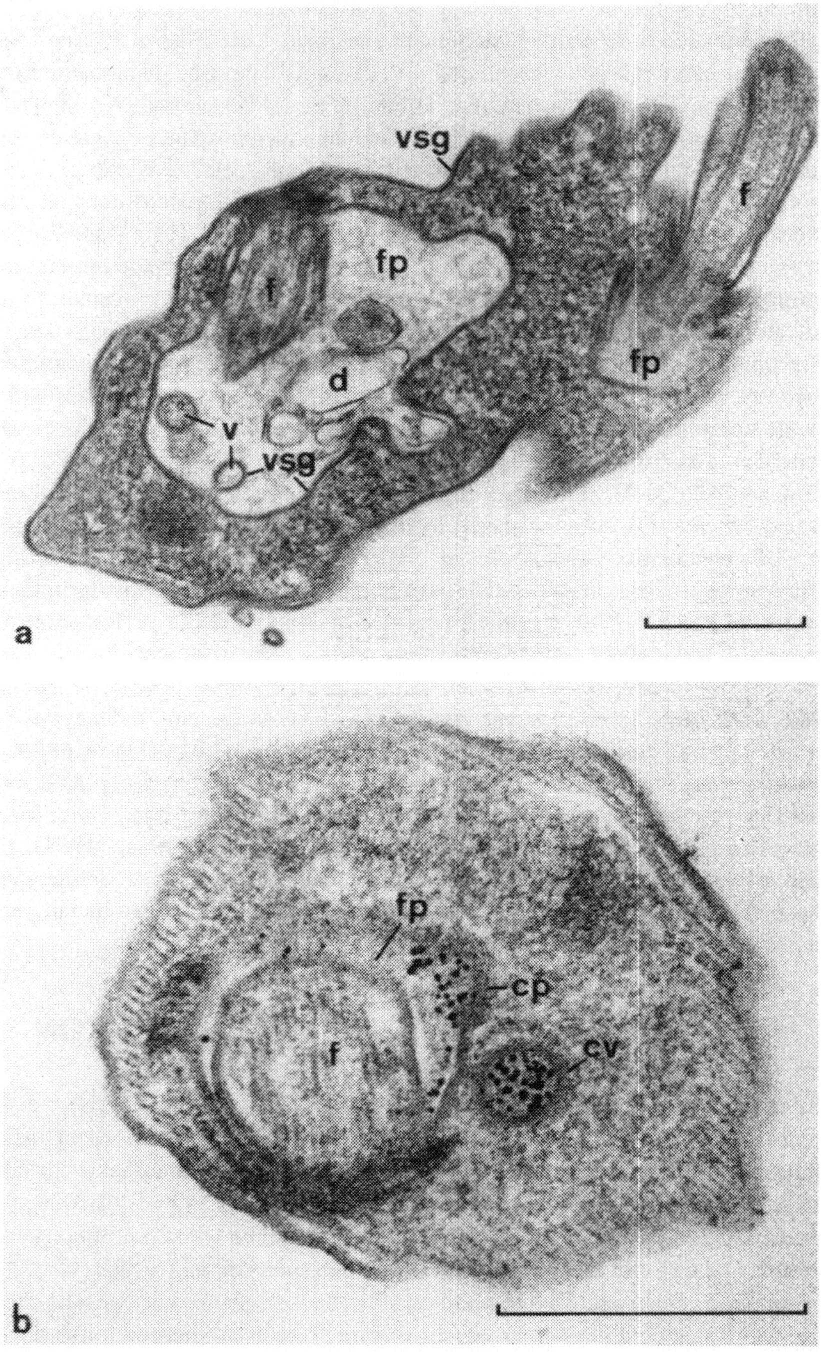
vsg
f
fp
f
d
v
vsg
fp
a
fp
cp
f
cv
b

have no admission to the scenery of this microdomain. Nevertheless, the flagellar pocket is also capable of retaining certain macromolecules in its lumen, including lysosomal hydrolases like acid phosphatase (Langreth and Balber, 1975; see below), and a fuzzy content is visible in electron micrographs (Fig. 4).

Other characteristics of the flagellar pocket include the insertion of the flagellum, which is essential for movement of the parasite and may also have a transport function along the flagellar adhesion zone (i.e., the "undulating membrane" of light microscopy terminology). Antibodies binding to surface VSG are found to rapidly accumulate along the line of flagellar attachment and to become transported to the flagellar pocket (*T. congolense*, Frevert and Reinwald, 1988a, 1990; *T. brucei*, A. Seyfang, unpublished results). The same transport function of the attached flagellum may apply for nutrients bound to trypanosome receptors. However, studies on the distribution of specific receptors are limited by the low relative concentration outside the coated pit region, where receptor-mediated endocytosis is initiated by local receptor accumulation.

Recently, flagellar pocket-specific glycoproteins have been demonstrated in *T. brucei* bloodstream forms (Brickman and Balber, 1990). They could be specifically labeled by the lectin ricin that recognizes *N*-acetyl galactosamine and terminal galactose of the glycosylation sites. Upon trypsinization, ricin-binding glycoproteins were also localized in the flagellar adhesion zone. In *T. brucei* procyclic forms three flagellar adhesion zone-specific glycoproteins were detected by monoclonal antibodies (Woods et al., 1989a, b). These findings further demonstrate the specificity of the flagellar pocket and the flagellar adhesion zone.

The flagellar pocket is the only surface membrane area without the underlying cytoskeleton of a continuous network of microtubules, and the only area where endocytosis can occur. All other surface membrane is stabilized by a subpellicular corset of microtubules, four of which are characteristically attached to the rER along the flagellar adhesion zone. Specifically in this area calcium depletion of the cells led to the polymerization of additional microtubules and to a retraction of the rER extension from its usual position (Selzer et al., 1991).

Figure 4. Electron micrographs of *T. b. brucei* bloodstream form. (**a**) Longitudinal section through the flagellar pocket region of a dividing cell. The flagellar pocket (fp) is coated by a VSG coat (vsg) like the pellicle membrane and filled with exocytic debris (d), possibly exocytic vesicle with VSG surface coat (v) and fuzzy material. (**b**) Cross section through the flagellar pocket region illustrating the endocytic pathway of VSG, 2 min after labeling of surface VSG with immunogold (anti-VSG 117/7nm-gold, preembedding labeling), and subsequent incubation at 37°C without immunogold (gelatin embedding, finally embedded in Epon resin). Coated pit (cp) and coated vesicle (cv; about 130 nm in diameter) with labeled VSG form exclusively along the flagellar pocket (fp). f = flagellum. (a) Unpublished micrograph by A. Seyfang. (b) reproduced from Seyfang et al. (1990) by permission of the Society of Protozoologists. Bars: 0.5 μm.

IV. THE ENDOCYTIC PATHWAY

A. Endocytosis and Turnover of VSG

Endocytosis of plasma membrane segments occurs by vesicle formation of specialized membrane areas, coated on their cytoplasmic surface with a number of distinct and characteristic filamentoid proteins. The best studied sample is the clathrin-coated vesicles decorated by a basket-like clathrin coat that appears as a bristle coat in cross sections. Other vesicle types include caveolae with a striated or spiral membrane coat of caveolin (Rothberg et al., 1992) or plasmalemmal vesicles (Palade and Bruns, 1968). Although the latter two types of nonclathrin-coated vesicles are prominent on the surface of endothelial cells, smooth muscle cells, and fibroblasts, they are thought to be ubiquitous in a wide range of cells (Rothberg et al., 1992). Interestingly, clathrin-coated pits seem to act as molecular filters, concentrating some proteins (e.g., the LDL [low-density lipoprotein] receptor), but excluding others like Thy-1 and H63 antigens (Bretscher et al., 1980). There is cumulative evidence now that GPI-anchored proteins, like the 5-methyltetrahydrofolate receptor (Rothberg et al., 1990), T-cell activating protein (Ly-6A.2), and Thy-1 antigen (Bamezai et al., 1992), as well as a GPI-anchored form of CD4 (CD4DAF fusion protein)—in contrast to its transmembrane form—(Keller et al., 1992) are all excluded from coated pits and are internalized by nonclathrin-coated membrane invaginations, which may involve formation of caveolae (Rothberg et al., 1990). However, there are also reports of GPI-anchored proteins that are clustered in clathrin-coated vesicles, like placental alkaline phosphatase (Makiya et al., 1992), VSG (see below), and the *T. brucei* transferrin receptor (Webster and Grab, 1988; Schell et al., 1991). Thus, the GPI anchor appears to be no strict sorting signal that exclusively links protein internalization to caveolae-mediated endocytosis. This is also true for the GPI-anchored VSG, which is clearly a constituent of clathrin-like coated vesicles in trypanosomes, as demonstrated in electron micrographs and detected by immunolocalization (Figs. 3–5). Although uncoated endocytic vesicles were described in *T. brucei* procyclic insect forms (Langreth and Balber, 1975), caveolae or other nonclathrin-coated vesicles have not yet been studied in trypanosomes.

Figure 5. Electron micrographs of *T. b. brucei* bloodstream form after labeling of surface VSG as described for Fig. 4b. (**a**) After 2 min of incubation: coated pits (cp) and coated vesicles (cv; here, about 100 nm in diameter) form exclusively along the flagellar pocket and rapidly fuse with lysosomes (l, arrows). (**b**, **c**) After 90 min of incubation: lysosome-like vesicles and secondary lysosomes (as judged by their contents of debris) have concentrated immunogold within their tubular lysosomal structures (tl). f = flagellum, fp = flagellar pocket, g = glycosome. Micrographs reproduced from Seyfang et al. (1990), by permission of the Society of Protozoologists. Bars: 0.5 μm.

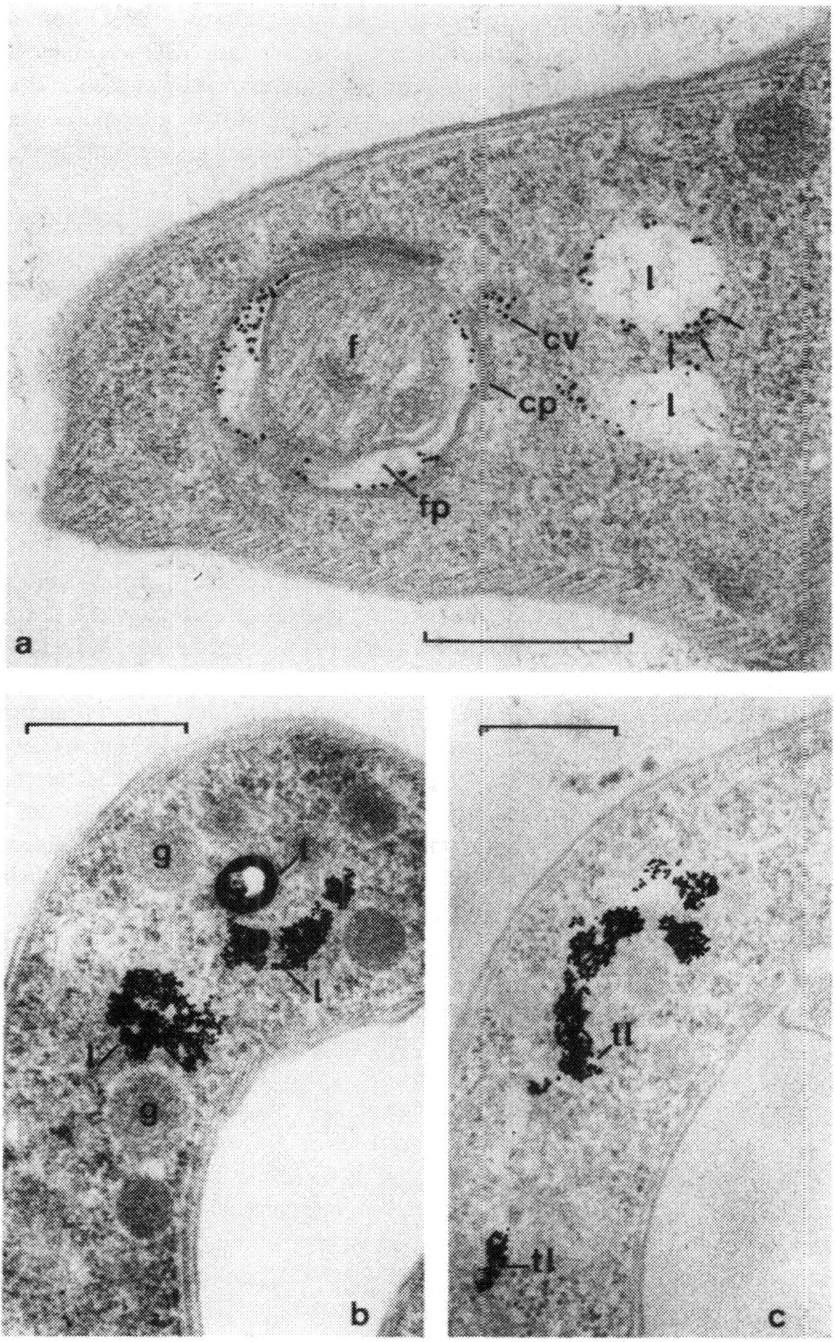
f
cv
cp
fp
l
l
a
g
l
l
g
b
tl
tl
c

In order to unambiguously separate the endocytic pathway of VSG from the molecule's exocytic pathway, specific labeling of cell surface VSG was employed on living trypanosomes. Immunogold labeling of surface VSG was used to study VSG uptake by electron microscopy (Seyfang et al., 1990). Another independent approach was the use of radiochemical labeling techniques to determine the rate of VSG turnover (Bülow et al., 1989; Seyfang et al., 1990).

Immunogold labeling of living *T. brucei* bloodstream forms was performed to localize and evaluate VSG uptake and degradation by electron microscopy. Following direct labeling of surface VSG with colloidal gold–antibody conjugate at 4°C, a temperature which prevents endocytosis, cells were subsequently incubated at 37°C under physiological conditions. VSG uptake by endocytosis was found to occur exclusively along the flagellar pocket membrane (see Section III) and was related to the formation of coated pits and coated vesicles in this functionally discrete surface domain (Figs. 4 b and 5 a; Fig. 7: step 2a, b). These vesicles were surrounded by a unit membrane showing a luminal VSG coat as indicated by both the electron-dense structure and specific immunogold labeling (Fig. 4 b). Additionally a clathrin-like structure around the vesicles could be distinguished.

Clathrin-coated endocytic vesicles have been reported in the mammalian bloodstream form as well as in the epimastigote and metacyclic insect forms of *T. brucei, T. vivax,* and *T. congolense* (Vickerman, 1969, 1973; Steiger 1973a, b; Langreth and Balber, 1975; Vickerman and Preston, 1976; Webster, 1989; Webster and Fish, 1989) but appear to be absent in the procyclic insect form of *T. brucei* (Langreth and Balber, 1975). Trypanosomal coated vesicles are about 100–150 nm in diameter (Langreth and Balber, 1975; Shapiro and Webster, 1989) and larger than clathrin-coated vesicles purified from mammalian cells (about 50–100 nm in diameter; Pearse, 1975, 1982). The clathrin coat of these vesicles contained a protein with similar electrophoretic mobility as mammalian clathrin heavy chains but did not cross-react with antisera against rat or bovine clathrin (Shapiro and Webster, 1989; Webster and Shapiro, 1990). However, Balber (1990) reported two proteins from *T. brucei* bloodstream and procyclic forms that are immunologically related to human clathrin light chains. Morphologically, the typical bristle-coat appearance in cross sections can be distinguished as a basket-like covering in grazing sections of trypanosomal coated vesicles, displaying the characteristic honeycomb arrangement of clathrin-coated vesicles in other eukaryotic cells (see Fig. 9, p. 55, in Vickerman and Preston, 1976).

Our results showed a high frequency of VSG uptake during endocytosis as indicated by a large number of gold-labeled vesicles present 2 min after immunogold labeling. This observation is in good agreement with an earlier study on receptor-mediated endocytosis in *T. brucei* where Coppens et al. (1987) reported an endocytosis of about 250 vesicles/hr per trypanosome (see also Section IV.C). We could show the rapid fusion of VSG-labeled and clathrin-coated vesicles with lysosomes in less than 2 min, and labeled VSG was found along the luminal side of the lysosomal membrane (Fig. 5 a, arrows; see also Fig. 7: step 2c).

In the course of lysosomal degradation, gold particles became concentrated and eventually filled the whole lysosome-like structure or secondary lysosomes (Fig. 5 b, c). However, the electron-dense VSG coat that initially covered the luminal membrane could no longer be detected at that stage, and it could not be distinguished whether gold particles were still associated with VSG molecules or if gold–antibody conjugate had been digested and gold particles were stored as debris.

Kinetic studies of radiochemically labeled VSG were used to determine the turnover rate of surface VSG. Therefore, direct surface labeling was performed with [^{125}I]-Bolton-Hunter reagent (Seyfang et al., 1990). The other approach was biosynthetic labeling of total protein with [^{35}S]-methionine followed by immunological or electrophoretical separation of labeled VSG (Bülow et al., 1989; Seyfang et al., 1990). All methods followed the fate of labeled VSG during subsequent incubation of labeled trypanosomes at physiological conditions for up to 48 hr.

Both labeling methods showed similar results, leading to a VSG degradation rate of only 0.3%/hr compared with a much higher VSG shedding rate of 2.2%/hr into the culture medium by actively dividing parasites (Table 1). Performing Western blot analysis with antiserum against cytosolic proteins, we could confirm that VSG release was not due to cell lysis of degenerating parasites (Seyfang et al., 1990).

Comparing the rather low VSG degradation rate (0.3%/hr) and the endocytosis rate of 9.4%/hr, it appears that more than 95% of the endocytosed surface VSG is recycled back to the plasma membrane (Table 1).

Table 1. Endocytosis and VSG Turnover in *Trypanosoma brucei* Bloodstream Form

	Percent/hr	*Half-time (hr)*	*Reference*
Endocytosis of surface membrane	9.4	7.0	Coppens et al., 1987
VSG degradation	0.37 ± 0.05	189 ± 24	Bülow et al., 1989
	0.29 ± 0.06	237 ± 45	Seyfang et al., 1990
mean	0.33	213	
VSG shedding	2.1 ± 0.2	32 ± 3	Bülow et al., 1989
	2.2 ± 0.6	32.6 ± 9.0	Seyfang et al., 1990
mean	2.2	32	
VSG recycling	9.1	7.3	Seyfang et al., 1990

Notes: Endocytosis of surface membrane was calculated from morphometric analysis of electron micrographs and clearance data of radiochemically labeled serum albumin, using [^{3}H]-inulin exclusion to determine the total cell volume.
VSG degradation and VSG shedding were measured by radiochemical labeling of living cells ([^{125}I] surface labeling and [^{35}S]-methionine biosynthetic labeling). VSG recycling was calculated from the difference between the rates of VSG endocytosis and VSG degradation.

B. Fate of Endocytosed Material

The high frequency of endocytosis in the *T. brucei* bloodstream form rises the question about the fate of this high amount of endocytosed material. Although we could estimate the recycling of more than 95% of this material, the exact intracellular recycling route of endocytosed material is yet unknown. However, lysosomal sequestration appears to follow the typical endosomal pathway of other eukaryotic cells, as illustrated in Fig. 5 a–c for *T. brucei*. Similar results for the endocytic pathway were observed in *T. congolense* (Frevert and Reinwald, 1988b; Webster, 1989; Webster and Fish, 1989).

Endocytic coated vesicles rapidly fuse with lysosome-like vesicles that sometimes could be identified as secondary lysosomes by their contents of debris (Fig. 5 a). Immunogold-labeled VSG could be followed as an electron-dense luminal surface coat until fusion with the lysosomal compartment (Fig. 5 a, arrows). Subsequent recycling or degradation was measured using kinetic studies of radiochemically labeled VSG (Table 1). Immunocytochemical labeling, however, reached its limitation at this stage since it could not be distinguished whether gold particles were still associated with VSG molecules or if they were just stored as debris after digestion of gold–antibody conjugate. Although Webster and Grab (1988) demonstrated an intracellular colocalization of endocytosed transferrin-gold and intracellular VSG, they could not assign immunogold-labeled VSG to either one, the exocytic or endocytic pathway. However, endocytic gold particles from transferrin-gold label (Webster and Grab, 1988), VSG-gold label (Seyfang et al., 1990), or pure endocytosed ferritin (Langreth and Balber, 1975) all accumulated in tubular vesicles of a collecting membrane system (Fig. 5 b, c; see also Fig. 7: step 2e), which is hence assumed to be part of the endosomal compartment. This is further supported by acid phosphatase staining of this tubulovesicular membrane system (Langreth and Balber, 1975; Fig. 6 a). It is unclear if these membranes are still covered with VSG molecules. Although no electron-dense luminal VSG coat can be seen in this compartment, intact VSG molecules may still be present in a less dense distribution.

Figure 6. Electron micrographs of *T. b. brucei* bloodstream form. (**a**) Localization of the lysosomal marker enzyme, acid phosphatase, in fixed cells. Reaction product is found in lysosomes, tubular lysosomal structures, and the flagellar pocket. (**b**) Control with nonspecific immunogold (anti-VSG 118/7nm-gold with trypanosomes of VSG 117-clone) showed only very little nonspecific labeling after 90 min of incubation (arrowheads), compared with specific labeling in Fig. 5b, c. f = flagellum, fp = flagellar pocket, g = glycosome, k = kinetoplast with kDNA, l = lysosome, tl = tubular lysosomal structure, n = nucleus, rer = rough ER. (a) Unpublished micrograph by A. Seyfang. (b) Reproduced from Seyfang et al. (1990) by permission of the Society of Protozoologists. Bars: 0.5 µm.

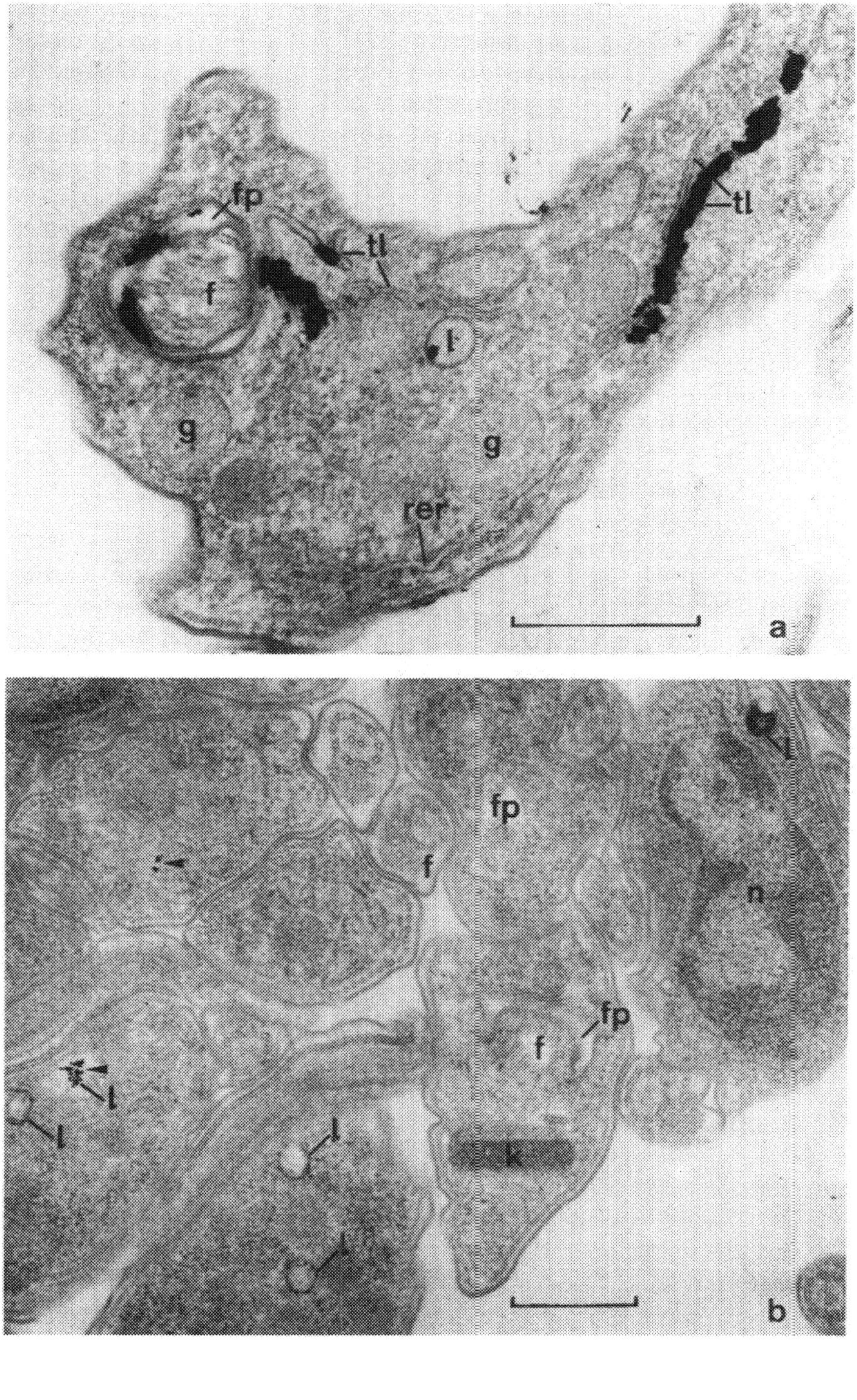
fp
tl
tl
f
l
g
g
rer
a
l
fp
f
n
fp
f
l
l
l
k
l
b

Likewise, endocytosed material may end up as debris in secondary lysosomes filled with very electron-dense material (Fig. 5 b; see also Fig. 7: step 2f), clearly distinguished from the thinner, less electron-dense, and homogeneous VSG surface coat (Figs. 4 b and 5 a). After accumulation of gold particles in tubular lysosomal vesicles, no exocytosis of gold particles into the flagellar pocket could be detected within 90 min of incubation in our immunogold labeling studies of surface VSG. This is in good agreement with Webster's study (1989), which did not observe any exocytosis of endocytosed transferrin-gold in *T. brucei* and *T. congolense* within 410 min. Taking into account the results of a very low VSG degradation rate, we interpret these data as consistent with a lysosomal cleavage of gold–antibody conjugates from the antigen. The resulting unlabeled VSG could be recycled to the cell membrane, whereas gold particles would remain as debris in residual bodies of the lysosomal compartment, as described for other eukaryotic systems (De Robertis and De Robertis, 1987).

C. Vesicle Recycling and Recycling of VSG

There is clear evidence for the recycling of large amounts of endocytosed VSG, as described above for the kinetic studies of radiochemically labeled VSG (Section IV.A and Table 1). This may be accomplished by exocytic lysosomal vesicles during membrane fusion along the flagellar pocket region and release of debris (Fig. 4 a; see also Fig. 7: step 2d). It is unlikely that VSG is transported within the cell directly through the cytosol as individual molecules since membrane form VSG includes the hydrophobic diacylglycerol with its two fatty acids. Instead, VSG is most likely associated with intracellular membrane systems throughout its entire intracellular trafficking as an intact molecule. This can be achieved by either an electron-dense VSG surface coat as found in endocytic clathrin-coated vesicles, or it can be accomplished by a less dense VSG packing as found in the ER and Golgi cisternae.

Vesicles with an electron-dense luminal VSG coat are also found in close proximity to Golgi stacks and TGN-like tubules (Fig. 7: step 2g), which may indicate the affiliation of the endocytic lysosomal pathway with the exocytic biosynthetic pathway of VSG. Webster and Grab (1988) found a similar association for endocytic transferrin-gold in tubules that were in close proximity to the Golgi apparatus and large vesicular structures.

D. Trypanosomal Receptors and Unspecific Endocytosis

Trypanosomes as parasites depend on nutrients supplied from the host's system and mechanisms for their uptake. This includes specific trypanosomal receptors as well as unspecific endocytosis.

Receptor-mediated endocytosis is of particular importance for the rapidly growing parasite. Substrates like carbohydrates, which are abundant in the body fluids, can be taken up easily by specific transmembrane transporters. However, some vital components such as cholesterol and metals are tightly bound to carrier proteins

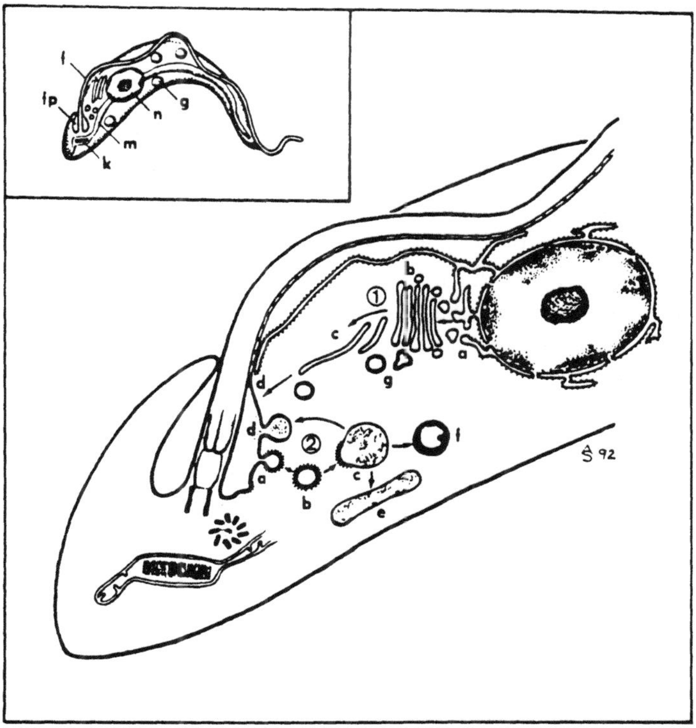

Figure 7. Longitudinal section through the flagellar pocket region of *T. brucei* bloodstream form, illustrating exocytic (**1**) and endocytic membrane traffic (**2**) of variant surface glycoprotein (VSG), based on electron micrographs. Pellicle membrane, flagellum, and flagellar pocket are covered by a dense VSG surface coat. Biosynthesis of VSG occurs at ribosomes of the rough ER. Initial *N*-glycosylation is carried out cotranslationally on the luminal side of the ER, where the GPI-membrane anchor is presumably also added in less than 1 min after translation (**1a**). Budding of vesicles at the ER and fusion with the Golgi apparatus is followed by transport through all Golgi cisternae (**1b**), where VSG glycosylation is completed. Subsequent transport to the cell surface follows tubular structures (**1c**) that may represent the trans-Golgi network (TGN). Finally, mature VSG is inserted into the surface coat (**1d**), presumably along the flagellar pocket. This exocytic pathway of VSG has a half-time of about 15 min.

The endocytic pathway of VSG starts with the formation of coated pits (**2a**) and coated vesicles (**2b**, about 100–150 nm in diameter) exclusively along the flagellar pocket. They are coated by a clathrin-like bristle coat and a luminal electron-dense VSG coat. Coated vesicles fuse with lysosomes (**2c**) in less than 2 min. Lysosomal debris is released into the flagellar pocket (**2d**). More than 95% of endocytosed VSG is recycled to the cell surface and may follow the same route as lysosomal debris. However, lysomal debris can also accumulate in tubular lysosomal structures (**2e**), or end up in lysosomes filled with a dark and inhomogeneous material (**2f**) like residual bodies. Finally, vesicles with a luminal VSG coat are also found in close proximity to Golgi cisternae and TGN-like tubules (**2g**), which may indicate the affiliation of the endocytic lysosomal pathway with the exocytic biosynthetic pathway of VSG.

Inset: *T. brucei* bloodstream form, overview. f = flagellum, fp = flagellar pocket, k = kinetoplast with mitochondrial kDNA, m = mitochondrion, n = nucleus with nucleolus, g = glycosome. Length including flagellum about 20–40 µm.

in the mammalian host and do not freely circulate. Specific receptors and receptor-mediated endocytosis can overcome this essential problem for the protozoan parasite. This was first demonstrated in the *T. brucei* bloodstream form for LDL and transferrin uptake, the parasite's cholesterol and iron sources, respectively (Coppens et al., 1987; Webster and Grab, 1988). Recently, both trypanosomal receptors have been purified, the LDL receptor by Coppens et al. (1988) and the transferrin receptor by Schell et al. (1991). However, since the trypanosomal LDL receptor cross-reacts immunologically with the human LDL receptor, the transferrin receptor appears to be a better target for a possible control of the parasite (Borst, 1991b). Like VSG, the latter contains a GPI membrane anchor and is genetically associated with the same gene expression site (ESAG = expression site associated gene), both features that differ from the human transferrin receptor. Interestingly, the trypanosome lytic factor(s) of human HDL (high-density lipoprotein), which specifically kills *T. brucei brucei*, is also thought to be taken up by receptor-mediated endocytosis (Hajduk et al., 1992).

Unspecific endocytosis, also called pinocytosis or fluid-phase endocytosis, is much slower than the receptor-mediated process and has a clearance of two to three orders of magnitude lower for serum albumin and total plasma proteins compared with the receptor-mediated LDL or transferrin uptake (0.1 μl/hr vs. 20–200 μl/hr per mg cell protein [Coppens et al., 1987]). Nevertheless, it is of importance for the unspecific uptake of macromolecules like many proteins which are too large for a membrane transporter. Other studies on endocytosis of exogenous macromolecules include ferritin (Brown et al., 1965; Langreth and Balber, 1975), polyvinylpyrrolidone (Fairlamb and Bowman, 1980), and Triton WR-1339 (Opperdoes and Van Roy, 1982).

The cell volume of trypanosomes was estimated by [^{3}H]-inulin exclusion and compared with the clearance data above to determine the amount of endocytosis (Coppens et al., 1987). Assuming a cell volume of 58 μm^3 for *T. brucei* bloodstream forms, unspecific endocytosis corresponds to 1.2–1.7%/hr of the cell volume. Morphometric analysis performed on electron micrographs in the same study revealed a flagellar pocket volume of roughly 0.2% of the total cell volume, implying that fluid-phase endocytosis results in the uptake of the equivalent of the flagellar pocket 6–8 times per hour. Furthermore, an endocytosis rate of about 250 vesicles/hr per trypanosome implies that the equivalent of the flagellar pocket membrane area is internalized every 2 min. Thus, the flagellar pocket membrane has to be renewed about 30 times per hour (Coppens et al., 1987), and more than 95% of this process seems to be performed by recycling (Seyfang et al., 1990).

Finally, unspecific endocytosis of serum albumin and other plasma proteins is of significant importance for chemotherapeutic considerations. The trypanocidal drug suramin shows anionic charge at blood pH and binds readily to plasma proteins like albumin. Thus, drug uptake is related to uptake of plasma proteins, including unspecific endocytosis of the dominant serum albumin. This explains that suramin does not pass the blood–brain barrier and is ineffective for treatment of sleeping sickness with involvement of the central nervous system. Nevertheless, suramin is very effective in the treatment of equine trypanosomiasis and early cases of sleeping sickness, where it produces a high rate of cure in both Gambian and Rhodesian forms of the disease (Raether, 1988). Accordingly, fluid-phase endocytosis is of importance for the uptake of hydrophobic fatty acids that also bind to serum albumin, including myristic acid—obviously essential for the GPI anchor formation (Masterson et al., 1990; Doering et al., 1991).

V. VSG TURNOVER AND ANTIGENIC VARIATION

Mammalian trypanosomes of Old World origin can change the antigenic character of their surface and therefore evade the host's immune response by expressing different antigen genes (Vickerman, 1978; Borst and Cross, 1982; Donelson, 1989; Cross, 1990). This antigenic variation has been described for salivarian trypanosomes from Africa like *T. brucei* and *T. congolense* (Boothroyd, 1985), as well as for *T. vivax* (Gardiner, 1989) and *T. evansi* (Mahmoud and Gray, 1980), which are both found from Africa through tropical Asia to South America. The cosmopolitan equine parasite *T. equiperdum* also displays antigenic variation (Cantrell, 1958) but is on the verge of extinction outside the laboratories (Vickerman, 1978).

The VSG repertoire may be restricted to only two in the cosmopolitan rodent parasites *T. lewisi* and *T. musculi* (Targett and Viens, 1979; Schmidt and Roberts, 1981). It was measured, however, to exceed over 100 serologically distinct antigens in *T. equiperdum* (Capbern et al., 1977) and seems to include up to several 1000 in *T. brucei* (Borst, 1991a). Antigenic variation in the latter and best studied species occurs with an antigen switching rate of about 2×10^{-7} to 3×10^{-6}/generation per cell (Lamont et al., 1986) and is independent of the host's specific immune response. However, this switching rate can increase to at least 1×10^{-2}/generation per cell in postmetacyclic trypanosomes in the early bloodstream infection for wild-type *T. brucei*-isolates, which are not adapted to the laboratory (Turner and Barry, 1989). During antigen switching, cells have to exchange the old VSG surface coat for a new, immunologically distinct, VSG coat before the host's specific immune response can recognize the cell. Since uncoated trypanosomes are easily recognized by the mammalian host's nonspecific defense mechanisms (Mosser and Roberts, 1982; Ferrante and Allison, 1983), a gradual exchange of old VSG by new VSG seems to be necessary. Such a simultaneous expression of both pre- and post-switch VSG on individual cells was demonstrated for *T. brucei* (Esser and Schoenbechler, 1985) and *T. equiperdum* (Baltz et al., 1986).

In the infected host, successive expression of immunologically distinct VSGs is paralleled by successive waves of parasitemia usually every 7–14 days (Ross and Thomson, 1910; Donelson and Rice-Ficht, 1985). Hence, it is essential for the parasite to remove the old VSG of switched (i.e., double expressor) cells before immune selection could eliminate the entire switched cell population. Elimination of old VSG could be performed by either intracellular VSG degradation or by VSG shedding. Degradation of VSG is relatively low, with a half-time of more than 210 hr (Table 1), and it would take about 9 days to degrade just half the old VSG coat. On the other hand, 2.2% of VSG is shed per hour and it would only take 32 hr to release half the old surface coat. Seyfang et al. (1990) calculated that merely about 1×10^{-6} % of pre-switch VSG remains on the cell 7 days after a VSG gene switching event, assuming a doubling time of 8 hr for cells and total VSG. Since about 1×10^7 VSG molecules constitute the surface coat of one cell (Cross, 1975; Jackson et al., 1985), about two pre-switch VSG molecules remain on an individual cell 6 days after a switching event, whereas after 7 days only every eighth cell has one "old" VSG molecule left. Following this calculation, after 7 days 87% and after 8 days 99% of the switched cell population has released all pre-switch VSG and is protected by a new, serologically distinct, surface coat (Seyfang et al., 1990). These results indicate that shedding of VSG at a regular turnover rate is sufficient to remove the old VSG coat within one week, and no increase of the VSG turnover rate seems to be necessary during antigenic variation.

Furthermore, released VSG could serve as an exoantigen, as demonstrated in the blood of infected animals (Weitz, 1960; Lanham and Taylor, 1972; Diffley et al., 1980). Hence, shedding of VSG could be responsible for the immune dysfunctions observed during the course of African trypanosomiases, including hypergammaglobulinemia (Houba et al., 1969), nonspecific immunosuppression (Hudson et al., 1976; Jayawardena and Waksman, 1977), and mitogenic effects of VSG as described for murine splenic T-lymphocytes (Diffley et al., 1980).

VI. CONCLUSIONS

The African hemoflagellate *Trypanosoma brucei* has been well studied for its major surface antigen, the variant surface glycoprotein (VSG), protecting the bloodstream forms of this parasite from the immune system of its mammalian host. Structure, biosynthesis, and turnover of VSG have been examined in detail, using *T. brucei* (and to a minor extent *T. congolense* and *T. equiperdum*) as trypanosome model systems. VSG studies have had an important impact in cell biology and related fields, including the discovery of the GPI anchor for membrane proteins and a better understanding of antigenic variation in infectious agents.

Biosynthesis of VSG accounts for about 10% of total cellular protein synthesis, and intracellular transport of VSG to the cell surface has a half-time of about 15 min. Membrane traffic of newly formed VSG follows the endoplasmic reticulum (ER), where the GPI anchor is rapidly added posttranslationally, the Golgi apparatus, and delivery to the flagellar pocket, where it is integrated into the surface coat. Although the GPI anchor is added in the ER, VSG is nevertheless subsequently transported along the classical intracellular route for glycoproteins. Hence, the GPI anchor appears to be no sorting signal for the exocytic pathway of VSG, neither does it significantly affect its lateral diffusion rate. Since the discovery of this novel GPI anchorage in trypanosomes in 1985, over 100 examples of GPI-anchored proteins have been described from many stages of eukaryotic evolution, including plants and man. Nevertheless, the function of this alternative membrane anchor and its phospholipase-specific cleavage remains a mystery, and trypanosomes, once again, may represent a model system when it comes to an answer.

The flagellar pocket is a functionally discrete surface domain of the trypanosomes. It is the insertion of the flagellum, the only surface membrane area without the underlying cytoskeleton of a continuous network of microtubules, and thus the only area where endocytosis and exocytosis can occur. Macromolecules like serum proteins and antibodies can enter the flagellar pocket, but whole cells such as macrophages appear to have no access. Nevertheless, the flagellar pocket is also capable to retain certain macromolecules in its lumen, including lysosomal hydrolases like acid phosphatase.

Endocytosis and lysosomal sequestration of VSG appear to follow the typical endosomal pathway of other eukaryotic cells. Essential macromolecules such as iron/transferrin and cholesterol/LDL can be taken up by receptor-mediated endocytosis at an efficiency of two to three orders of magnitude higher than for unspecific fluid-phase endocytosis. Coated pits and coated vesicles form along the flagellar pocket membrane and rapidly fuse with lysosome-like vesicles and secondary lysosomes in less than 2 min. Trypanosomal coated vesicles have a bristle coat with a clathrin-like structure and additionally contain a luminal VSG coat, as indicated by both its electron-dense structure and specific immunogold labeling. Subsequent lysosomal sequestration can lead to storage as debris in tubular lysosomal structures and residual bodies filled with a very electron-dense material. However, kinetic studies of radiochemically labeled VSG have revealed a high VSG recycling rate of more than 95% of the endocytosed material. About 9.4% of the surface membrane is endocytosed per hour during the formation of about 250 vesicles/hr per trypanosome. Turnover and removal of old VSG are predominantly performed by VSG shedding at a rate of 2.2%/hr, but only to a minor extent by VSG degradation of about 0.3%/hr.

During the course of antigenic variation, bloodstream form trypanosomes can change the antigenic character of their VSG coat and therefore evade the host's specific immune response, resulting in successive peaks of parasitemia usually every 7–14 days. Antigen switching follows the expression of different (up to several 1000) antigen genes. Cells have to exchange the old VSG surface coat for a new, immunologically distinct, VSG coat before the host's specific immune response can recognize the switched cell. The VSG turnover studies revealed that after 7 days 87% and after 8 days 99% of the switched cell population has released all pre-switch VSG and is protected by a new and antigenically distinct surface coat. Hence it was demonstrated that shedding of VSG at a regular turnover rate is sufficient to remove the old VSG coat within one week, and no increase of the VSG turnover rate seems to be necessary during antigenic variation.

ABBREVIATIONS

GPI:	glycosyl-phosphatidylinositol
LDL:	low-density lipoprotein
PLC:	phospholipase C
rER:	rough endoplasmic reticulum
TGN:	trans-Golgi network
VSG:	variant surface glycoprotein
mfVSG:	membrane form VSG
pVSG:	precursor VSG
sVSG:	soluble VSG.

ACKNOWLEDGMENT

We are grateful to G. A. M. Cross for critical reading of the manuscript and to D. Mecke for stimulating discussions. The experimental work in our laboratory was supported by grants from the Deutsche Forschungsgemeinschaft (DFG).

REFERENCES

Allan, G., Gurnett, L. P., and Cross, G. A. M. (1982). Complete amino acid sequence of a variant surface glycoprotein (117) from *Trypanosoma brucei*. *J. Mol. Biol.* 157: 527–546.

Balber, A. E. (1990). The pellicle and the membrane of the flagellum, flagellar adhesion zone, and flagellar pocket: functionally discrete surface domains of the bloodstream form of African trypanosomes. *Crit. Rev. Immunol.* 10: 177–201.

Baltz, T., Giroud, C., Baltz, D., Roth, C., Raibaud, A., and Eisen, H. (1986). Stable expression of two variable surface glycoproteins by cloned *Trypanosoma equiperdum*. *Nature* 319: 602–604.

Bamezai, A., Goldmacher, V. S., and Rock, K. L. (1992). Internalization of glycosyl-phosphatidylinositol (GPI)-anchored lymphocyte proteins. II. GPI-anchored and transmembrane molecules internalize through distinct pathways. *Eur. J. Immunol.* 22: 15–21.

Bangs, J. D., Doering, T. L., Englund, P. T., and Hart, G. W. (1988). Biosynthesis of a variant surface glycoprotein of *Trypanosoma brucei*: processing of the glycolipid membrane anchor and *N*-linked oligosaccharides. *J. Biol. Chem.* 263: 17697–17705.

Bangs, J. D., Hereld, D., Krakow, J. L., Hart, G. W., and Englund, P. T. (1985). Rapid processing of the carboxyl terminus of a trypanosome variant surface glycoprotein. *Proc. Natl. Acad. Sci. U.S.A.* 82: 3207–3211.

Bomsel, M. and Mostov, K. (1991). Sorting of plasma membrane proteins in epithelial cells. *Curr. Opinion Cell Biol.* 3: 647–653.

Boothroyd, J. C. (1985). Antigenic variation in African trypanosomes. *Annu. Rev. Microbiol.* 39: 475–502.

Boothroyd, J. C., Paynter, C. A., Coleman, S. L., and Cross, G. A. M. (1982). Complete nucleotide sequence of complementary DNA coding for a variant surface glycoprotein from *Trypanosoma brucei*. *J. Mol. Biol.* 157: 547–556.

Borst, P. (1986). Discontinuous transcription and antigenic variation in trypanosomes. *Annu. Rev. Biochem.* 55: 701–732.

Borst, P. (1991a). Molecular genetics of antigenic variation. In: *Immunoparasitology Today* (Ash, C. and Gallagher, R. B., eds.), pp. A29–A33, Elsevier Trends Journals, Cambridge.

Borst, P. (1991b). Transferrin receptor, antigenic variation and the prospect of a trypanosome vaccine. *Trends Genet.* 7: 307–309.

Borst, P. and Cross, G. A. M. (1982). Molecular basis for trypanosome antigenic variation. *Cell* 29: 291–303.

Breitfeld, P. P., McKinnon, W. C., and Mostov, K. E. (1990). Effect of nocodazole on vesicular traffic to the apical and basolateral surfaces of polarized MDCK cells. *J. Cell Biol.* 111: 2365–2373.

Bretscher, M. S., Thomson, J. N., and Pearse, B. M. F. (1980). Coated pits act as molecular filters. *Proc. Natl. Acad. Sci. U.S.A.* 77: 4156–4159.

Brickman, M. J. and Balber, A. E. (1990). *Trypanosoma brucei rhodesiense* bloodstream forms: surface ricin-binding glycoproteins are localized exclusively in the flagellar pocket and the flagellar adhesion zone. *J. Protozool.* 37: 219–224.

Brown, K. N., Armstrong, J. A., and Valentine, R. C. (1965). The ingestion of protein molecules by blood forms of *Trypanosoma rhodesiense*. *Exp. Cell Res.* 39: 129–135.

Bühling, O., Kremoser, C., Bernhagen, J., Landgraf, R., and Duszenko, M. (1993). *In vitro* translation and fatty acid labelling of variant surface glycoprotein (VSG) in lysates of *Trypanosoma brucei*. *Eur. J. Cell Biol.* 60 (Suppl. 37): 114.

Bülow, R., Nonnengässer, C., and Overath, P. (1989). Release of the variant surface glycoprotein during differentiation of bloodstream to procyclic forms of *Trypanosoma brucei*. *Mol. Biochem. Parasitol.* 32: 85–92.

Bülow, R. and Overath, P. (1986). Purification and characterization of the membrane form variant surface glycoprotein hydrolase of *Trypanosoma brucei*. *J. Biol. Chem.* 261: 11918–11923.

Bülow, R., Overath, P., and Davoust, J. (1988). Rapid lateral diffusion of the variant surface glycoprotein in the coat of *Trypanosoma brucei*. *Biochemistry* 27: 2384–2388.

Cantrell, W. (1958). Mutation rate and antigenic variation in *Trypanosoma equiperdum*. *J. Infect. Dis.* 103: 263–271.

Capbern, A., Giroud, C., Baltz, T., and Mattern, P. (1977). *Trypanosoma equiperdum*: Étude des variations antigéniques au cours de la trypanosomose expérimentelle de lapin. *Exp. Parasitol.* 42: 6–13.

Cardoso de Almeida, M. L., and Turner, M. L. (1983). The membrane form of variant surface glycoprotein of *Trypanosoma brucei*. *Nature* 302: 349–352.

Coppens, I., Baudhuin, P., Opperdoes, F. R., and Courtoy, P. J. (1988). Receptors for the host low density lipoproteins on the hemoflagellate *Trypanosoma brucei*: purification and involvement in the growth of the parasite. *Proc. Natl. Acad. Sci. U.S.A.* 85: 6753–6757.

Coppens, I., Opperdoes, F. R., Courtoy, P. J., and Baudhuin, P. (1987). Receptor-mediated endocytosis in the bloodstream form of *Trypanosoma brucei*. *J. Protozool.* 34: 465–473.

Cross, G. A. M. (1975). Identification, purification and properties of clone-specific glycoprotein antigens constituting the surface coat of *Trypanosoma brucei*. *Parasitology* 71: 393–417.

Cross, G. A. M. (1984). Structure of the variant surface glycoproteins and surface coat of *Trypanosoma brucei*. *Phil. Trans. R. Soc. Lond.* B307: 3–13.

Cross, G. A. M. (1990a). Cellular and genetic aspects of antigenic variation in trypanosomes. *Annu. Rev. Immunol.* 8: 83–110.

Cross, G. A. M. (1990b). Glycolipid anchoring of plasma membrane proteins. *Annu. Rev. Cell Biol.* 6: 1–39.

Cruse, J. M. and Lewis, R. E. (eds.) (1987). Antigenic Variation: Molecular and Genetic Mechanisms of Relapsing Disease. (Contributions to Microbiology and Immunology, vol. 8), Karger, Basel.

De Robertis, E. D. P. and De Robertis, E. M. F. (1987). Cell and Molecular Biology. Chapter 10: Lysosomes, endocytosis, coated vesicles, endosomes, and peroxisomes; pp. 267–290, Lea and Febiger, Philadelphia.

Diffley, P., Strickler, J. E., Patton, C. L., and Waksman, B. H. (1980). Detection and quantification of variant specific antigen in the plasma of rats and mice infected with *Trypanosoma brucei brucei*. *J. Parasitol.* 66: 185–191.

Doering, T. L., Masterson, W. J., Hart, G. W., and Englund, P. T. (1990). Biosynthesis of glycosyl-phosphatidylinositol membrane anchors. *J. Biol. Chem.* 265: 611–614.

Doering, T. L., Raper, J., Buxbaum, L. U., Adams, S. P., Gordon, J. I., Hart, G. W., and Englund, P. T. (1991). An analog of myristic acid with selective toxicity for African trypanosomes. *Science* 252: 1851–1854.

Donelson, J. E. (1989). DNA rearrangements and antigenic variation in African trypanosomiasis. In: *Contributions to Microbiology and Immunology,* vol. 8, Antigenic Variation: Molecular and Genetic Mechanisms of Relapsing Disease. (Cruse, J. M. and Lewis, R. E., eds.), pp. 138–175, Karger, Basel.

Donelson, J. E. and Rice-Ficht, A. C. (1985). Molecular biology of trypanosome antigenic variation. *Microbiol. Rev.* 49: 107–125.

Duszenko, M., Ivanov, I. E., Ferguson, M. A. J., Plesken, H., and Cross, G. A. M. (1988). Intracellular transport of a variant surface glycoprotein in *Trypanosoma brucei. J. Cell Biol.* 106: 77–86.

Edidin, M. and Stroynowski, I. (1991). Differences between the lateral organization of conventional and inositol phospholipid-anchored membrane proteins: a further definition of micrometer scale membrane domains. *J. Cell Biol.* 112: 1143–1150.

Esser, K. M. and Schoenbechler, M. J. (1985). Expression of two variant surface glycoproteins on individual African trypanosomes during antigen switching. *Science* 229: 190–193.

Fairlamb, A. H. and Bowman, I. B. R. (1980). *Trypanosoma brucei:* maintenance of concentrated suspensions of bloodstream trypomastigotes in vitro using continuous analysis for measurement of endocytosis. *Exp. Parasitol.* 49: 366–380.

Farquhar, M. G. (1985). Progress in unraveling pathways of Golgi traffic. *Annu. Rev. Cell Biol.* 1: 447–488.

Ferguson, M. A. J. (1991). Lipid anchors on membrane proteins. *Curr. Opinion Struct. Biol.* 1: 522–529.

Ferguson, M. A. J. and Cross, G. A. M. (1984). Myristylation of the membrane form of a *Trypanosoma brucei* variant surface glycoprotein. *J. Biol. Chem.* 259: 3011–3015.

Ferguson, M. A. J., Duszenko, M., Lamont, G. S., Overath, P., and Cross, G. A. M. (1986). Biosynthesis of *Trypanosoma brucei* variant surface glycoproteins: N-glycosylation and addition of a phosphatidylinositol membrane anchor. *J. Biol. Chem.* 261: 356–362.

Ferguson, M. A. J., Homans, S. W., Dwek, R. A., and Rademacher, T. W.(1988). Glycosyl-phosphatidylinositol moiety that anchors *Trypanosoma brucei* variant surface glycoprotein to the membrane. *Science* 239: 753–759.

Ferguson, M. A. J., Low, M. G., and Cross, G. A. M. (1985). Glycosyl-sn-1,2-dimyristylphosphatidylinositol is covalently linked to *Trypanosoma brucei* variant surface glycoprotein. *J. Biol. Chem.* 260: 14547–14555.

Ferguson, M. A. J. and Williams, A. F. (1988). Cell-surface anchoring of proteins via glycosyl-phosphatidylinositol structures. *Annu. Rev. Biochem.* 57: 285–320.

Ferrante, A. and Allison, A. C. (1983). Alternative pathway activation of complement by African trypanosomes lacking a glycoprotein coat. *Parasite Immun.* 5: 491–498.

Fox, J. A., Duszenko, M., Ferguson, M. A. J., Low, M. G., and Cross, G. A. M. (1986). Purification and characterization of a novel glycan-phosphatidylinositol-specific phospholipase C from *Trypanosoma brucei. J. Biol. Chem.* 261: 15767–15771.

Franke, E. (1903). Therapeutische Versuche bei Trypanosomenerkrankung. Doctoral thesis, University of Giessen, Germany (Reprinted 1905 at Gustav-Fischer-Verlag, Jena, Germany).

Franke, E. (1905). Ueber Trypanosomentherapie. *Münchener Med. Wochenschr.* 42: 2059–2060.

Frevert, U. and Reinwald, E. (1988a). Formation of filopodia in *Trypanosoma congolense* by crosslinking the variant surface antigen. *J. Ultrastruct. Mol. Struct. Res.* 99: 124–136.

Frevert, U. and Reinwald, E. (1988b). Endocytosis and intracellular occurrence of the variant surface glycoprotein in *Trypanosoma congolense. J. Ultrastruct. Mol. Struct. Res.* 99: 137–149.

Frevert, U. and Reinwald, E. (1990). *Trypanosoma congolense* bloodstream forms evade complement lysis in vitro by shedding of immune complexes. *Eur. J. Cell Biol.* 52: 264–269.

Gardiner, P. R. (1989). Recent studies of the biology of *Trypanosoma vivax. Adv. Parasitol.* 28: 229–317.

Gilbert, T., Le Bivic, A., Quaroni, A., and Rodriguez-Boulan, E. (1991). Microtubular organization and its involvement in the biogenetic pathways of plasma membrane proteins in Caco-2 intestinal epithelial cells. *J. Cell Biol.* 113: 275–288.

Goud, B. and McCaffrey, M. (1991). Small GTP-binding proteins and their role in transport. *Curr. Opinion Cell Biol.* 3: 626–633.

Hajduk, S. L., Hager, K., and Esko, J. D. (1992). High-density lipoprotein-mediated lysis of trypanosomes. *Parasitol. Today* 8: 95–98.

Hereld, D., Krakow, J. L., Bangs, J. D., Hart, G. W., and Englund, P. T. (1986). A phospholipase C from *Trypanosoma brucei* which specifically cleaves the glycolipid on the variant surface glycoprotein. *J. Biol. Chem.* 261: 13813–13819.

Holder, A. A. (1983). Carbohydrate is linked through ethanolamine to the C-terminal amino acid of *Trypanosoma brucei* variant surface glycoprotein. *Biochem. J.* 209: 261–262.

Holder, A.A. (1985). Glycosylation of the variant surface antigens of *Trypanosoma brucei. Curr. Top. Microbiol. Immunol.* 117: 57–75.

Houba, V., Brown, K. N., and Allison, A. C. (1969). Heterophile antibodies, M-antiglobulins, and immunoglobulins in experimental trypanosomiasis. *Clin. Exp. Immunol.* 4: 113–123.

Hudson, K. M., Byner, C., Freeman, J., and Terry, R. J. (1976). Immunodepression, high IgM levels and evasion of the immune response in murine trypanosomiasis. *Nature* 264: 256–258.

Ishihara, A., Hou, Y., and Jacobsen, K. (1987). The Thy-1 antigen exhibits rapid lateral diffusion in the plasma membrane of rodent lymphoic cells and fibroblasts. *Proc. Natl. Acad. Sci. U.S A.* 84: 1290–1293.

Jackson, D. G., Owen, M. J., and Voorheis, H. P. (1985). A new method for the rapid purification of both the membrane-bound and released forms of the variant surface glycoprotein from *Trypanosoma brucei. Biochem. J.* 230: 195–202.

Jayawardena, A. N. and Waksman, B. H. (1977). Suppressor cells in experimental trypanosomiasis. *Nature* 265: 539–541.

Keller, G.-A., Siegel, M. W., and Caras, I. W. (1992). Endocytosis of glycophospholipid-anchored and transmembrane forms of CD4 by different endocytic pathways. *EMBO J.* 11: 863–874.

Kodukula, K., Micanovic, R., Gerber, L., Tamburrini, M., Brink, L. and Udenfried, S. (1991). Biosynthesis of phosphatidylinositol glycan-anchored membrane proteins—design of a simple protein substrate to characterize the enzyme that cleaves the COOH-terminal signal peptide. *J. Biol. Chem.* 266: 4464–4470.

Krakow, J. L., Hereld, D., Bangs, J. D., Hart, G. W., and Englund, P. T. (1986). Identification of a glycolipid precursor of the *Trypanosoma brucei* variant surface glycoprotein. *J. Biol. Chem.* 261: 12147–12153.

Lamont, G. S., Tucker, R. S., and Cross, G. A. M. (1986). Analysis of antigen switching rates in *Trypanosoma brucei. Parasitology* 92: 355–367.

Langreth, S. G. and Balber, A. E. (1975). Protein uptake and digestion in bloodstream and culture forms of *Trypanosoma brucei. J. Protozool.* 22: 40–53.

Lanham, S. M. and Taylor, A. E. R. (1972). Some properties of the immunogens (protective antigens) of a single variant of *Trypanosoma brucei brucei. J. Gen. Microbiol.* 72: 101–116.

Lisanti, M. P., Caras, I.W., Davitz, M. A., and Rodriguez-Boulan, E. (1989). A glycophospholipid membrane anchor acts as an apical targeting signal in polarized epithelial cells. *J. Cell Biol.* 109: 2145–2156.

Lisanti, M. P., Sargiacomo, M., Graeve, L., Saltiel, A. R., and Rodriguez-Boulan, E. (1988). Polarized apical distribution of glycosyl-phosphatidylinositol-anchored proteins in a renal epithelial cell line. *Proc. Natl. Acad. Sci. U.S.A.* 85: 9557–9561.

Mahmoud, M. M. and Gray, A. R. (1980). Trypanosomiasis due to *Trypanosoma evansi* (Steel, 1885) Balbiani, 1888: A review of recent research. *Trop. Anim. Hlth. Prod.* 12: 35–47.

Makiya, R., Thornell, L.-E., and Stigbrand, T. (1992). Placental alkaline phosphatase, a GPI-anchored protein, is clustered in clathrin-coated vesicles. *Biochem. Biophys. Res. Commun.* 183: 803–808.

Masterson, W. J., Raper, J., Doering, T. L., Hart, G. W., and Englund, P. T. (1990). Fatty acid remodeling: a novel reaction sequence in the biosynthesis of trypanosome glycosyl-phosphatidylinositol membrane anchor. *Cell* 62: 73–80.

Mayor, S., Menon, A. K., and Cross, G. A. M. (1991). Transfer of glycosyl phosphatidylinositol membrane anchors to polypeptide acceptors in a cell-free system. *J. Cell Biol.* 114: 61–71.

Mayor, S., Menon, A. K., and Cross, G. A. M. (1992). Galactose-containing glycosyl-phosphatidylinositols in *Trypanosoma brucei*. *J. Biol. Chem.* 267: 754–761.

McCloskey, M. A. and Poo, M. (1986). Rates of membrane-associated reactions: reduction of dimensionality revisited. *J. Cell Biol.* 102: 88–96.

McConnell, J., Cordingley, J. S., and Turner, M. J. (1982). The biosynthesis of *Trypanosoma brucei* variant surface glycoproteins—in vitro processing of signal peptide and glycosylation using heterologous rough endoplasmic reticulum vesicles. *Mol. Biochem. Parasitol.* 6: 161–174.

McConnell, J., Gurnett, A. M., Cordingley, J. S., Walker, J. E., and Turner, M. J. (1981). Biosynthesis of *Trypanosoma brucei* variant surface glycoprotein. I. Synthesis, size, and processing of an N-terminal signal peptide. *Parasitology* 4: 225–242.

Menon, A. K., Mayor, S., Ferguson, M. A. J., Duszenko, M., and Cross, G. A. M. (1988). Candidate glycophospholipid precursor for the glycosylphosphatidylinositol membrane anchor of *Trypanosoma brucei* variant surface glycoproteins. *J. Biol. Chem.* 263: 1970–1977.

Menon, A. K., Mayor, S., and Schwarz, R. T. (1990a). Biosynthesis of glycosyl-phosphatidylinositol lipids in *Trypanosoma brucei*: involvement of mannosyl-phosphoryldolichol as the mannose donor. *EMBO J.* 9: 4249–4258.

Menon, A. K., Schwarz, R. T., Mayor, S., and Cross, G. A. M. (1990b). Cell-free synthesis of glycosyl phosphatidylinositol precursors for the glycolipid membrane anchor of *Trypanosoma brucei* variant surface glycoprotein. *J. Biol. Chem.* 265: 9033–9042.

Mosser, D. M. and Roberts, J. F. (1982). *Trypanosoma brucei*: recognition in vitro of two developmental forms by murine macrophages. *Exp. Parasitol.* 54: 310–317.

Noda, M., Yoon, K., Rodan, G. A., and Koppel, D. E. (1987). High lateral mobility of endogenous and transfected alkaline phosphatase: a phosphatidylinositol-anchored membrane protein. *J. Cell Biol.* 105: 1671–1677.

Opperdoes, F. R. and Van Roy, J. (1982). Involvement of lysosomes in the uptake of macromolecular material by bloodstream forms of *Trypanosoma brucei*. *Mol. Biochem. Parasitol.* 6: 181–190.

Palade, G. (1975). Intracellular aspects of the process of protein secretion. *Science* 189: 347–358.

Palade, G. E. and Bruns, R. R. (1968). Structural modulation of plasmalemmal vesicles. *J. Cell Biol.* 37: 633–649.

Pearse, B. M. F. (1975). Coated vesicles from pig brain: purification and biochemical characterization. *J. Mol. Biol.* 97: 93–98.

Pearse, B. M. F. (1982). Coated vesicles from human placenta carrying ferritin, transferrin and immunoglobulin G. *Proc. Natl. Acad. Sci. U.S.A.* 79: 451–455.

Pelham, H. R. B. (1989). Control of protein exit from the endoplasmic reticulum. *Annu. Rev. Cell Biol.* 5: 1–23.

Pelham, H. R. B. (1991). Recycling of proteins between the endoplasmic reticulum and Golgi complex. *Curr. Opinion Cell Biol.* 3: 585–591.

Phelps, B. M., Primakoff, P., Koppel, D. E., Low, M. G., and Myles, D. G. (1988). Restricted lateral diffusion of PH-20, a PI anchored sperm membrane protein. *Science* 240: 1780–1782.

Pingel, S. and Duszenko, M. (1992). Identification of two distinct galactosyltransferase activities acting on the variant surface glycoprotein of *Trypanosoma brucei*. *Biochem. J.* 283: 479–485.

Raether, W. (1988). Chemotherapy and other control measures of parasitic diseases in domestic animals and man. In: *Parasitology in Focus: Facts and Trends* (Mehlhorn, H., ed.), pp. 739–866, Springer-Verlag, Berlin.

Rose, J. K. and Doms, R. W. (1988). Regulation of protein export from the endoplasmic reticulum. *Annu. Rev. Cell Biol.* 4: 257–288.

Ross, R. and Thomson, D. (1910). A case of sleeping sickness studied by precise enumerative methods: regular periodical increase of the parasites disclosed. *Proc. R. Soc. Lond.* B82: 411–415.

Rothberg, K. G., Heuser, J. E., Donzell, W. C., Ying, Y.-S., Glenney, J. R., and Anderson, R. G. W. (1992). Caveolin, a protein component of caveolae membrane coats. *Cell* 68: 673–682.

Rothberg, K. G., Ying, Y.-S., Kolhouse, J. F., Kamen, B. A., and Anderson, R. G. W. (1990). The glycophospholipid-linked folate receptor internalizes folate without entering the clathrin-coated pit endocytic pathway. *J. Cell Biol.* 110: 637–649.

Schell, D., Evers, R., Preis, D., Ziegelbauer, K., Kiefer, H., Lottspeich, F., Cornelissen, A. W. C. A., and Overath, P. (1991). A transferrin-binding protein of *Trypanosoma brucei* is encoded by one of the genes in the variant surface glycoprotein gene expression site. *EMBO J.* 10: 1061–1066.

Schmidt, G. D. and Roberts, L. S. (1981). Foundations of Parasitology. Chapter 4: Order *Kinetoplastida*: trypanosomes and their kin; pp. 55–82, C. V. Mosby Company, St. Louis.

Selzer, P. M., Webster, P., and Duszenko, M. (1991). Influence of Ca^{2+} depletion on cytoskeleton and nucleolus morphology in *Trypanosoma brucei*. *Eur. J. Cell Biol.* 56: 104–112.

Seyfang, A., Mecke, D., and Duszenko, M. (1990). Degradation, recycling, and shedding of *Trypanosoma brucei* variant surface glycoprotein. *J. Protozool.* 37: 546–552.

Shapiro, S. Z. and Webster, P. (1989). Coated vesicles from the protozoan parasite *Trypanosoma brucei*: purification and characterization. *J. Protozool.* 36: 344–349.

Steiger, R. (1973a). On the ultrastructure of *Trypanosoma (Trypanozoon) brucei* in the course of its life cycle and some related aspects. *Acta Trop.* 30: 64–168.

Steiger, R. (1973b). Ultracytochemistry of the surface coat/pellicle complex in *Trypanosoma brucei*. *Acta Trop.* 32: 152–158.

Targett, G. A. T. and Viens, P. (1979). Immunity to *Trypanosoma (Herpetosoma)* infections in rodents. In: *Biology of the Kinetoplastida*, vol. 2 (Lumsden, W. H. R. and Evans, D. A., eds.), pp. 461–479, Academic Press, London.

Thomas, J., Webb, W., Davitz, M. A., and Nussenzweig, V. (1987). Decay accelerating factor diffuses rapidly on HeLa cell surfaces. *Biophys. J.* 51: 522a (abstract).

Turner, C. M. R. and Barry, J. D. (1989). High frequency of antigenic variation in *Trypanosoma brucei rhodesiense* infections. *Parasitology* 99: 67–75.

Van der Ploeg, L. H. T., Valerio, D., de Lange, T., Bernards, A., Borst, P., and Grosveld, F. G. (1982). An analysis of cosmid clones of nuclear DNA from *Trypanosoma brucei* shows that the genes for VSGs are clustered in the genome. *Nucl. Acids Res.* 10: 5905–5923.

Van Meer, G. (1989). Lipid traffic in animal cells. *Annu. Rev. Cell Biol.* 5: 247–275.

Vickerman, K. (1969). The fine structure of *Trypanosoma congolense* in its bloodstream phase. *J. Protozool.* 16: 54–69.

Vickerman, K. (1973). The mode of attachment of *Trypanosoma vivax* in the proboscis of the tsetse fly *Glossina fuscipes*. *J. Protozool.* 20: 394–404.

Vickerman, K. (1978). Antigenic variation in trypanosomes. *Nature* 273: 613–617.

Vickerman, K. and Preston, T. M. (1976). Comparative cell biology of the kinetoplastid flagellates. In: *Biology of the Kinetoplastida*, vol. 1 (Lumsden, W. H. R. and Evans, D. A., eds.), pp. 35–130, Academic Press, London.

Von Figura, K. (1991). Molecular recognition and targeting of lysosomal proteins. *Curr. Opinion Cell Biol.* 3: 642–646.

Waters, M. G., Griff, I. C., and Rothman, J. E. (1991). Proteins involved in vesicular transport and membrane fusion. *Curr. Opinion Cell Biol.* 3: 615–620.

Webster, P. (1989). Endocytosis by African trypanosomes. I. Three-dimensional structure of the endocytic organelles in *Trypanosoma brucei* and *T. congolense*. *Eur. J. Cell Biol.* 49: 295–302.

Webster, P. and Fish, W. R. (1989). Endocytosis by African trypanosomes. II. Occurrence in different life-cycle stages and intracellular sorting. *Eur. J. Cell Biol.* 49: 303–310.

Webster, P. and Grab, D. J. (1988). Intracellular colocalization of variant surface glycoprotein and transferrin-gold in *Trypanosoma brucei*. *J. Cell Biol.* 106: 279–288.

Webster, P. and Shapiro, S. Z. (1990). *Trypanosoma brucei*: A membrane-associated protein in coated endocytotic vesicles. *Exp. Parasiol.* 70: 154–163.

Weitz, B. G. F. (1960). A soluble protective antigen of *Trypanosoma brucei*. *Nature* 185: 788–789.

Wieland, F. T., Gleason, M. L., Serafini, T. A., and Rothman, J. E. (1987). The rate of bulk flow from the endoplasmic reticulum to the cell surface. *Cell* 50: 289–300.

Woods, A., Baines, A. J., and Gull, K. (1989a). Evidence for a M_r 88000 glycoprotein with a transmembrane association to a unique flagellum attachment region in *Trypanosoma brucei*. *J. Cell Sci.* 93: 501–508.

Woods, A., Sherwin, T., Sasse, R., MacRae, T. H., Baines, A. J., and Gull, K. (1989b). Definition of individual components within the cytoskeleton of *Trypanosoma brucei* by a library of monoclonal antibodies. *J. Cell Sci.* 93: 491–500.

Zamze, S. E., Ashford, D. A., Wooten, E. W., Rademacher, T. W., and Dwek, R. A. (1991). Structural characterization of the asparagine-linked oligosaccharides from *Trypanosoma brucei* type II and type III variant surface glycoproteins. *J. Biol. Chem.* 266: 20244–20261.

Zamze, S. E., Wooten, E. W., Ashford, D. A., Ferguson, M. A. J., Dwek, R. A., and Rademacher, T. W. (1990). Characterization of the asparagine-linked oligosaccharides from *Trypanosoma brucei* type-I variant surface glycoproteins. *Eur. J. Biochem.* 187: 657–663.

Ziegelbauer, K., Multhaup, G., and Overath, P. (1992). Molecular characterization of two invariant surface glycoproteins specific for the bloodstream stage of *Trypanosoma brucei*. *J. Biol. Chem.* 267: 10797–10803.

A COMPARATIVE SURVEY ON PHAGOSOME FORMATION IN PROTOZOA

Klaus Hausmann and Renate Radek

I. INTRODUCTION

The formation of phagosomes is a central requirement for heterotrophic protists in order to gain substrates for energy production. In general, extracellular material has to be internalized via a focal enlargement of the plasma membrane, which is succeeded by the formation of a vacuole. This vacuole is eventually pinched off from the plasma membrane and transported into the cytoplasm.

Advances in Cell and Molecular Biology of Membranes, Volume 2B
Membrane Traffic in Protozoa, pages 259–282

ISBN: 1-55938-628-2

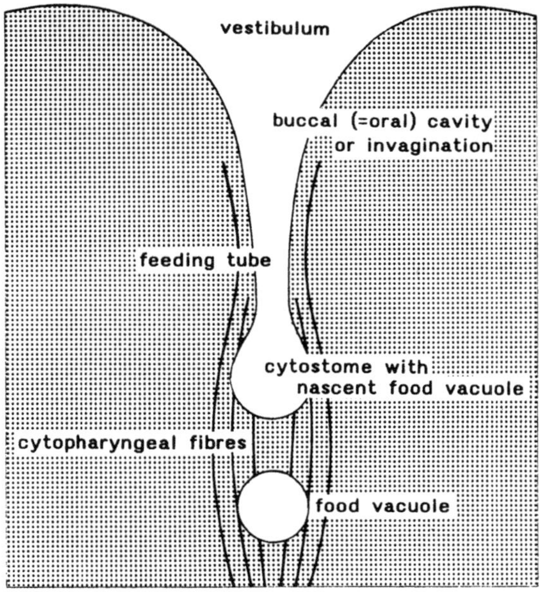

Figure 1. Definition of the terms used to describe the components of the buccal (= oral = feeding) apparatuses.

This chapter provides a survey of these phenomena in the different protists, focusing interest on the source of the additional membrane which is needed for food vacuole growth and on the force generating structures for food vacuole internalization. We will discuss the situations in the order of the classical protistan taxa: Mastigophora, Sarcodina, Apicomplexa, Ciliophora, and Myxozoa.

The terms used to describe the components of the oral apparatuses have no uniform meaning in the literature. We use them in the following sense (Fig. 1): an invagination into which food particles are drawn and which is supported by cytoskeletal elements is generally called "buccal (= oral) invagination" or "cavity," or "feeding tube" if channel-like. In ciliates the buccal invagination may be the continuation of a shallow depression, the "vestibulum." At its closed end the "cytostome" whose membrane is deprived of pellicular structures pinches off "food vacuoles." Microtubular or filamentous fibrils may conduct the food vacuole deeper into the cell. They form a "cytopharynx." Frequently, the fibrils are continuous with the cytoskeletal elements bordering the buccal invagination.

II. MASTIGOPHORA

Little is definitely known about the mechanism of phagosome formation in flagellates. So far, only ultrastructural details give hints to possible motors for force generation and membrane growth. Physiological experiments are largely missing. Two types of endocytosis may be morphologically distinguished, pinocytosis and phagocytosis.

Dissolved nutrients are internalized via a vesicular and/or tubular kind of pinocytosis. Coated pits and coated vesicles are thought to be responsible for the uptake of selective nutrients by a receptor-mediated pinocytosis. The type of pinocytosis may depend on the developmental stage of the protist. The sites of pinocytosis may be randomly distributed or may be situated at characteristic places, for example, at the base of pellicular folds of the Opalinida (Münch, 1970), at the membrane of the euglenoid reservoir (Kivic and Vesk, 1974), or at depressions rich in intramembranous particles in the trypanosome flagellar pocket membrane (Yoshikawa et al., 1990). The coated vesicles arising from the pocket membrane of *Trypanosoma brucei* differ from clathrin-coated vesicles of other eukaryotes in containing an external and an inner coat of electron-dense material and in being larger (100–150 nm) (Shapiro and Webster, 1989). The inner coat corresponds to the parasite's variant glycocalyx. Serological cross reactivity of protein components with clathrin could not be shown. A selective receptor-mediated and Ca^{2+}-dependent uptake of substances such as radioactively labeled low-density lipoprotein (LDL) and transferrin could, however, be proven, in addition to an unspecific uptake with low clearance rate (Coppens et al., 1987).

Particulate material is engulfed by numerous species. Phagocytosis may take place randomly on large areas of the cell surface or via specialized cytostomal complexes. Uptake all along the surface is only possible in protists lacking rigid cortical structures. In these cases, food may be preferentially engulfed laterally and at the posterior cell pole (Trichomonadida, Hypermastigida, Oxymonadida) or at the anterior end (Chrysomonadida, Choanoflagellida). Hypermastigotes and oxymonads engulf food particles (cellulose) by submerging them into the body via the formation of deepenings of the plasma membrane without extension of pseudopods (Lavette, 1966, 1975; Radek et al., 1992). The same is described for the kinetoplastid *Colponema loxodes* (Mignot and Brugerolle, 1975). In these cases, the cytoplasm underlying the cellular membrane is conspicuously free of organelles at the site of ingestion. It contains microfibrils resembling the pseudopodial ultrastructure of amoebae. In *Trichomonas vaginalis*, where under certain circumstances a similar situation can be observed (Brugerolle et al., 1974), cytochalasin B and colchicine inhibit phagocytosis of bacteria, indicating an involvement of microfilaments and microtubules (Francioli et al., 1983; Juliano and Bandiera, 1985, data presented at the Int. Symp. on Trichomonads and Trichomoniasis, Prague, abstr. 52.).

Other flagellates extend pseudopods to enclose the food. In choanoflagellates, for instance, which filter food particles by the collar microvilli, lobopods arise at the base of the collar. They extend toward the trapped prey by moving up along the outside of the collar (Fig. 2; Leadbeater, 1983). The pseudopod fuses with the collar very close to the site of its origin, thus totally encircling the prey. Food is internalized into the cytoplasm beneath the microvilli. Also certain species of

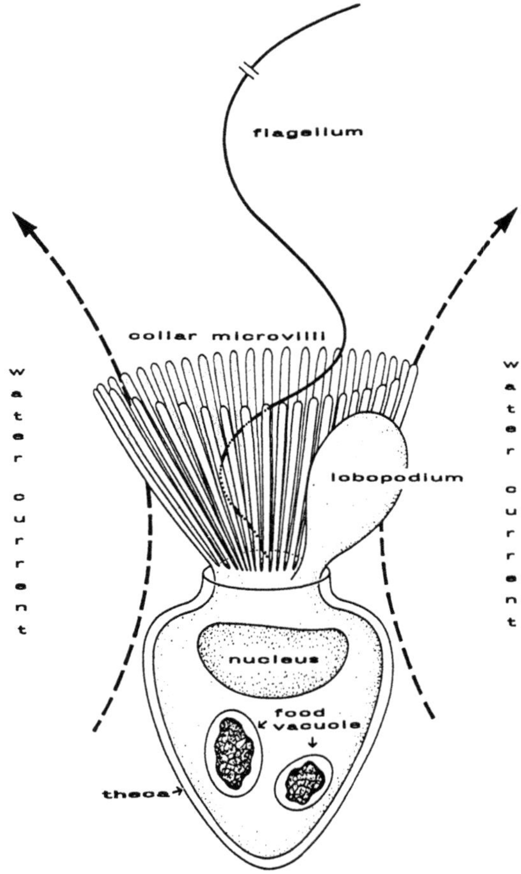

Figure 2. Feeding choanoflagellate *Salpingoeca pelagica.* A lobopod arises at the base of the collar microvilli entrapping food which has been filtered from the medium. After Laval (1971).

chrysomonads use pseudopods for food capture and ingestion. In this group a great variety of pseudopodial forms is reported: tentacle-like in *Brehmiella chrysohydra*, forming a collar in *Monas*, fine and anastomozing in *Derepyxis* and *Dinobryon* or like a pseudopodial veil in *Chrysarachnion* (Pringsheim, 1963). In the dinoflagellate *Gymnodinium helveticum* a large lobopod emerging near the longitudinal flagellar groove by dilating the sulcus is involved in phagocytosis (Popovsky, 1982). Many armored dinoflagellates enclose large prey cells by a pseudopod and digest them in a food vacuole outside the cell body proper, since the sulcal opening is too narrow for internalization of large food particles (Jacobson and Anderson, 1986). The pseudopod extends from the flagellar groove at a rate of 2–6 μm per second and encloses the food within one or few minutes. No information about the forces driving pseudopodial expansion and retraction is available.

The parasitic flagellate *Pirsonia guinardiae* forms a cell process near the apical cell pole which pierces the weakly silicified cell wall of the marine diatome *Guinardia flaccida* (Schnepf et al., 1990). Inside the host cell, the process develops into a feeding organelle consisting of distal pseudopods and a proximal digestion vacuole. The main body of *P. guinardiae* remains outside the host cell.

Long ramified pseudopods develop from the ventral groove of the colorless flagellate *Thaumatomonas lauterborni* and entangle food particles (Karpov and Makarenkova, 1989). A population of small vesicles on the ventral side is thought to be responsible for the quick enlargement of the filopodial surface and, thus, also for the food vacuole formation. These vesicles seem to be derived from coated vesicles which are pinched off from the dorsal surface and the flagellar pocket. They lose their coat on the way to the ventral cell region, where they are stored. In most cases of food vacuole formation by extending pseudopods, however, nothing is known about the origin of the membrane material required to enlarge the surface.

The phagotrophic apparatuses of flagellates exhibit a relatively common pattern in that a certain membrane domain is supported by microtubules with various fibrous connectives, in addition to the presence of a force-generating mechanism and associated vesicles (Willey et al., 1988). The specific ingestion mechanisms, however, are quite distinct, involving a wide spectrum of oral apparatuses to capture, concentrate, and engulf food. The buccal opening may be a funnel- or tube-like cavity or a shallow depression, supported by longitudinal microtubular arrays and fibrillar systems. In *Ochromonas* two bundles of microtubules lining the right margin of a ventral furrow of the cell seem to mediate the phagocytosis of bacteria. When a bacterium touches that region the right margin rises to form a 2 μm high, concave wall in which the food particle is trapped by enclosure (Fenchel, 1982). The entire process requires about 20 sec. The membrane of the food vacuole is thought to arise from the plasma membrane (Cole and Wynne, 1974). In many chrysomonads a cup-like feeding apparatus or food basket emanates near the basal bodies, if a particle captured between the flagella is accepted as food. The transitory food basket is formed by the dynamic interaction of flagellar root microtubules (Fig. 3; Andersen, 1990; Moestrup and Andersen, 1991). In the chrysomonad *Epipyxis pulchra* the f-microtubule of the R_3 root, which consists of 6 microtubules, has been shown to slide upward. By that movement the size of the f-microtubular loop under the short flagellum increases to form the enlarging rim of the cup (Andersen and Wetherbee, 1992). The f-tubule probably interacts with the a-tubule of the same root and with the R_4 root. As the loops of the a- and f-microtubules separate, two types of vesicles fuse with the membrane of the cup, thus enlarging its surface and increasing its depth. The striated fibers accompanying the a-tubule and the periodically striated material coating the a- and f-tubules are supposed to be contractile, thus possibly playing a role in closing the feeding cup. Prey which previously was trapped and moved into the cup by the action of the two flagella thus becomes enclosed in a food vacuole (Wetherbee and Andersen, 1992).

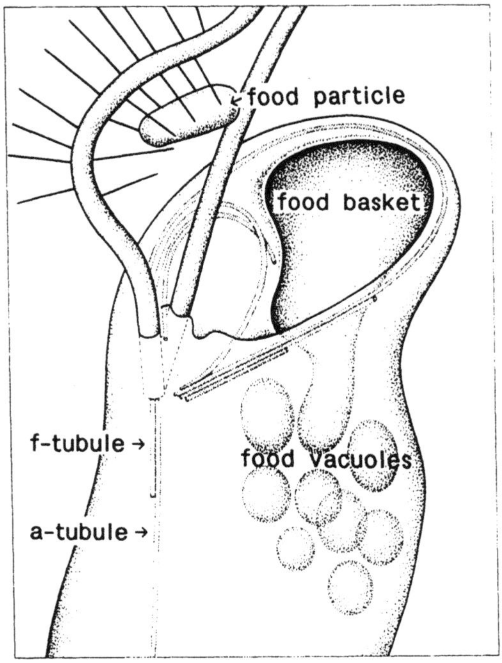

Figure 3. Cup-like feeding apparatus (= food basket) of the chrysomonad *Epipyxis pulchra* emanating near the basal bodies. After Andersen and Wetherbee (1992).

Other oral apparatuses can be protruded and retracted to engulf food, for example, in some Euglenida. The mechanism of force production may be based on an antagonistic system of microtubular rods and attached lamella and on cytoplasmic pressure (*Peranema trichophorum*; Nisbet, 1974). *Peranema trichophorum* is thus thought to "suck" food into the cytostomal sack by withdrawal of the protruded feeding organelle. In *Entosiphon sulcatum* an elongate invagination of the plasma membrane extends downward into the tube-like feeding apparatus (siphon) consisting of four striated vanes arising from three microtubular bundles (Fig. 4). During the protrusion of the feeding organelle a cap covering the opening of the invagination is withdrawn and the invagination is extended to a large cavity engulfing food particles. The model of sliding microtubules within the microtubular bundle as the generator of the driving force, as proposed by Triemer and Fritz (1987), seems to be inadequate since the siphon does not change in length during feeding (Lackey, 1929).

In the dinoflagellate *Noctiluca* the cytostome opens like a slit during ingestion. After cytoplasm has aggregated toward the cytostome by vigorous cytoplasmic streaming, the membrane of the cytostome invaginates, thus enclosing the food (Nawata and Sibaoka, 1983). The movement seems to be triggered by a rise in the cytoplasmic calcium concentration. During the initiation of feeding about 5% of

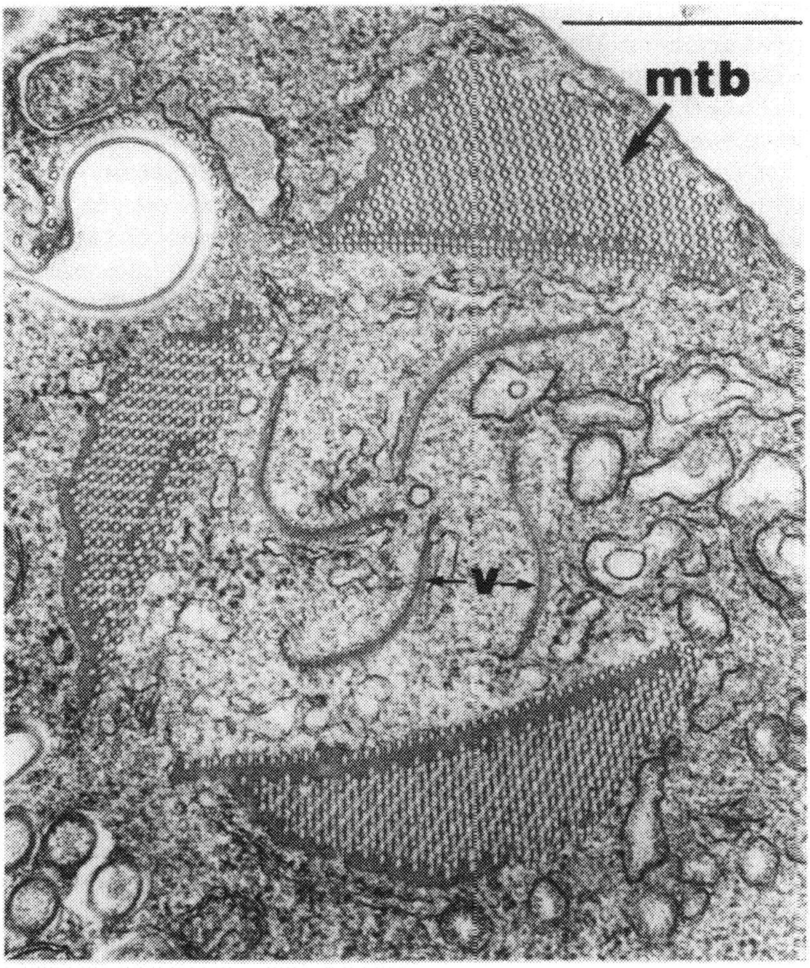

Figure 4. Cross section of the tube-like feeding apparatus of *Entosiphon sulcatum*. mtb = microtubular bundle, v = vane. Original D. J. Patterson, Sydney. Bar: 0.5 µm.

the inward transmembrane ion currents are carried by Ca^{2+} (Nawata and Sibaoka, 1987). The wide spectrum of feeding mechanisms realized by dinoflagellates was recently reviewed by Elbrächter (1991) and Schnepf and Elbrächter (1992).

Food uptake is called myzocytosis when the prey is pierced and its contents "sucked out" (Schnepf and Deichgräber, 1984). *Paulsenella*, for example, extends a feeding tube which penetrates into the host cytoplasm (Schnepf et al., 1985). Longitudinal sliding of microtubular ribbons may be responsible for the extension of the tube. The host cytoplasm flows into the tube as described for Suctoria (Ciliophora; Bardele, 1974). Two types of vesicles, namely, rod-shaped bodies with electron-dense contents and electron translucent vacuoles, disappear during the emergence of the tube. They may provide proteolytic enzymes, substances which affix the parasite to the host cell, or membrane material for the elongation of the feeding tube, its inner channel, and the food vacuole (Schnepf et al., 1985). The membrane lining the feeding tube channel of the parasite is not connected to the longitudinal microtubules in the tube wall. Thus, food uptake is probably not driven by microtubule–membrane interactions, as postulated, for example, for Suctoria (Bardele, 1974). Schnepf et al. (1985) discussed the hypothesis of a hydrostatic gradient caused by rhythmic ion pumping to deliver the driving force for the uptake of food by suction (Schnepf et al., 1985). It has been proposed that actin or actomyosin, but not microtubules, may be involved in the uptake of host material. However, the true nature of the "sucking" force remains speculative (Schnepf and Deichgräber, 1984) and convincing experimental proof is still missing.

A mechanism similar to suctorian food uptake may be realized also in some kinetoplastids (Burzell, 1973). *Rhynchomonas* possesses a mobile proboscis consisting of two lobes. When food particles are ingested, the proboscis becomes shorter and broadens, sucking in the food into a membrane-limited lumen (Swale, 1973). Two rings of microtubules which are continuous with the cytopharyngeal microtubules surround the lumen. Burzell (1973) speculated that during the process of ingestion the inner ring of microtubules may move to the outer ring, thus widening the lumen. In other kinetoplastids, for example, in *Bodo designis* and *B. curvifilus*, a compact, prismatic microtubular rod is associated with the buccal invagination. Eyden (1977) discussed the hypothesis that the buccal invagination may be extended either by displacement of the rod-organelle as a whole or by sliding of its microtubular rows. Some of the microtubules are connected to the buccal membrane and may promote membrane movement during ingestion. Microtubule–membrane interactions also occur in the colonial kinetoplastid *Cephalothamnium cyclopum* (Hitchen, 1974). It has not been clarified whether the abundant oral vesicles of this organism contribute membrane material or digestive enzymes to the enlarging cytostome, or whether they are in the process of exocytosis.

III. SARCODINA

In amoebae of the *proteus*-group pinocytosis has been investigated in great detail (Chapman-Andresen, 1962; Stockem and Wohlfarth-Bottermann, 1969; Stockem and Klein, 1988). The phenomena are to some extent similar to those being reported for flagellates in this chapter.

In this type of amoebae, phagocytosis begins with pseudopods surrounding a prey cell, thus forming trap-like structures. This mode of food vacuole origin is possible only in those sarcodines where a certain surplus of plasma membrane is available (Chapman-Andresen, 1973). Since these aspects are covered separately in this book (see chapters by Prusch and by Jeon) we will concentrate merely on some more exotic, not so well-known sarcodines.

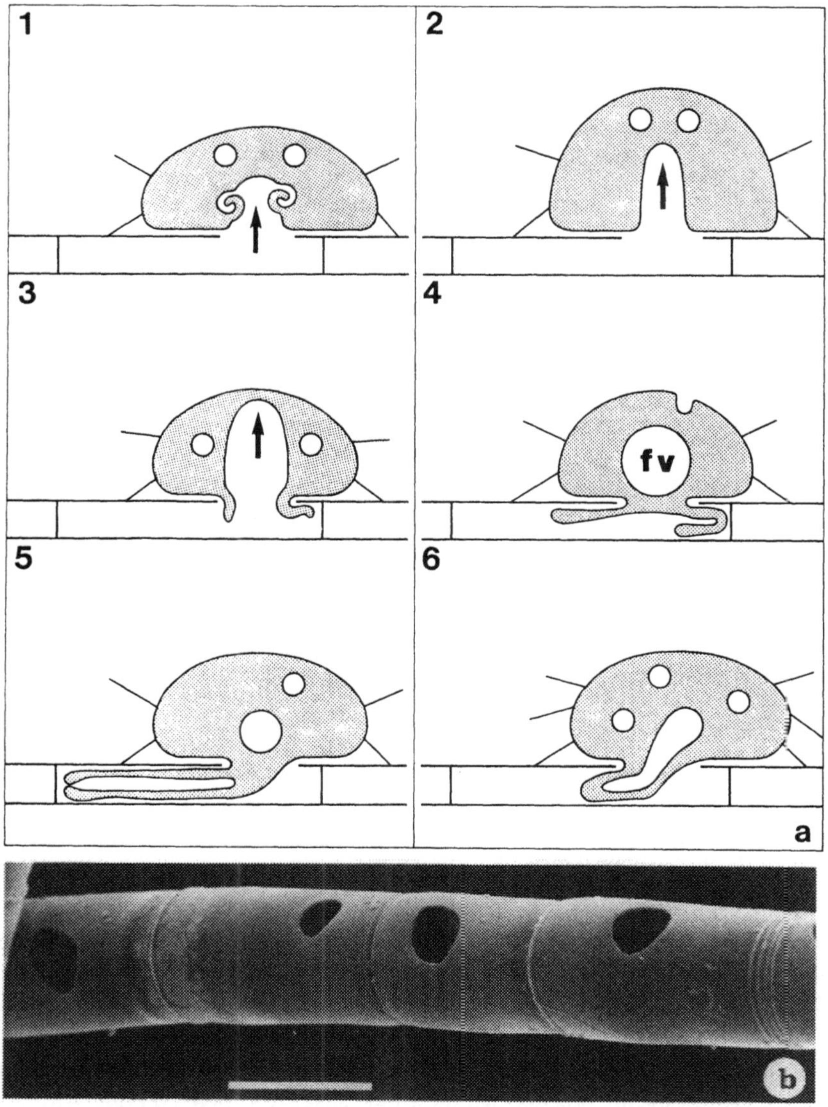

Figure 5. (**a**) Schematic representation of the formation of food vacuoles (fv) in a vampyrellid. Steps 1–4 occur within a few seconds. The arrows indicate the explosive release of algal protoplasm. (**b**) A filament of the green alga *Oedogonium* with holes in the cell walls which are caused by the enzymatic activity of vampyrellids. Original N. Hülsmann, Berlin. Bar: 25 μm.

Compared with the lobopodial amoebae the situation differs greatly in those sarcodines, which exhibit a rather rapid mode of food uptake, for example, vampyrellids, or in those organisms which obviously do not have additional plasma membrane available, as it is true for instance in heliozoa (Page and Siemensma, 1991).

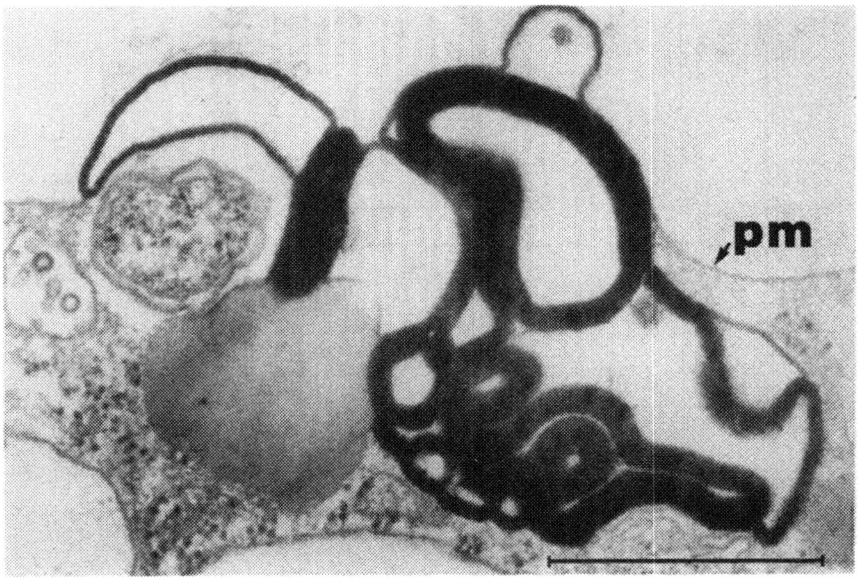

Figure 6. Myelinic body of a vampyrellid in the process of intussusception into the plasma membrane (pm). Original N. Hülsmann, Berlin. Bar: 0.5 μm.

In the vampyrellids as well as in the heliozoa, membrane stores preexist which can be used rapidly for the formation of food vacuoles, providing mainly the lipid phase of the membranes. In the vampyrellids an extremely rapid formation of food vacuoles is reported (Hülsmann, 1985). These protists are able to open, for example, cell walls of filamentous and unicellular green algae of the genera *Oedogonium, Spirogyra,* or *Closterium* by enzymatic activity. Nothing is known about the chemistry of this process. Due to the turgor of the algal cell, the opening of the cell wall leads to a sudden, almost explosive release of at least a part of the algal cytoplasm (Fig. 5). The vampyrellid immediately encloses the food by intussusception of lipid structures (myelinic bodies) into the plasma membrane (Fig. 6). Owing to this process, within a few seconds a food vacuole is formed which has a membrane area of more than 4000 μm^2. The further steps of food vacuole formation are characterized by the additional incorporation of myelinic bodies and by the gradual formation of a cup-shaped feeding pseudopodium of so far unexplained origin (Fig. 5; Hülsmann, 1985).

In heliozoa, for example, in *Actinophrys sol* and in *Echinosphaerium nucleofilum,* food uptake might be initiated by an incidental contact with a potential prey organism like a flagellate, a ciliate, or even a rotifer. When the ciliate *Colpidium* is given as food, the formation of a cup-shaped pseudopodium begins which encloses the prey completely in about 15 min (Fig. 7; Patterson and Hausmann, 1981).

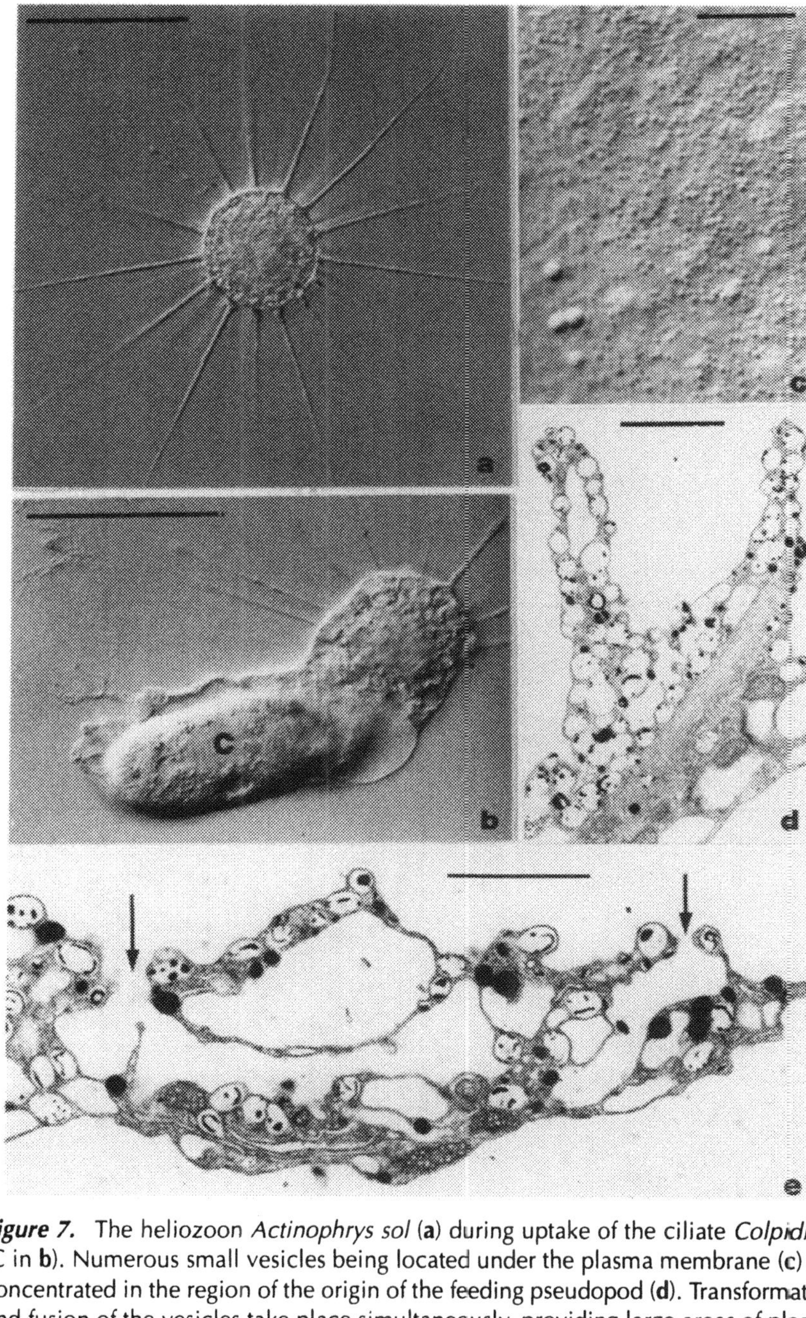

Figure 7. The heliozoon *Actinophrys sol* (**a**) during uptake of the ciliate *Colpidium* (C in **b**). Numerous small vesicles being located under the plasma membrane (**c**) are concentrated in the region of the origin of the feeding pseudopod (**d**). Transformation and fusion of the vesicles take place simultaneously, providing large areas of plasma membrane (**e**, arrows). After Patterson and Hausmann (1981) and Hausmann and Patterson (1982). Bars: 50 µm (a, b); 10 µm (c); 2.5 µm (d, e).

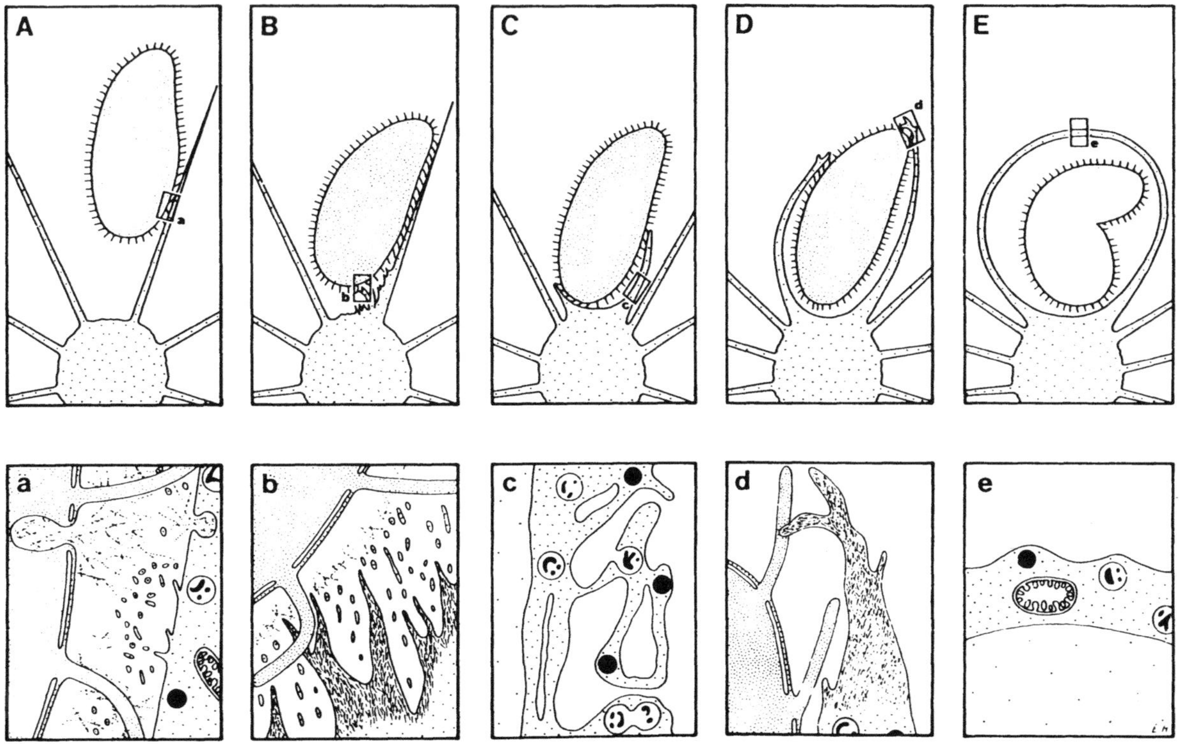

Figure 8. Schematic representation of food uptake in *A. sol* on the light microscopical (A–E) and electron microscopical level (**a–e**). For more details, see text. After Hausmann and Patterson (1982).

Ultrastructural analyses of this type of prey capture and food ingestion in actinophryid heliozoa revealed the following details (Fig. 8; Suzaki et al., 1980; Hausmann and Patterson, 1982; see also chapter by Suzaki and Shigenaka in this volume): The first contact of predator and prey leads to a rather stable connection between these two organisms. The association is mediated by the excretion of an obviously adhesive material of unknown chemical nature by the predator. Similar situations have been reported for other groups of actinopodial sarcodines (Bardele, 1976). Additionally, long narrow tubes are formed by the plasma membrane. These tubes make close contact with the prey's surface. Some axopods of the heliozoon might be retracted. Finally, fine pseudopods appear which entangle cellular appendages of the prey, for example, flagella or cilia. At this stage the connection between the predator and the prey is already extremely strong. An escape of the prey is almost impossible.

At the same moment that the prey is arrested, the growth of a cup-shaped feeding pseudopod commences. The plasma membrane of this pseudopod is provided by granule-like extrusomes of 1 μm diameter. They are located immediately beneath the plasma membrane in great numbers. These granules, which become concentrated close to the region of the continuously growing food vacuole, are highly osmiophilic structures in their condensed (resting) stage. During pseudopod formation they dilate considerably and fuse with each other and finally with the plasma membrane. By this, large areas of membrane are provided (Fig. 7). The presumed

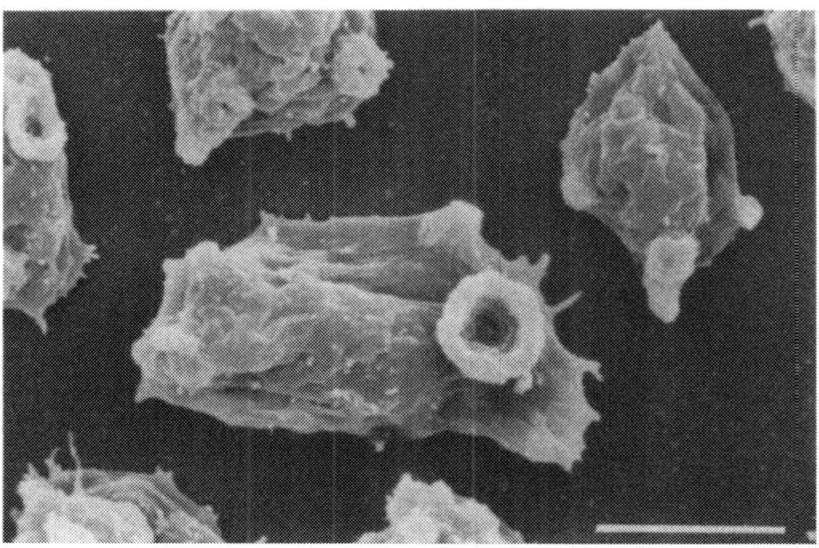

Figure 9. Formation of amoebostomes in *Naegleria fowleri*. From John et al. (1985) by permission of Allen Press, Lawrence. Bar: 10 μm.

lipid nature of these granules has been demonstrated using the freeze-fracture technique (Mulisch and Hausmann, 1991).

Calculations reveal that about 8000 μm^2 of membrane are necessary to form a food vacuole containing, for instance, one *Colpidium* cell. Under certain instances more might be needed. About 2000 μm^2 of membrane might become available from the plasma membrane as a result of the retraction of some axopods. As over 10000 of the dense granules are used up in the process of food uptake (Patterson and Hausmann, 1981), an adequate quantity of membrane can be supplied.

Although in sarcodines no permanent oral apparatuses for food ingestion are known, an at least transitional, obviously circular ingestion area arises. This is indicated by the round holes in the cell walls of food algae created by the vampyrellids (Fig. 5), and by the more or less round cup-shaped feeding pseudopods of heliozoa (Fig. 7). Moreover, circular feeding structures, called amoebostomes (Fig. 9), can be induced at least experimentally in certain schizopyrenid sarcodines, which normally feed on small bacteria (John et al., 1985). The origin of these structures and their biological significance are so far obscure.

IV. APICOMPLEXA

Only little is known about the mechanisms of uptake of external material in Apicomplexa. Micropores (ultracytostomes) are the typical sites of endocytosis (Fig. 10). The inner pellicular complex is interrupted at this site of the pellicle so that the plasma membrane can be invaginated. The inner pellicular structures turn inward for a short distance, thus forming a cylindrical structure at the neck of the invagination. Even in a single sporozoan genus the diameter of the invaginations varies from 50 to 200 nm (Scholtyseck and Mehlhorn, 1970) and the depth ranges from 50 to 350 nm (Scholtyseck, 1973). A large and deep opening is interpreted as an indication of endocytotic activity. Material enters the cavity, while the micropore deepens. Finally, a vesicle is pinched off. Alternatively, ingesting micropores may increase in length and turn into long tubes in *Besnoitia* (Sheffield, 1967; Sénaud et al., 1976). Pinocytotic vesicles deriving from such tubes have been described in *Plasmodium chabaudi* and *P. berghei* (Slomianny et al., 1983, 1985). The intraerythrocytic stages of *P. chabaudi* and *P. berghei* take up hemoglobin by their micropores. Two membranes lying closely together border the micropores. One of them is the membrane of the parasitophorous vacuole which enters the pore. When the invagination reaches a certain size, a food vacuole bordered by two membranes pinches off. Slomianny et al. (1983) showed that later on the inner (parasitophorous) vacuole membrane disappears in *P. chabaudi*. No information is available on the origin of the enlarging micropore membrane and the driving force for invagination and the pinching off of vesicles. For endocytotic processes in Apicomplexa, see also chapter by Duszenko and Seyfang in this volume.

Gregarines and some developmental stages of coccidia and piroplasmia do not possess typical micropores. Nevertheless, endocytotic processes are proposed, in

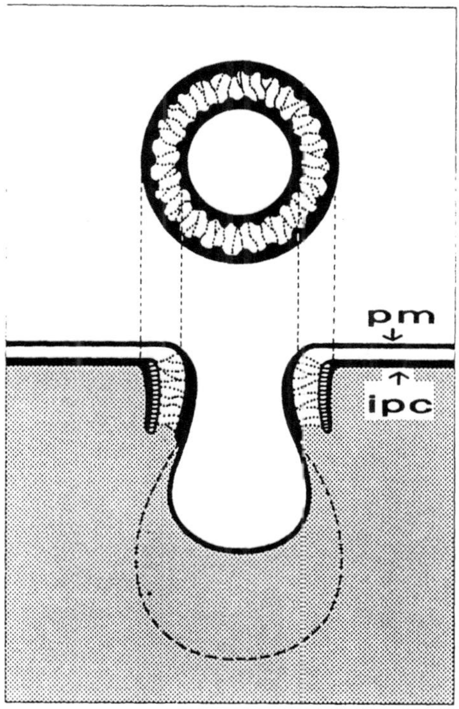

Figure 10. Diagrammatic representation of a sporozoan micropore in longitudinal and tangential sections. The broken line depicts the possible enlargement of the cytostomal invagination. ipc = inner pellicular membrane complex, pm = plasma membrane. After Mehlhorn et al. (1988).

addition to osmotrophic nutrition. The mode of feeding may change in the environment of different hosts and depending on the developmental stage of the parasite (Rudzinska et al., 1984). Based on electron micrographs Vivier (1968) postulated that gregarines form endocytotic vesicles by invagination of the three pellicular membranes. Later on, the inner two membranes disappear. Intraerythrocytic forms of *Babesia microti* pinocytose via narrow channels emerging from the entire cortex, as visualized by added ferritin (Langreth, 1976).

Young stages of the malaria parasite *Plasmodium* form micropinocytotic vesicles at the periphery of the cell body. Older stages, however, ingest host cell cytoplasm by formation of deep tubes which can break up into vesicles or pinch off vesicles from their terminal portion (Slomianny, 1990).

V. CILIOPHORA

Ciliophora internalize food using the most complex oral apparatuses known in protists (Figs. 11–13). Frequently, special devices for food capture are developed

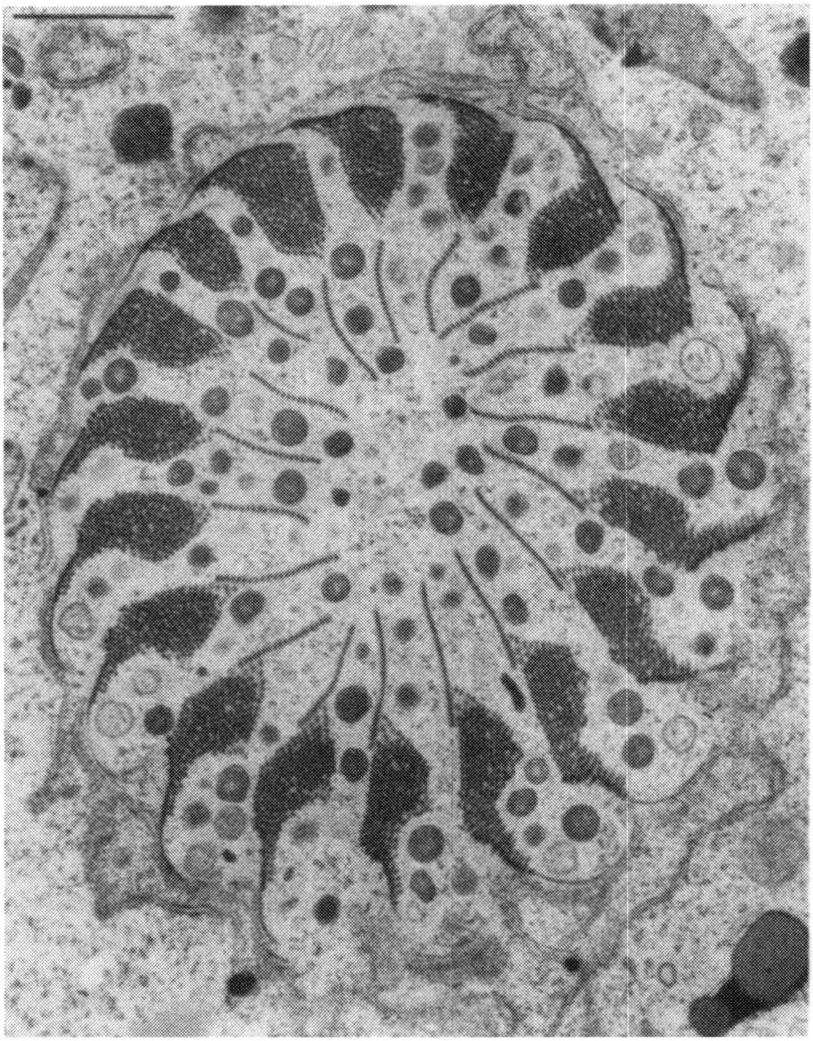

Figure 11. Cross section through the cytopharyngeal basket of the ciliate *Pseudomicrothorax dubius.* K. Hausmann, unpublished micrograph. For more details, see text. Bar: 1 μm.

in the vicinity or even inside the oral apparatus, for example, special sets of cilia for filter feeding or different types of extrusomes for prey capture (Hausmann, 1978). In few cases a special ritual has been detected before food uptake starts (Hausmann and Eisenbarth, 1977; Peck, 1985; Peck and Duborgel, 1985). The observations indicate that chemical recognition is involved in the initiation of the process of phagocytosis.

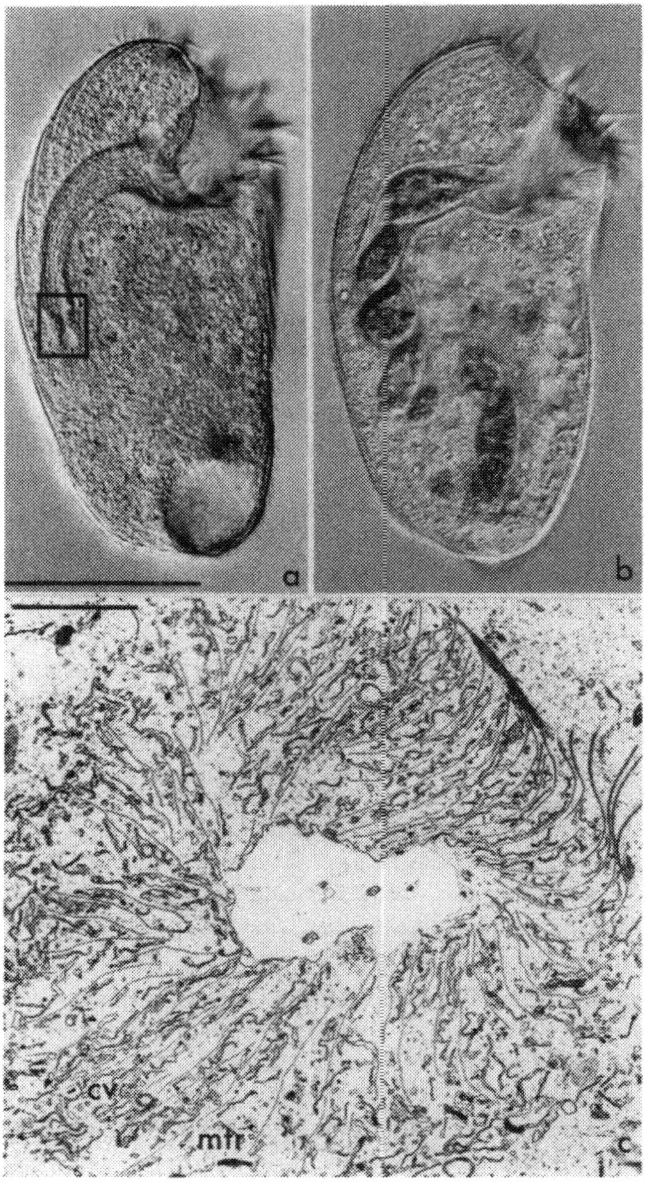

Figure 12. The ciliate *Climacostomum virens* (**a**) ingests unicellular algae via its feeding tube (**b**). A cross section of the area of food vacuole growth (**c**) reveals numerous microtubular ribbons (mtr) and cytostomal vesicles (cv). The rectangle in (a) depicts the part of the feeding tube which is shown schematically in Fig. 13. From Fischer-Defoy and Hausmann (1981) by permission of Springer Verlag (Heidelberg). Bars: 50 μm (a, b), 2.5 μm (c).

It would be far too excessive to review the different structural and functional variations of the oral apparatuses of ciliated protozoa. Consequently it should only be generally stated that the formation of food vacuoles occurs almost exclusively in the cytostomal region of the oral apparatus. This region of the plasma membrane is deprived of underlaying cortex structures such as membranous alveoli, microtubules, bundles of microfilaments, and the root elements associated with the basal bodies of the cilia (= infraciliature).

As has been demonstrated by ultrastructural investigations of a great variety of ciliates, in general a nascent food vacuole gains its membrane from preexisting membrane stores by the fusion of small vesicles of different sizes and shapes (Fig. 12; for literature see Fischer-Defoy and Hausmann, 1981). So far the only exception to this general path has been found in the marine ciliate genus *Kentrophoros* (Raikov, 1971, 1974; Fenchel and Finlay, 1989). The ciliates carry a dense layer of sulfur bacteria on their dorsal surface which accounts for about half the biomass of the symbiotic consortium. It has been demonstrated that the ciliates phagocytose the symbionts by invagination of the dorsal plasma membrane, which is reported to be free of cortical structures, especially free of alveoli. Owing to problems with a proper fixation for electron microscopy of this marine organism the latter statement needs further investigations.

No protein biosynthesis is required during the actual food vacuole growth (Weidenbach and Thompson, 1974; Ricketts and Rappitt, 1975; Fischer-Defoy and Hausmann, 1981; Eichenlaub-Ritter and Ruthmann, 1983). The vesicles are synthetized continuously in all parts of the cell and guided to the cytostomal region by ribbons of microtubules (Figs. 12, 13). No conclusive information is available on the driving forces for this transport. Similarly, the mechanism which is responsible for pinching off food vacuoles has not yet been revealed. Of course, for both requirements motor proteins such as dynein, kinesin, and actomyosin are believed to be potential candidates. However, these proposals are still a matter of speculation.

The speed by which membrane for the growth of food vacuoles is provided via the fusion of vesicles with the cytostomal membrane differs in the diverse groups of ciliates. The mode of food uptake differs as well. Rough estimates show that, for instance, in the filter feeder *Paramecium*, which feeds by concentrating mainly bacteria and small unicellular algae from the surrounding medium and ingesting them, 15 μm^2 of membrane per second are provided. In the gulper *Chilodonella* feeding on pennate diatoms, maximally 300 μm^2 of membrane are needed per second. In *Pseudomicrothorax*, which is highly specialized on a rapid ingestion of filamentous blue-green algae of the genus *Oscillatoria*, the values are also in the range of 300 μm^2 membrane per second.

Concerning the content of the food vacuole forming vesicles, at least two alternatives are known. In *Paramecium*, for instance, these vesicles are reported to be free of any special content such as digestive enzymes (see chapters by Allen and Fok as well as by Fok and Allen in this volume). In other ciliates, however, it has been demonstrated by cytochemistry that at least a certain population of these vesicles is filled with digestive enzymes. The enzymes are already secreted into

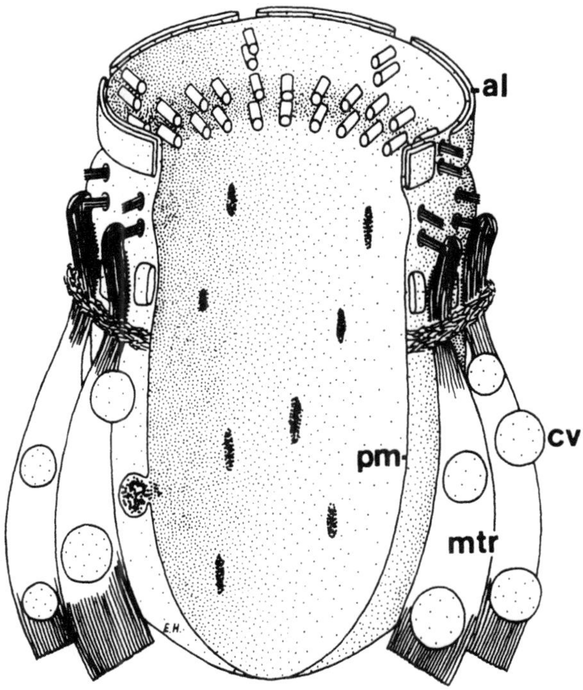

Figure 13. At the end of the feeding tube of *Climacostomum virens* the plasmalemma is free of underlying alveoli (al) and infraciliary structures. Alongside microtubular ribbons (mtr) cytostomal vesicles (cv) are guided to the plasma membrane (pm), where fusion occurs. After Fischer-Defoy and Hausmann (1981).

the nascent food vacuole by the fusion of the concerned vesicles (Peck and Hausmann, 1980). In *Pseudomicrothorax dubius*, which exclusively feeds on filamentous blue-green algae, this first set of digestive enzymes is obviously responsible for the extremely rapid degradation of the algal cell walls, lasting only some 10 sec. In certain other ciliates it has been observed ultrastructurally that some distinct material is released into the growing food vacuole (Fischer-Defoy and Hausmann, 1981; Mulisch et al., 1981; Mulisch and Hausmann, 1988). The chemical nature of this material is so far unknown.

There exists only rare information about the forces responsible for internalization of food vacuoles in ciliates. Microfilaments as well as microtubules seem to be important prerequisites and some structural evidence exists that microtubular arms might be involved in this process (Tucker, 1972, 1978; Bardele, 1974; Hausmann and Peck, 1979). In *P. dubius* ATPase activity is associated with the arm-bearing microtubules (Hauser and Hausmann, 1982). Using immunocytochemistry actin was found to be a substantial component of the feeding apparatus of *P. dubius* (Hauser et al., 1980). However, until now, no convincing experimentally proved model exists that explains the incorporation of food vacuoles against the internal pressure of the cell.

As in flagellates and sarcodines, pinocytosis also occurs in ciliates. It takes place at the parasomal sacs. These are deep invaginations of the plasma membrane that are located close to the basal bodies of cilia, where the alveolar system is interrupted. They remind us to some extent of the apicomplexan micropores (see Section IV). A chapter by R. D. Prusch in this volume deals with the endocytotic activity of these plasma membrane domains of ciliates in detail.

VI. MYXOZOA

Different modes of nutrition are reported for Myxozoa. Pinocytosis has been detected in various histozoic species, for example, *Henneguya psorospermica, Myxobolus* sp. (Lom and Puytorac, 1965), *Myxosoma funduli* (Current et al., 1979), and *Henneguya adiposa* (Current, 1979). Numerous plasmalemmal invaginations perpendicular to the parasites' surface were seen to pinch off vesicles, internalizing material from the intercellular (host–parasite) space. Fibrils may be associated with the pinocytotic canals (Lom and Puytorac, 1965). Coelozoic myxosporidia may form pinocytotic vesicles at the base of their microvilli which enlarge the absorbing area (Uspenskaya, 1982).

Phagocytosis is another common process of food internalization (Uspenskaya, 1984). *Myxosoma cerebralis*, for example, phagocytoses cartilage cells, *Kudoa quadratum* takes up muscle protofibrils (Uspenskaya, 1979, 1982). *Hoferellus gilsoni* has evolved a type of phagocytosis that is rather unique among coelozoic Myxozoa. It attaches to the surface of urinary tract epithelial cells and draws evaginations of the host cell into invaginations of the own cell (Lom et al., 1986). These invaginations are eventually pinched off and become food vacuoles.

In general, there is no explanation for the origin of the endosomal membranes in Myxozoa. Similarly, nothing is known about the force generation for endocytosis.

VII. CONCLUSIONS AND OUTLOOK

Internalization of dissolved and particulate nutrients occurs in protozoa mainly by endocytosis, that is, by the formation of endosomes. There exists a great variety of feeding devices, and different modes of food vacuole formation are reported. On the other hand there is one basic similarity insofar as phagosomes are generated by preexisting membrane stores such as plasmalemma, cytostomal vesicles, myelinic bodies, and lipid granules. In many cases this membrane material is guided to the region of the nascent food vacuole, where it is incorporated simply by fusion or after a preceding transformation. Protein biosynthesis is not required during the actual process of phagosome formation.

Further investigations should concentrate on still open questions such as: What are the recognition processes which lead to the initiation of food vacuolar growth? By what mechanism are the membrane stores selected and guided to the area of phagosome formation? How can the rapid transformations of, for instance, lipid granules or myelinic bodies be explained? To what extent and purpose are enzymes secreted into the nascent food vacuole? What are the driving forces for translocation of food vacuoles into the cytoplasm?

REFERENCES

Andersen, R. A. (1990). The three-dimensional structure of the flagellar apparatus of *Chrysosphaerella brevispina* (Chrysophyceae) as viewed by high voltage electron microscopy stereo pairs. *Phycologia* 29: 86–97.

Andersen, R. A. and Wetherbee, R. (1992). Microtubules of the flagellar apparatus are active during prey capture in the chrysophycean alga *Epipyxis pulchra*. *Protoplasma* 166: 8–20.

Bardele, C. F. (1974). Transport of materials in the suctorian tentacle. In: *Transport at the Cellular Level*, Society of Experimental Biology Symposium, Vol. 28 (Sleigh, M. A. and Jennings, D. H., eds.), pp. 191–208, Cambridge University Press, Cambridge.

Bardele, C. F. (1976). Particle movement in heliozoan axopods associated with lateral displacement of highly ordered membrane domains. *Z. Naturforsch.* 31c: 190–194.

Brugerolle, G., Gobert, J. G., and Savel, J. (1974). Étude ultrastructurale des lésions viscérales provoquées par l'injection intra-péritonéale de *Trichomonas vaginalis* chez la souris *Ann. Parasitol.* 49: 301–318.

Burzell, L. A. (1973). Observations on the proboscis–cytopharynx complex and flagella of *Rhynchomonas metabolita* Pshenin, 1964 (Zoomastigophorea: Bodonidae). *J. Protozool.* 20: 385–393.

Chapman-Andresen, C. (1962). Studies on pinocytosis in amoeba. *Trav. Lab. Carlsberg* 33: 73–264.

Chapman-Andresen, C. (1973). Endocytic processes. In: *The Biology of Amoeba* (Jeon, K. W., ed.), pp. 319–348, Academic Press, New York.

Cole, G. T. and Wynne, M. J. (1974). Endocytosis of *Microcystis aeruginosa* by *Ochromonas danica*. *J. Phycol.* 10: 397–410.

Coppens, I., Opperdoes, F. R., Courtoy, P. J., and Baudhuin, P. (1987). Receptor-mediated endocytosis in the bloodstream form of *Trypanosoma brucei*. *J. Protozool.* 34: 465–473.

Current, W. L. (1979). *Henneguya adiposa* Minchew (Myxosporidia) in the channel catfish: ultrastructure of the plasmodium wall and sporogenesis. *J. Protozool.* 26: 209–217.

Current, W. L., Janovy, J., and Knight, S. A. (1979). *Myxosoma funduli* in *Fundulus kansae*: ultrastructure of plasmodium wall and sporogenesis. *J. Protozool.* 26: 574–583.

Eichenlaub-Ritter, U. and Ruthmann, A. (1983). The oral apparatus of the ciliate *Nyctotherus ovalis* Leidy: possible involvement of microtubules in the structural support of a cortical domain and the translocation of vesicles. *Differentiation* 24: 97–110.

Elbrächter, M. (1991). Food uptake mechanisms in phagotrophic dinoflagellates and classification. In: *The Biology of Free-living Heterotrophic Flagellates*. The Systematics Association, Special Vol. 45 (Patterson, D. J. and Larsen, J., eds.), pp. 303–312, Clarendon Press, Oxford.

Eyden, B. P. (1977). Morphology and ultrastructure of *Bodo designis* Skuja 1948. *Protistologica* 13: 169–179.

Fenchel, T. (1982). Ecology of heterotrophic microflagellates. I. Some important forms and their functional morphology. *Mar. Ecol. Prog. Ser.* 8: 211–223.

Fenchel, T. and Finlay, B. J. (1989). *Kentrophoros*: a mouthless ciliate with a symbiotic kitchen garden. *Ophelia* 30: 75–93.

Fischer-Defoy, D. and Hausmann, K. (1981). Microtubules, microfilaments, and membranes in phagocytosis: structure and function of the oral apparatus of the ciliate *Climacostomum virens*. *Differentiation* 20: 141–151.
Francioli, P., Shio, H., Roberts, R. B., and Müller, M. (1983). Phagocytosis and killing of *Neisseria gonorrhoeae* by *Trichomonas vaginalis*. *J. Infect. Dis.* 147: 87–94.
Hauser, M. and Hausmann, K. (1982). Electron microscopic localization of ATPase activity in the cytopharyngeal basket of the ciliate *Pseudomicrothorax dubius*. *Differentiation* 22: 67–72.
Hauser, M., Hausmann, K., and Jockusch, B. M. (1980). Demonstration of tubulin, actin and alpha-actinin by immunofluorescence in the microtuble–microfilament complex of the cytopharyngeal basket of the ciliate *Pseudomicrothorax dubius*. *Exp. Cell Res.* 125: 265–274.
Hausmann, K. (1978). Extrusive organelles in protists. *Int. Rev. Cytol.* 53: 197–276.
Hausmann, K. and Eisenbarth, A. (1977). Phagocytoseverhalten von *Pseudomicrothorax dubius* unter Normalbedingungen und nach Applikation von Colchicin sowie Cytochalasin B. *Protoplasma* 92: 269–279.
Hausmann, K. and Patterson, D. J. (1982). Pseudopod formation and membrane production during prey capture by a heliozoon: feeding by *Actinophrys*, II. *Cell Motil.* 2: 9–24.
Hausmann, K. and Peck, R. K. (1979). The mode of function of the cytopharyngeal basket of the ciliate *Pseudomicrothorax dubius*. *Differentiation* 14: 147–158.
Hitchen, E. T. (1974). The fine structure of the colonial kinetoplastid flagellate *Cephalothamnium cyclopum* Stein. *J. Protozool.* 21: 221–231.
Hülsmann, N. (1985). Entwicklung und Ernährungsweise von *Vampyrella lateritia* (Rhizopoda). Publ. Wiss. Film. (Göttingen), Sekt. Biol., Ser. 17, 16: 1–23.
Jacobson, D. M. and Anderson, D. M. (1986). Thecate heterotrophic dinoflagellates: feeding behavior and mechanisms. *J. Phycol.* 22: 249–258.
John, D. T., Cole, T. B.,and Bruner, R. A. (1985). Amebostomes of *Naegleria fowleri*. *J. Protozool.* 32: 12–19.
Karpov, S. A. and Makarenkova, E. P. (1989). Phagocytosis and endocytosis in a colorless flagellate *Thaumatomonas lauterborni* (in Russian). *Tsitologiya* 31: 347–349.
Kivic, P. A. and Vesk, M. (1974). Pinocytotic uptake of protein from the reservoir in *Euglena*. *Arch. Microbiol.* 96: 155–159.
Lackey, J. B. (1929). Studies in the life histories of Euglenida. I. The cytology of *Entosiphon sulcatum* (Duj.) Stein. *Arch. Protistenk.* 66: 176–200.
Langreth, S. G. (1976). Feeding mechanisms in extracellular *Babesia microti* and *Plasmodium lophurae*. *J. Protozool.* 23: 215–223.
Laval, M. (1971). Ultrastructure et mode de nutrition du choanoflagellé *Salpingoeca pelagica*, sp. nov. Comparaison avec les choanocytes des spongiaires. *Protistologica* 7: 325–336.
Lavette, A. (1966). Le problème du vacuome chez quelques flagellés termiticole et la prise du bois par ces derniers. *C. R. Acad. Sci. Paris* 263 D: 1739–1742.
Lavette, A. (1975). Nouvelles recherches sur l'ultrastructure et la biologie de quelques flagellés symbiotiques de termites. *Ann. Sci. Nat., Zool.* 12e Sér., 17: 179–214.
Leadbeater, B. S. C. (1983). Distribution and chemistry of microfilaments in choanoflagellates, with special reference to the collar and other tentacle systems. *Protistologica* 19: 157–166.
Lom, J. and Puytorac, P. (1965). Observations sur l'ultrastructure des trophozoites de myxosporidies. *C. R. Acad. Sci. Paris* 260 D: 2588–2590.
Lom, J., Molnár, K., and Dyková, I. (1986). *Hoferellus gilsoni* (Debaisieux, 1925) comb. n. (Myxozoa, Myxosporea): redescription and mode of attachment to the epithelium of the urinary bladder of its host, the european eel. *Protistologica* 22: 405–413.
Mehlhorn, H., Dubremetz, J. F., Franz, M., Gustafsson, M., Peters, W., Taraschewski, H., Walldorf, V., and Voigt, W. P. (1988). Morphology, Protozoa. In: *Parasitology in Focus* (Mehlhorn, H., ed.), pp. 161–196, Springer, Berlin, Heidelberg.
Mignot, J. P. and Brugerolle, G. (1975). Étude ultrastructurale du flagellé phagotrophe *Colponema loxodes* Stein. *Protistologica* 11: 429–444.

Moestrup, O. and Andersen, R. A. (1991). Organization of heterotrophic heterokonts. In: *The Biology of Free-living Heterotrophic Flagellates.* The Systematics Association, Special Vol. 45. (Patterson, D. J. and Larsen, J., eds.), pp. 333–360, Clarendon Press, Oxford.

Mulisch, M. and Hausmann, K. (1988). Transmission electron microscopical observations on stomatogenesis during metamorphosis of *Eufolliculina uhligi* (Ciliophora: Heterotrichida). *J. Protozool.* 35: 450–458.

Mulisch, M. and Hausmann, K. (1991). Rapid lipid intussusception into the plasma membrane of the heliozoon *Actinophrys sol* during phagocytosis. *Naturwissenschaften* 78: 232–233.

Mulisch, M., Barthlott, W., and Hausmann, K. (1981). Struktur und Ultrastruktur von *Eufolliculina* spec. Schwärmer und sessiles Stadium. *Protistologica* 17: 285–312.

Münch, R. (1970). Endocytosenachweis bei *Opalina ranarum. Cytobiologie* 2: 108–122.

Nawata, T. and Sibaoka, T. (1983). Experimental induction of feeding behavior in *Noctiluca miliaris. Protoplasma* 115: 34–42.

Nawata, T. and Sibaoka, T. (1987). Local ion currents controlling the localized cytoplasmic movement associated with feeding initiation of *Noctiluca miliaris. Protoplasma* 137: 125–133.

Nisbet, B. (1974). An ultrastructural study of the feeding apparatus of *Peranema trichophorum. J. Protozool.* 21: 39–48.

Page, F. C. and Siemensma, F. J. (1991). Nackte Rhizopoda und Heliozoa. Protozoenfauna (Matthes, D., Hrsg.), Band 2. Gustav Fischer Verlag, Stuttgart.

Patterson, D. J. and Hausmann, K. (1981). Feeding by *Actinophrys sol* (Protista, Heliozoa): 1. Light microscopy. *Microbios* 31: 39–55.

Peck, R. K. (1985). Feeding behavior in the ciliate *Pseudomicrothorax dubius* is a series of morphologically and physiologically distinct events. *J. Protozool.* 33: 492–501.

Peck, R. K. and Duborgel, F. (1985). Effects of cations on phagocytosis in the ciliate *Pseudomicrothorax dubius. J. Protozool.* 32: 501–508.

Peck, R. K. and Hausmann, K. (1980). Primary lysosomes of the ciliate *Pseudomicrothorax dubius:* cytochemical identification and role in phagocytosis. *J. Protozool.* 27: 401–409.

Popovsky, J. (1982). Another case of phagotrophy by *Gymnodinium helveticum* Penard f. *achroum* Skuja. *Arch. Protistenk.* 125: 73–78.

Pringsheim, E. G. (1963). Farblose Algen. Gustav Fischer Verlag, Stuttgart.

Radek, R., Hausmann, K., and Breunig, A. (1992). Ectobiotic and endocytobiotic bacteria associated with the termite flagellate *Joenia annectens. Acta Protozoologica* 31: 93–107.

Raikov, I. B. (1971). Bactéries épizoiques et mode de nutrition du cilié psammophile *Kentrophoros fistulosum* Fauré-Fremiet (étude au microscope électronique). *Protistologica* 7: 365–378.

Raikov, I. B. (1974). Étude ultrastructurale des bactéries épizoiques et endozoiques de *Kentrophoros latum* Raikov, cilié holotriche mésopsammique. *Cah. Biol. Mar.* 15: 379–393.

Ricketts, T. R. and Rappitt, A. F. (1975). The effect of puromycin and cycloheximide on vacuole formation and exocytosis in *Tetrahymena pyriformis* G1-9. *Arch. Microbiol.* 102: 1–8.

Rudzinska, M., Spielman, A., Lewengrub, S., Piesman, J., and Karakashian, S. (1984). The sequence of developmental events of *Babesia microti* in the gut of *Ixodes dammini. Protistologica* 20: 649–663.

Schnepf, E. and Deichgräber, G. (1984). "Myzocytosis," a kind of endocytosis with implications to compartmentation in endosymbiosis: observations in *Paulsenella* (Dinophyta). *Naturwissenschaften* 71: 218–219.

Schnepf, E. and Elbrächter, M. (1992). Nutritional strategies in dinoflagellates: a review with emphasis on cell biological aspects. *Eur. J. Protistol.* 28: 3–24.

Schnepf, E., Deichgräber, G., and Drebes, G. (1985). Food uptake and the fine structure of the dinophyte *Paulsenella* sp., an ectoparasite of marine diatoms. *Protoplasma* 124: 188–204.

Schnepf, E., Drebes, G., and Elbrächter, M. (1990). *Pirsonia guinardiae*, gen. et spec. nov: a parasitic flagellate in the marine diatom *Guinardia flaccida* with an unusual mode of food uptake. *Helgol. Meeresunters.* 44: 275–293.

Scholtyseck, E. (1973). Ultrastructure. In: *The Coccidia: Eimeria, Isospora, Toxoplasma, and Related Genera* (Hammond, D. M. and Long, P. L., eds.), pp. 81–144, University Park Press, Baltimore.

Scholtyseck, E. and Mehlhorn, H. (1970). Ultrastructural study of characteristic organelles (paired organelles, micronemes, micropores) of Sporozoa and related organisms. *Z. Parasitenk.* 34: 97–127.

Sénaud, J., Chobotar, B., and Scholtyseck, E. (1976). Role of the micropore of the Sporozoa: ultrastructural study of *Plasmodium cathemerium, Eimeria ferrisi, E. stidai, Besnoitia jellisoni,* and *Frenkelia* sp. *Tropenmed. Parasitol.* 27: 145–159.

Shapiro, S. Z. and Webster, P. (1989). Coated vesicles from the protozoan parasite *Trypanosoma brucei*: Purification and characterization. *J. Protozool.* 36: 344–349.

Sheffield, H. G. (1967). The function of the micropyle in the cyst organism of *Besnoitia jellisoni. J. Parasitol.* 53: 888–889.

Slomianny, C. (1990). Three-dimensional reconstruction of the feeding process of the malaria parasite. *Blood Cells* 16: 369–378.

Slomianny, C., Charet, P., and Prensier, G. (1983). Ultrastructural localization of enzymes involved in the feeding process on *Plasmodium chabaudi* and *Babesia hylomysci. J. Protozool.* 30: 376–382.

Slomianny, C., Prensier, G., and Charet, P. (1985). Étude ultrastructurale comparée du processus de dégradation de l'hémoglobine par *P. berghei* (Vincke et Lips, 1948) en fonction de l'état de maturité de la cellule hôte. *J. Protozool.* 32: 1–5.

Stockem, W. and Klein, H. P. (1988). Pinocytosis and locomotion of amoebae. XVII. Influence of different cations on induced pincytosis in *Amoeba proteus. Eur. J. Protistol.* 23: 317–326.

Stockem, W. and Wohlfarth-Bottermann, K. E. (1969). Pinocytosis (endocytosis). In: *Handbook of Molecular Cytology* (Lima de Faria, A., ed.), pp. 1373–1400, North-Holland, Amsterdam.

Suzaki, T., Shigenaka, V., Watanabe, S., and Toyohara, A. (1980). Food capture and ingestion in the large heliozoa, *Echinosphaerium nucleofilum. J. Cell Sci.* 42: 61–79.

Swale, E. M. F. (1973). A study of the colourless flagellate *Rhynchomonas nasuta* (Stokes) Klebs. *Biol. J. Linn. Soc.* 5: 255–264.

Triemer, R. E. and Fritz, L. (1987). Structure and operation of the feeding apparatus in a colorless euglenoid, *Entosiphon sulcatum. J. Protozool.* 34: 39–47.

Tucker, J. B. (1972). Microtubule-arms and propulsion of food particles inside a large feeding organelle in the ciliate *Phascolodon vorticella. J. Cell Sci.* 10: 883–903.

Tucker, J. B. (1978). Endocytosis and streaming of highly gelated cytoplasm alongside rows of arm-bearing microtubules in the ciliate *Nassula. J. Cell Sci.* 29: 213–232.

Uspenskaya, A. V. (1979). The mode of nutrition of vegetative stages of *Myxosoma cerebralis* Hofer 1903 (Plehn, 1905) (Myxosporidia). *Parasitologiya* (in Russian) 13: 169–171.

Uspenskaya, A. V. (1982). New data on the life cycle and biology of myxosporidia. *Arch. Protistenk.* 126: 309–338.

Uspenskaya, A. V. (1984). *The Cytology of Myxosporidia* (in Russian). Nauka Publishing House, Leningrad.

Vivier, E. (1968). L'organisation ultrastructurale corticale de la grégarine *Lecudina pellucida*: ses rapports avec l'alimentation et la locomotion. *J. Protozool.* 15: 230–246.

Weidenbach, A. L. S. and Thompson, G. A. (1974). Studies of membrane formation in *Tetrahymena pyriformis*. VIII. On the origin of membranes surrounding food vacuoles. *J. Protozool.* 21: 745–751.

Wetherbee, R. and Andersen, R. A. (1992). Flagella of a chrysophycean alga play an active role in prey capture and selection: direct observation on *Epipyxis pulchra* using enhanced video microscopy. *Protoplasma* 166: 1–7.

Willey, R. L., Walne, P. L., and Kivic, P. (1988). Phagotrophy and the origins of the euglenoid flagellates. *CRC Crit. Rev. Plant Sci.* 7: 303–340.

Yoshikawa, H., Furuki, J., Takahashi, Y., Morioka, H., and Yoshida, Y. (1990). Freeze-fracture study of the bloodstream form of *Trypanosoma brucei gambiense. J. Protozool.* 37: 27–32.

ENDOSOMAL MEMBRANE TRAFFIC OF CILIATES

Richard D. Allen and Agnes K. Fok

I. INTRODUCTION

Endocytosis in ciliates has been generally neglected as an area of investigation, as only a few papers have been devoted to this subject in the past 25 years. Most of our current understanding of the endocytic process in a protozoan comes from

Advances in Cell and Molecular Biology of Membranes, Volume 2B
Membrane Traffic in Protozoa, pages 283–309

ISBN: 1-55938-628-2

studies of pinocytosis in amoeba, both from the earlier classical studies (reviewed in Chapman-Andresen, 1962; Stockem and Wohlfarth-Bottermann, 1969; Stockem, 1977) and more recent studies such as those from Stockem's and Prusch's laboratories (see corresponding chapters in this volume) and of Grębecki (1991). The extensive studies on mammalian cells, both tissue culture cells and cultured peritoneal macrophages (reviewed by Mellman et al., 1987; Courtoy, 1989; Gruenberg and Howell, 1989; Van Deurs et al., 1989), have provided a comparative basis for our understanding of the endocytic processes in ciliates. The lack of endocytic studies using ciliates probably stems from a paucity of information about what ciliates may need to acquire or concentrate from their environment either by way of receptor-mediated endocytosis or by some other less specific means. This lack may also be related to the problems of visualizing this process either at the light or electron-microscopic levels. On the other hand, since only limited areas on the ciliates' surface are potential sites for endocytic uptake as most of the surface is covered internally by an extensive system of alveoli and underlying fibrous epiplasm, these cells may provide the best model system for certain types of studies on endocytosis.

As one of the most extensive endocytic studies in ciliates has recently been carried out on *Paramecium*, this review will feature information derived from this ciliate supplemented by additional information from other ciliates. At the end of the chapter we will compare the endocytic system of ciliates with that in mammalian cells, where the endocytic processes are more fully understood.

II. SITES OF ENDOCYTOSIS AT THE CELL SURFACE

The cell surface of ciliates is covered with a pellicle consisting of the outer plasma membrane, a mosaic of underlying interconnected flattened sacs, called the alveoli, and a subtending fibrous epiplasm. This system is interrupted only at the cytopharynx, the cytoproct, contractile vacuole pores, and along the junctions of the abutting units of alveolar membrane sacs. Cilia, extrusomes, and indentations of the plasma membrane penetrate the pellicle along these alveolar junctions or septa. Thus the actual surface sites where endocytosis can easily occur are fairly limited as well as highly predictable in location. This expedites the search for endocytic ports and makes it possible to identify them using various electron-microscopic techniques. In cells such as amoeba and those metazoans which lack such a complex surface architecture, it is much more difficult to pinpoint the precise location where endocytosis will likely occur until some visible change in the membrane topography or the attachment of a specialized armature (e.g., a clathrin cage) signals that endocytosis will follow at these sites.

Not all regions covered by a single membrane in ciliates are sites of endocytic uptake. Digestive-vacuole membrane retrieval does occur at the cytoproct (Allen and Wolf, 1974, 1979) and extrusome membrane at the extrusome fusion sites (Hausmann and Allen, 1976). However, the cell appears to process these two groups of retrieved membranes in ways differing from primary endocytic vesicles. Phagocytosis occurs at the cytopharynx or homologous single-membrane-limited oral regions but so far there is no strong evidence for submicron-size vesicles arising from this region of the cell. Thus only the punctate indentations of the plasma membrane, called parasomal sacs (refer to scheme at the end of this chapter), which are regularly associated with basal bodies, or pellicular pores which are not associated with basal bodies, are known to be potential ports of endocytic entry of all fluid phase and putative receptor-mediated endocytosed macromolecules in these cells. Such indentations have been shown by thin-section electron microscopy to be coated on their cytosolic sides (Fig. 1) (Allen, 1967; Noirot-Timothée, 1968; Franke, 1971). We now see, using deep-etch rotary-shadowed images of quick-frozen *Paramecium*, that this armature has a morphology identical to the clathrin-coated cages typical of mammalian cells (Allen et al., 1992). A polyhedral arrangement of macromolecular assemblies covers the cytosolic side of these indentations (Figs. 2, 3). The sizes of these coated pits vary but are around 200 nm in diameter. In *Paramecium multimicronucleatum* they average 188 nm in diameter at their bulbous ends and are connected to ~100 nm wide necks. The size of the coated pit varies in accordance with the time of coated vesicle release, being largest just before release and smallest immediately following release. The endocytic sites in many ciliates are permanent and do not disappear upon removal of the coated vesicles, as usually occurs in metazoan cells. The plasma membrane is continuously indented through the pellicle to the level of the epiplasm, as this is the level below which the coated vesicle enlarges and pinches from the pit.

It is appropriate to give these permanent necks leading to coated pits specialized names in ciliates, though one name would be adequate to cover the two apparently homologous structures. Pellicular pore is the more general term, as this term does not suggest that the pore or sac needs to be located next to a basal body. However, the term *parasomal sac* is in such wide use that it would be difficult to change the terminology at this date. Thus in this review we will use both terms as defined by Corliss (1979) and Corliss and Lom (1985). In addition, we will use *coated pit* to refer only to the coated sac that is continuous with the plasma membrane at the bottom of the pellicular pore or parasomal sac as well as coated indentations of the plasma membrane found in ciliates lacking permanent pores.

Parasomal sacs are usually found in regular association with the basal bodies in the somatic kinetids as well as the oral polykinetids of most ciliates. One characteristic of the subphylum Postciliodesmatophora (Small and Lynn, 1985) is the absence of parasomal sacs. Because such an absence would mean that these ciliates might be incapable of endocytosis, we have taken another look at the published literature on members of this group, *Blepharisma*, *Spirostomum*, *Gastrostyla*, *Climacostomum*, *Oxytricha*, and *Halteria*, and at our own electron micrographs of *Blepharisma* and *Halteria* (Table 1). In *Blepharisma* coated pits are found in abundance near the basal bodies of the kineties as pits of the plasma membrane (Fig. 4). In *Halteria* coated pits are found interspersed with the basal bodies of the

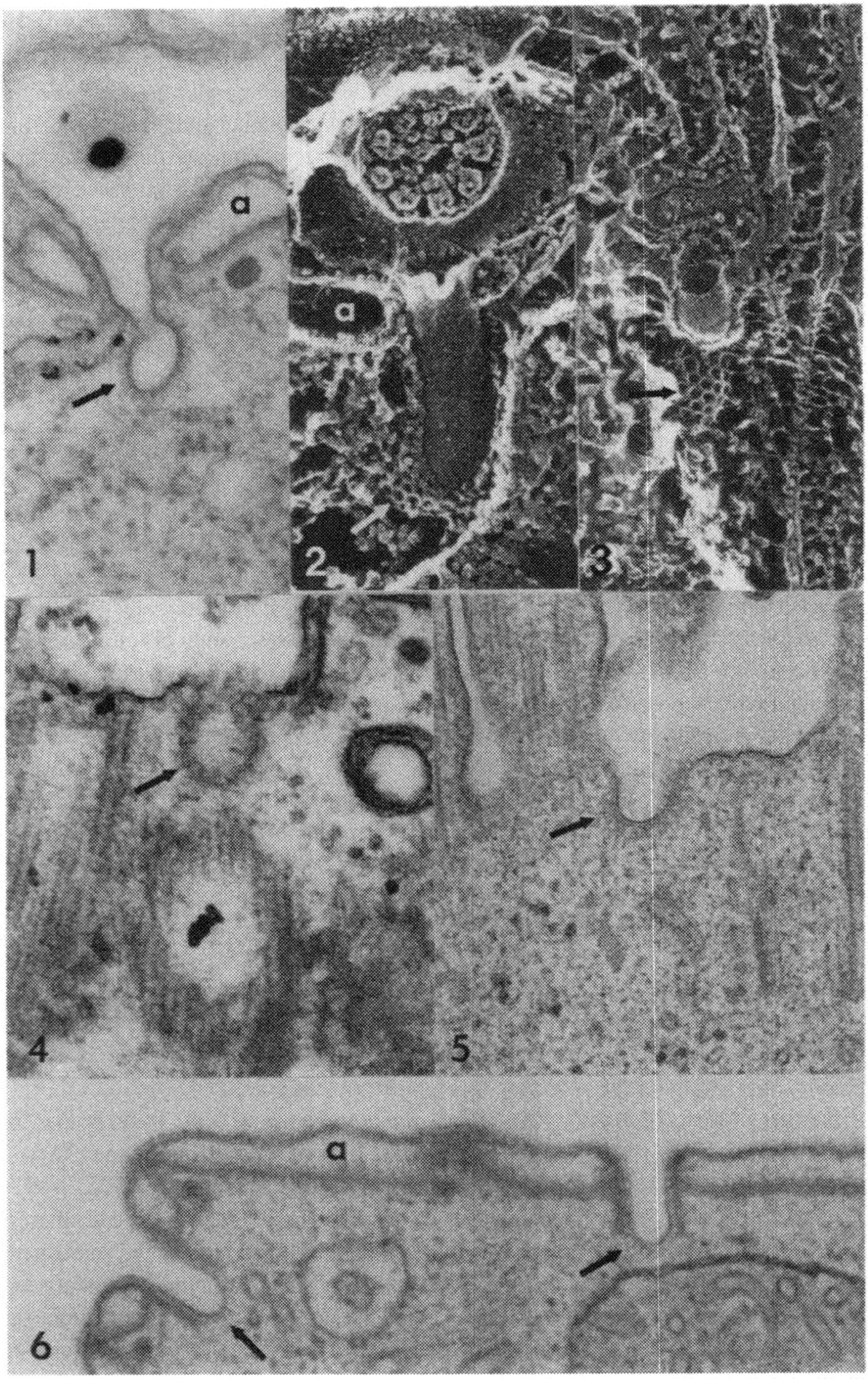
a
a
1
2
3
4
5
a
6

polykinetids. However, these coated pits are shallow and located at the cell surface; permanent pores are, indeed, found to be lacking (Fig. 5). There are no extensions of an alveolar system interspersed between or near the basal bodies of the kinetids and polykinetids of these ciliates as in ciliates of the other two subphyla, Rhabdophora and Cyrtophora. We conclude that coated pits are characteristic of the subphylum Postciliodesmatophora as in other ciliates, but that permanent pores are not present where an alveolar system is absent. However, the region immediately adjacent to basal bodies, even in this subphylum, appears to be the preferred and possibly exclusive site of coated pit formation. Functionally, it is the coated pit that plays the vital role in endocytosis and that occurs in all ciliates. The permanent pores are present only when there are alveolar sacs through which the plasma membrane must penetrate.

In nonciliated regions of ciliates such as the peritrichs, pellicular pores are found arranged in the grooves formed by the pellicular folds (Fig. 6) and encircling the scopula, where they may function primarily as sites for exocytosis of stalk material (Suchard and Goode, 1982). In *Tokophrya*, a suctorian, these pores are arranged over the feeding tentacles as well as the body (Rudzinska, 1980). The specific arrangements of the pores suggest a possible role for endocytosis in such cellular activities as myonemal contractility, ciliary beating, ciliary reversal, and/or chemotaxis. However, how endocytosis would regulate these activities remains only a matter of speculation.

We have estimated that *Paramecium multimicronucleatum*, which is ~200 μm long and 50 μm wide, has roughly 8000 pores on its somatic surface and another 500 concentrated in its buccal cavity. If each pore has an attached coated pit of 188 nm in diameter, this would account for a total coated pit surface area of over 900 μm^2 at the cell surface of one cell, which is presumably available for concentrating external macromolecules for endocytosis. Various markers have been shown to label these pits. These include thorotrast (Noirot-Timothée, 1968; Rudzinska,

---

Figures 1–6. Coated pits (arrows) in various ciliates. All figures are ×80,000. Figure **1**: A parasomal sac of *Paramecium multimicronucleatum* penetrates through the alveolus (a) and has a clathrin-like coat on the cytosolic surface of its coated pit. Figures **2**, **3**: Deep-etch rotary-shadowed images of *Paramecium multimicronucleatum* illustrate the polygonal arrangements of the elements that form the clathrin-like cage (arrows). The E-fracture face of the membrane of the parasomal sacs/coated pits is also partially exposed. A cross-fractured cilium is exposed in Fig. 2. a = alveolus. Figures **4**, **5**: Coated pits in *Blepharisma* (Fig. 4) and *Halteria* (Fig. 5); both are members of subphylum Postciliodesmatophora. Though permanent parasomal sacs are lacking in this subphylum, coated pits are found, as in other ciliates, next to basal bodies. Figure **6**: Pellicular pores penetrating the alveoli (a) in the pellicle of the peritrich *Opercularia coarctata*. Pellicular pores are homologous to parasomal sacs and bear coated pits (arrows). Figures 1–3 reproduced from Allen et al. (1992) by permission of The Company of Biologists (Cambridge, GB).

Table 1. Endosome-like Compartments in Ciliates

Ciliates[a]	Coated Pits[b]	Preendosomal Vesicles[b]	Early Endosomes[b]	References
Postciliodesmatophora				
Karyorelictea				
Loxodes striatus	14	8, 14	7, 14	Putyorac and Njine, 1970
Spirotrichea				
Blepharisma japonicum	5, 6	2	1a, 4c	Ishida et al., 1991
Blepharisma undulans	4, 5	5, 16	22	Kennedy, 1965
Climacostomun virens	19, 20	15, 18, 24	23	Peck et al., 1975
Condylostoma magnum	15, 18	7	13, 14, 22	Bohatier, 1979
Spirostomum ambiguum	1, 2, 4	3, 4	4	Daniel and Mattern, 1965
Spirostomum ambiguum	14	13, 14	13, 14	Finley et al., 1964
Stentor coeruleus	15, 19	14, 15	14, 19	Paulin and Bussey, 1971
Eufolliculina sp.	34	34, 35, 37		Mulisch et al., 1981
Eufolliculina sp.			19, 20	Mulisch and Hausmann, 1983
Halteria grandinella	+[c]	+	+	R.D. Allen,, unpublished observations
Gastrostyla steinii	19, 35	15	14, 15	Grim, 1972
Oxytricha fallax	2	9	7	Grimes, 1972
Rabdophora				
Prostomatea				
Bursellopsis spaniopogon	16, 17	17	25	Hiller, 1991
Coleps amphacanthus	13	15	14	Huttenlauch and Bardele, 1987
Prorodon aklitolophon	9	11	9	Hiller and Bardele, 1988
Litostomatea				
Didinium nasutum	11	11		Lipscomb and Riordan, 1992
Enchelydium polynucleatum	28	28	30	Foissner and Foissner, 1985
Bryophyllum tegularum	2	9	21	Bohatier et al., 1978
Spathidium spathula	13, 15	13, 18		Williams et al., 1981
Lepidotrachelophyllum fornicis	5	1, 7, 10, 17	5	Lynn and Nicholls, 1985
Epidinium caudatum	30	29	22	Furness and Butler, 1983
Cochliatoxum periachtum	7, 14	12, 14	14, 15	Sénaud and Grain, 1972
Cyrtophora				
Phyllophorynea				
Chilodonella cyprini	9, 11	13		Hofmann, 1987
Trithigmostoma steini	6, 11	8, 11	6	Hofmann and Bardele, 1987
Brooklynella hostilis	12, 13	13, 14	29	Lom and Corliss, 1971
Ignotocoma sabellarum	12–15	13–16	17, 18	Lom and Kozloff, 1969
Chilodochona quennerstedti	8, 9	9, 17	10, 17	Grain and Batisse, 1974
Tokophrya infusionum	2, 7–16	14–16	9, 10, 17	Rudzinska, 1980
Spirochona gemmipara	18, 19	20–22	21, 24	Fahrni, 1982

(*continued*)

Table 1. (Continued)

Ciliates[a]	*Coated Pits*[b]	*Preendosomal Vesicles*[b]	*Early Endosomes*[b]	*References*
Nassophorea				
Nassula spp.	15			Tucker, 1971
Pseudomicrothorax dubius	23, 25–28			Hausmann, 1979
Leptopharyx macrostoma	4, 6			Njine and Didier, 1980
Paramecium multimicronucleatum	Type endosomal system[d]			Allen et al., 1992
Certesia quadrinucleata	5	4	5	Wicklow, 1983
Euplotes eurystomus	10	10, 13	13	Fauré-Fremiet and André, 1968
Oligohymenophorea				
Glaucoma chattoni	2, 6, 13	4	8	Peck, 1978
Tetrahymena pyriformis	1	11, 18	9, 18	Allen, 1967
Tetrahymena pyriformis	Type endosomal system[d]			Nilsson and van Deurs, 1983
Ichthyophthirius multifiliis	3	3, 5	1, 3, 4	Chapman and Kern, 1983
Dexiotricha media	10, 21	16	21, 24, 26	Peck, 1977
Uronema spp.	27, 28	33	40, 41	Kaneshiro and Holz, 1976
Conchophthirus curtus	2, 27	27	27	Antipa, 1971
Cyclidium glaucoma	4	4	4	Bardele, 1983
Opercularia coarctata	10, 15	6, 10, 15	10, 15	Ruiz and Anadon, 1986
Thuricola folliculata	4	2, 4	2	Eperon and Grain, 1983
Vorticella convallaria	+[c]	+	+	R.D. Allen, unpublished observations
Trichodinopsis paradoxa	1–3, 5	3, 4, 7, 8	3, 6	Noirot-Timothée, 1968
Hyalophysa chattoni	31	30	32	Bradbury, 1966
Hyalophysa chattoni	13	14, 16	15	Bradbury, 1989
Terebrospira chattoni	3	3, 5	3, 5, 6	Bradbury and Goyal, 1976
Plagiopyla minuta	6, 10	10, 24	4, 22	Berger and Lynn, 1984
Colpodea				
Cyrtolophosis mucicola	19, 22	22	22, 25	Didier et al., 1980
Grossglockneria acuta	9	19	20	Aescht et al., 1991
Bryophrya bavariensis	11	22	6	Grain et al., 1979
Colpoda maupasi	11, 16	15	7, 8, 12	Lynn, 1976b
Colpoda simulans	14	16	15	Lynn, 1976c
Colpoda steinii	19, 20	5	5, 9	Lynn, 1976a
Tillina magna	3, 17	18, 23	7, 17, 23	Lynn, 1976d

Notes: [a]Classification scheme according to Small and Lynn (1985).

[b]Numbers refer to the figure numbers in the papers referenced at the right, which we or the authors have interpreted to show the designated compartments.

[c]+ indicates the presence of the designated compartments in unpublished electron micrographs.

[d]The endosomal systems of *Paramecium* and *Tetrahymena* served as type systems for identifying homologous structures in other ciliates.

1980), ferritin (Rudzinska, 1977, 1980; Nilsson and van Deurs, 1983; Westcot et al., 1985), horseradish peroxidase (HRP) (Allen and Fok, 1980; Allen et al., 1992), and monoclonal antibodies raised against surface antigens, such as the 190 kD antigen C6 on *P. multimicronucleatum* (Allen et al., 1992). However, no information about what these pits incorporate in the wild is known. Under laboratory conditions it has been suggested that they could substitute in part as a pathway for nutrient uptake in mutant cells unable to form digestive vacuoles (Rasmussen and Orias, 1975). However, the fact that these pits have a clathrin-like coat strongly suggests that in the cell's natural environment they are involved in receptor-mediated endocytosis.

III. PREENDOSOMAL VESICLES

In a recent paper Hansen et al. (1991) have introduced the useful term *preendosomal vesicle* for those coated or uncoated primary vesicles that arise from the plasma membrane and that communicate with an early endosome compartment. The initial coated vesicles, formed by the release of coated pits from parasomal sacs or pellicular pores, fit this definition. In *Paramecium* these spherical-to-flattened vesicles are relatively large, measuring 100 to 300 nm in diameter (Figs. 7–9). In the peritrich, *Trichodinopsis*, coated vesicles of a similar size were seen near the pellicular pores (Noirot-Timothée, 1968). As both coated and uncoated vesicles of this size are present, it is assumed that such vesicles are at first coated and then quickly lose their coats. Such vesicles are found to contain markers such as HRP (Fig. 7) (Allen et al., 1992), thorotrast (Noirot-Timothée, 1968), or ferritin (Nilsson and van Deurs, 1983) when cells were exposed to these substances. In *Paramecium* they also contain the resident plasma membrane C6 antigen, which presumably enters the pit and coated vesicle from the cell surface (Allen et al., 1992).

Deep-etch rotary-shadowed images show coated vesicles to be surrounded at first by a clathrin-like cage, while other nearby vesicles of similar size are free of such a cage (Fig. 9) (Allen et al., 1992). The vesicles tend to be found near the basal

Figures 7–9. Coated pits (arrows) of *P. multimicronucleatum* give rise to preendosomal vesicles (large arrowheads) which dock at and fuse with early endosomes (ee). Figure 7; ×31,000; Figs. 8, 9; ×80,000. Figure **7**: The cell was exposed to HRP and trifluoperazine, which resulted in an accumulation of HRP-containing endosomal compartments in the cortex. The small arrowhead indicates an evagination of the early endosome. Figure **8**: The two preendosomal vesicles (large arrowheads) as well as the coated pit (arrow) are all much larger than the three smaller coated evaginations (small arrowheads) budding from the early endosome. Figure **9**: Deep-etch rotary-shadowed replicas of coated and uncoated preendosomal vesicles (arrowheads). Figures 7, 9 reproduced from Allen et al. (1992) by permission of The Company of Biologists (Cambridge, GB).

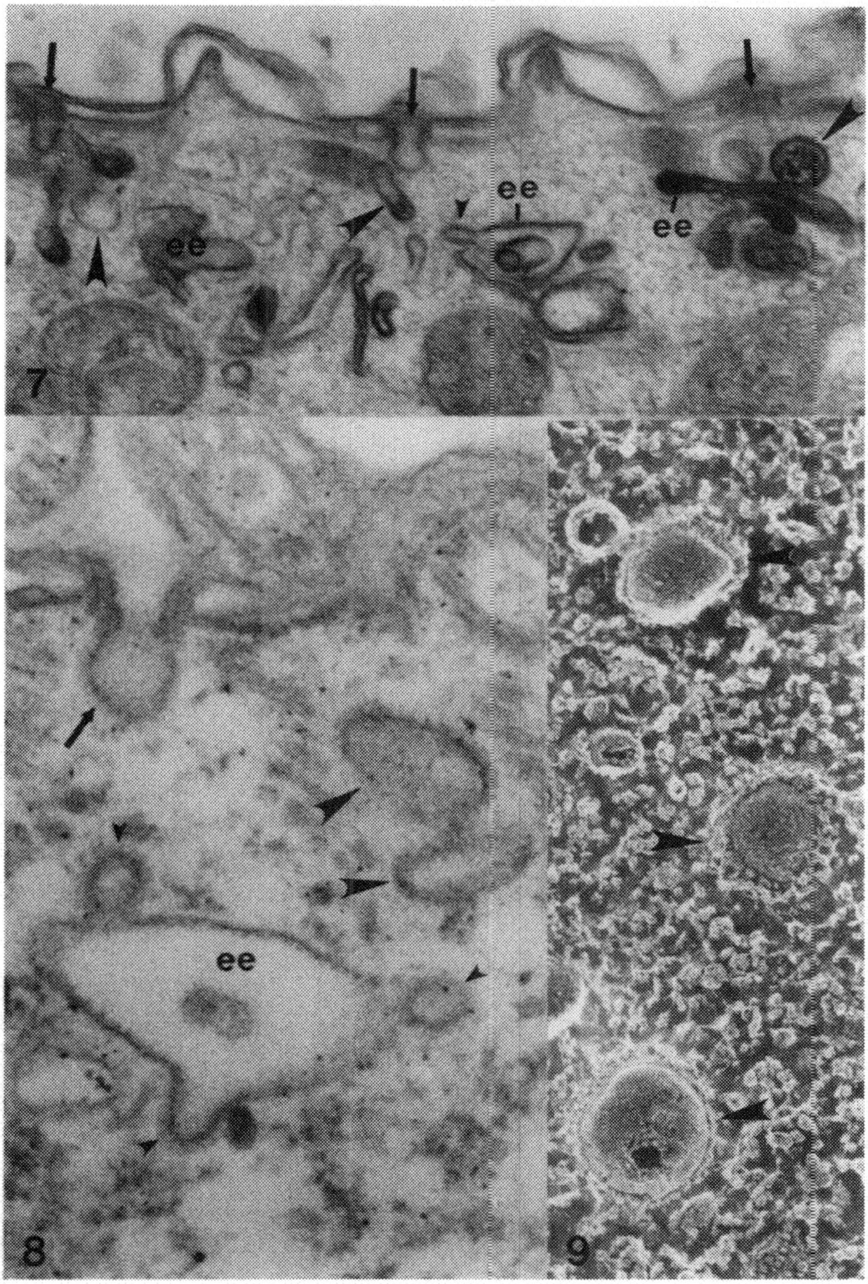
7
ee
ee
ee
8
ee
9

bodies and between the pellicle and a zone containing flattened cisternae located just below the basal bodies. However, such vesicles can also be found deeper in the endoplasm.

In order to deliver their cargo to a subsequent compartment, preendosomal vesicles must be able to recognize, dock at, and fuse with the appropriate receiving compartment. These processes must be efficiently regulated and presumably require mechanisms involving regulatory proteins such as the small GTP-binding rab proteins found in metazoan cells (Balch, 1990) or their close homologues found in yeast as well as fusion proteins (Wilson et al., 1991) analogous to the *N*-ethylmaleimide-sensitive fusion protein (NSF) (Glick and Rothman, 1987), and the soluble NSF attachment proteins (SNAPs) (Weidman et al., 1989; Wilson et al., 1992).

IV. EARLY ENDOSOMES

Preendosomal vesicles which have lost their clathrin-like cages transfer their contents to the cisternae that lie near the proximal ends of the basal bodies. These cisternae have appeared in many electron microscopic studies of various ciliates (Table 1) but their function has not been understood until very recently. Patterson (1978) believed such cisternae, which he called "cilisacs," might be involved in regulating the divalent cation concentration in the vicinity of cilia. However, Stelly et al. (1991) have provided compelling evidence that cortical alveoli more likely play this cation regulating role, a function originally proposed for alveolar sacs by Allen and Eckert (1969). Lobo-da-Cunha and Azevedo (1990a, 1990b) have observed similar cortical cisternae in *Ichthyophthirius multifiliis* filled with acid phosphatase-reaction product, which they consider to be similar to those of the Golgi complex. Rudzinska (1977, 1980) earlier came to this same conclusion for the flattened cisternae found in *Tokophrya* (a suctorian). Estève (1970) also found acid phosphatase-reaction product in cortical cisternae in *Paramecium*, which he identified as endoplasmic reticulum but which appears to us more like the cisternae discussed here.

We can now ascribe the role of early endosomes to all these cortically located ribosome-free cisternae in both the somatic and oral regions. The cisternae themselves do not appear to be interconnected in an extensive network but rather appear to be composed of many discrete units more or less discoidal in shape (Figs. 10, 11). These units have a tendency to curve into a bowl or crescent shape, and, in *Paramecium*, to be 0.2 to 0.7 μm in diameter (Allen et al., 1992). In all ciliates where they have been observed these cisternae appear to have a similar nonreticular shape. There is generally more than one cisterna near each basal body. In *Tokophrya* ferritin is found to be transferred from spherical vesicles arising from coated pits to flat vesicles of 380 nm in length by 40 nm wide which are frequently grouped into stacks of 3 to 5 or sometimes more. These were considered by the author to be vehicles for transporting macromolecules into the cytoplasm (Rudzinska, 1977, 1980). In our opinion, these are the suctorian's early endosomes.

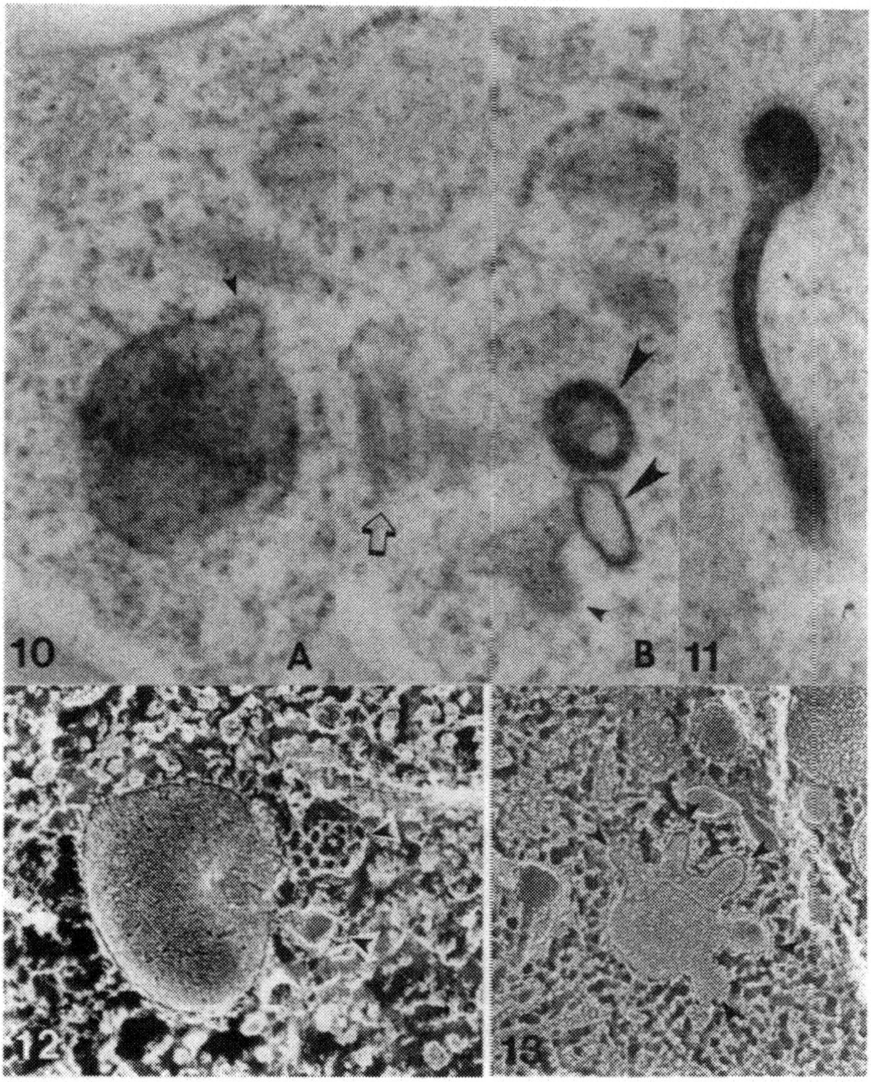

Figures 10–13. Early endosomes of *Paramecium*. All figures ×80,000. Figures **10A, B**: Two serial sections of an early endosome containing HRP reaction-product lying mostly in the plane of the section shown in Fig. 10A but having a small piece containing a coated evagination (small arrowhead) in Fig. 10B. This early endosome is thus discus shaped with two coated evaginations and has two preendosomal vesicles docked at its surface (large arrowheads, Fig. 10B). The edge of a basal body (open arrow) is present in Fig. 10B. Figure **11**: An early endosome sectioned in a plane perpendicular to the early endosome in Fig. 10. Judging from its large size, the bulbous end may be a fused preendosomal vesicle. Figure **12**: Deep-etch rotary-shadowed image of early endosome bearing two evaginations (small arrowheads) coated with clathrin-like cages. Figure **13**: A microtubule-affinity-purified early endosome-like vesicle bearing 5 evaginations. Early endosomes are thus apparently capable of binding to microtubules, which may ensure that they are transported to and/or remain in the cell cortex. Figs. 10–13 reproduced from Allen et al. (1992) by permission of The Company of Biologists (Cambridge, GB).

The early endosomes thus acquire markers from the coated pits via the preendosomal vesicles, although actual fusion between these two compartments is not easily recognized (see Fig. 11). As it takes some time for early endosomes to become labeled under normal laboratory growth conditions, the coated vesicles, at least in the somatic region, do not appear to pinch off from coated pits at a very rapid rate. It required up to 1 min of exposure to HRP for the first of these early endosome cisternae to become labeled in *Paramecium* (Allen et al., 1992). We have the impression that there are regions of the cell in which the coated pits are more active, for example, the anterior end of *Paramecium* seems to incorporate marker through the coated pits at a rate greater than other parts of the cell; also the equatorial region of a dividing cell appears to be more active in endocytosis. However, possibly the fastest uptake of marker occurs in the oral region along the cytopharynx because we have found marker in vesicles in this region within 3 sec of exposing *Paramecium* to HRP. We have observed that these cisternae are more intensely labeled in NH_4Cl-treated cells, as this weak base may raise the pH of these compartments. Also in cells treated with trifluoperazine, a calmodulin antagonist, these compartments appear to be retained and to accumulate in the cortical region (Fig. 7).

These cisternae have the remarkable characteristic of bearing one to several coated 80-nm-diameter evaginations extending from their membranes (Figs. 8, 10). In deep-etch rotary-shadowed images these evaginations, like the coated pits, were found to be coated by clathrin-like cages (Figs. 12, 13). Thus it appears that these early endosomes may be compartments for receiving exogenous macromolecules and for sorting these contents so that selected components will pass on into the cell, while others such as resident plasma membrane proteins are held back. This function has received support from our immunogold-labeling experiments of the early endosomes of *Paramecium*. By using the indirect immunogold-labeling technique, we have found that the C6 resident plasma membrane antigen is present in the body of the early endosome but not in its coated evaginations (Fig. 14). Thus the C6 antigen which resides on the cell surface, in preendosomal vesicles and in early endosomes, does not enter the coated evaginations of early endosomes and is not found normally in compartments deeper in the cell. We infer that this antigen is being sorted and may be recycled to the plasma membrane, although we have yet to prove the recycling step.

Based on the absence of the C6 antigen we conclude that the coated evaginations on early endosomes are not remnants of fusing preendosomal vesicles. When compared to these preendosomal vesicles they are also consistently smaller in diameter (Fig. 8). A similar difference in size of coated pits on the plasma membrane and those on early endosomes was also noted in macrophage (Rabinowitz et al., 1992). Finally, as it is generally accepted that coated vesicles must lose their clathrin coat before they can fuse with another membrane compartment (e.g., Petersen and van Deurs, 1983), remnants of preendosomal vesicles being incorporated into early endosomes would, for the most part, not be coated. In fact, preendosomal vesicles will probably flow into the early endosome membrane so

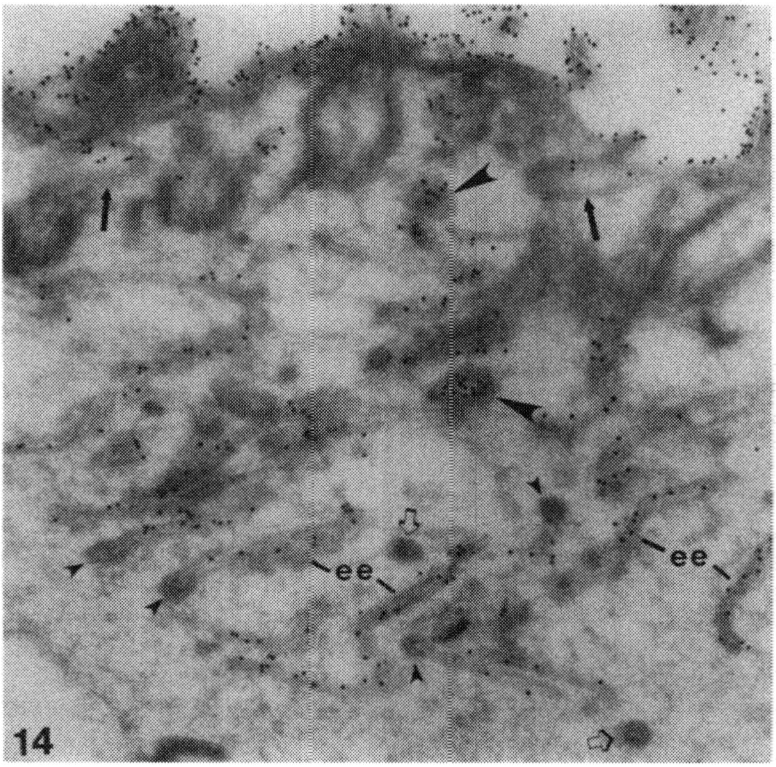

Figure 14. Immunogold labeling of the resident plasma membrane antigen C6 of *Paramecium multimicronucleatum*. This antigen is found on the cell surface, in coated pits (solid arrows), in preendosomal vesicles (large arrowheads), and in the bodies of early endosomes (ee). This antigen is not seen in the coated evaginations (small arrowheads) of these early endosomes or in "carrier" vesicles derived from these buds (open arrows). ×32,000. Reproduced from Allen et al. (1992) by permission of The Company of Biologists (Cambridge, GB).

quickly that they will not usually be recognizable as fused vesicles following preparation for electron microscopy, unless the cell has been subjected to the quick-freeze technique without prior chemical fixation. Quick freezing living cells is the only technique currently available that will adequately preserve intermediate steps in fast-acting biological processes. Though we have used this technique on *Paramecium*, we have not yet studied enough replicas to see the initial stages of preendosomal fusion.

Many examples of cortical cisternae in ciliates bearing coated evaginations can be found in the literature (Table 1). Such flattened vesicles are found in the tentacles as well as the body of *Tokophrya* (Rudzinska, 1980). In *Nassula* they accumulate in the pellicular ridges (Fig. 15). In ciliates with polykinetids such cisternae

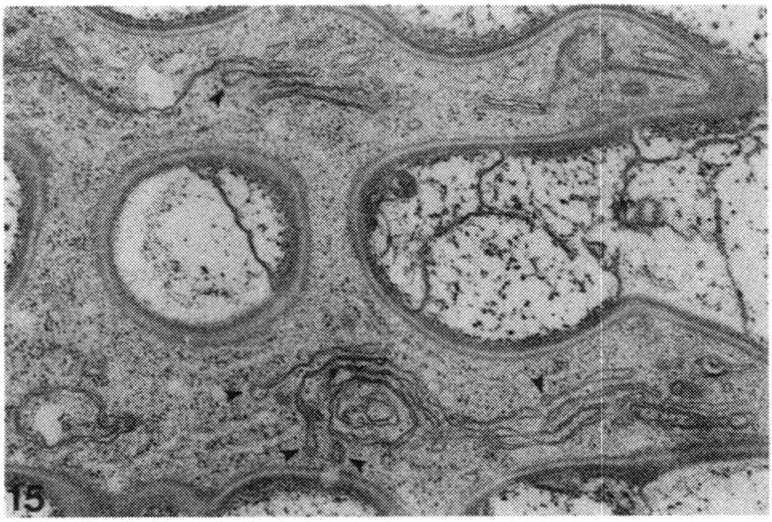

Figure 15. The ciliate *Nassula* sp. contains many early endosome-like compartments in their pellicular ridges which bear coated evaginations (arrowheads). Unpublished micrograph. ×32,000.

accumulate in large numbers just internal to the proximal ends of the basal bodies. This also includes members of the subphylum Postciliodesmatophora, such as *Blepharisma* and *Halteria*, which lack the typical parasomal sacs but have coated pits interspersed between the basal bodies of the polykinetids. We conclude that such cisternae are probably found in all ciliates and that all ciliates have both a preendosomal vesicle compartment as well as the early endosome compartment. By analogy with *Paramecium* and mammalian cells we assume this compartment is involved in sorting and recycling resident antigens back to the surface membrane. Some exogenous markers that are transferred to the early endosomes tend to be concentrated in the coated evaginations which are found at the rims of these early endosomes. This is true of endocytosed ferritin in *Tetrahymena* (Nilsson and van Deurs, 1983) and *Paramecium* (Westcot et al., 1985) and to some extent HRP in *Paramecium* (Allen et al., 1992). The flattened nature of these compartments may favor the movement and the concentration of the fluid contents into the bulbous buds, while receptor-bound ligands may accumulate in the buds through an active process of receptor flow in the plane of the membrane and ultimate entrapment in the buds by interactions between receptors and clathrin-binding proteins such as the clathrin-binding adaptins (reviewed by Keen, 1990). Other integral membrane proteins that do not bind to adaptin-like proteins will presumably not accumulate in the buds and will remain in the body of the early endosome. Such is the apparent fate of the C6 antigen in *Paramecium*.

Though usually found in the cell's cortex, early endosomes can also be found deeper in the endoplasm. It is possible to separate early endosomes from other components of cellular homogenates by making use of the fact that they bind to microtubules (Fig. 13) (Schroeder et al., 1990). Microtubules are typically found interspersed between the early endosome compartments and so may play a role in moving these structures to the cell's cortex and ensuring that they remain in the cortex. Early endosomes found deeper in the endoplasm may reflect stages in their development, in the event that they are not simply products of fusion of preendosomal vesicles, or they may be early endosomes that have become disassociated from the microtubules and so have been displaced into the endoplasm. The whole question of whether early endosomes are permanent preexisting compartments with their own resident proteins, a view favored by Griffiths and Gruenberg (1991), or transitory structures without their own resident proteins that mature into late endosomes, a view favored by Murphy (1991) and Dunn and Maxfield (1992), has not yet been settled. In ciliates, since there is some reason to believe that small carrier vesicles not destined to return to the plasma membrane arise from the early endosomes and then move deeper into the cell (see Section V), we tend to favor the idea that early endosomes are permanent compartments that do not themselves evolve into late endosomes. However, to confirm this hypothesis, resident proteins specific for early endosomes will need to be identified.

V. CARRIER VESICLES

The coated evaginations at the margins of early endosomes appear to pinch off to form a pool of ~100-nm vesicles (Fig. 16). Vesicles of this size with and without clathrin-like coats can be found near the early endosomes, usually on the side of the early endosomes toward the cell's center. Vesicles of a similar size have also been found aligned along microtubular ribbons of *Paramecium* following the quick-freeze deep-etch technique (Schroeder et al., 1990). Though the two vesicle populations have not yet been shown to be one and the same, on occasion endocytic marker has been found in the 100-nm vesicles next to the ribbons of microtubules arising from the cytopharynx region.

Our conclusion from these observations is that 100-nm carrier vesicles are pinching off the early endosomes and transporting their cargo from the sorting compartment to a later compartment (Allen et al., 1992). Because such vesicles are very small and become lost in the milieu of the ciliate's endoplasm, carrier vesicles have not been recognized in other ciliates as separate from the many other small vesicles which the cells normally harbor. However, the presence of coated evaginations on the cortical cisternae of a large number of ciliates as seen in thin sections strongly indicates the presence of such carrier vesicles in these ciliates. We venture the guess that 100-nm carrier vesicles will be found in all ciliates when appropriate markers are available to label them.

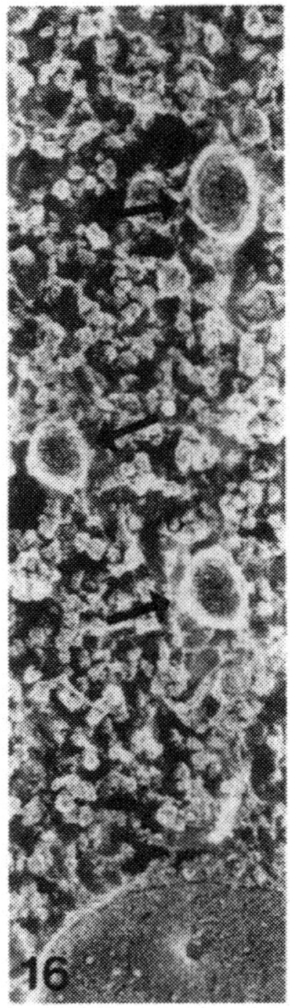

Figure 16. Small "carrier" vesicles (arrows), seen here in a deep-etch rotary-shadowed image, lie near the early endosomes but more often on the side of these cisternae closer to the cell's center. These vesicles may carry cargo from early to late endosomes as they do not appear to recycle to the cell's surface. ×80,000. Reproduced from Allen et al. (1992) by permission of The Company of Biologists (Cambridge, GB).

Where are the carrier vesicles going? It would be reasonable to conclude that these vesicles are recycling resident plasma membrane proteins back to the cell's surface. However, these vesicles do not contain the resident plasma membrane C6 antigen in *Paramecium*, so this possibility appears unlikely. The second possibility is that they are transporting endocytic cargo to a late endosome compartment or an intermediate endosome compartment where the cargo will be brought into contact with a strongly acidic environment and then with the battery of acid hydrolases. These hydrolases are transported to this putative late endosome compartment from the Golgi complex. At this time we favor this second possibility that these vesicles are moving cargo more deeply into the cell.

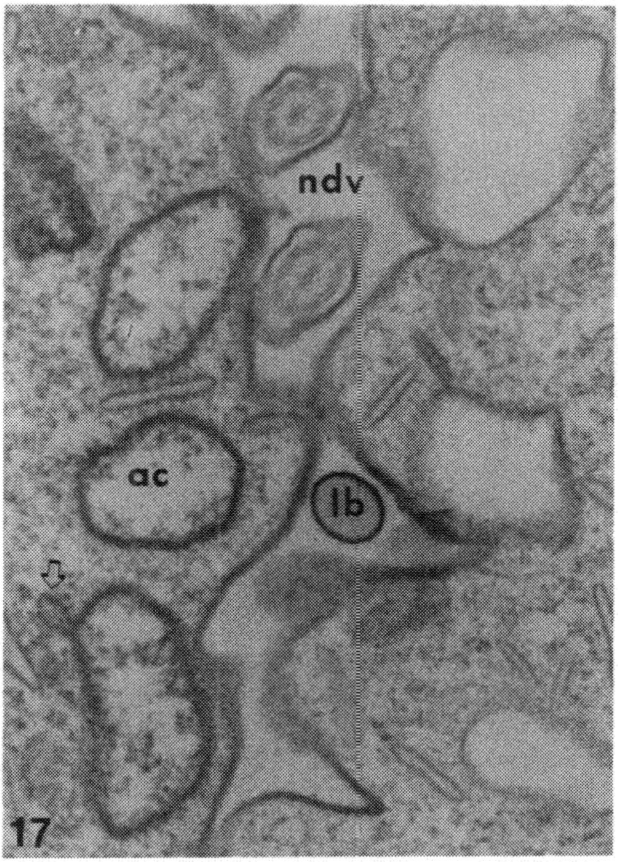

Figure 17. Acidosomes (ac) are large vesicles that bind to the nascent digestive vacuole membrane (ndv) and sometimes acquire exogenous marker, such as HRP, in a remarkably short time. In this case some acidosomes have become labeled with HRP after only 3 sec of exposure to this marker, while other acidosomes remain free of marker. Many examples of vesicles fusing or arising from these acidosomes (e.g., open arrow) have been found. Acidosomes are thus one of the possible destinations of the small carrier vesicles. Alternatively, preendosomal vesicles may be able to fuse with these acidosomes under some conditions. This nascent vacuole contains two sectioned cilia and one latex bead (lb). Unpublished micrograph. ×40,000.

VI. LATE ENDOSOMES

Following the analogy established in mammalian systems the endocytic cargo should eventually be delivered to a late endosome compartment. Such a compartment has yet to be identified positively in ciliates. Nilsson and van Deurs (1983) show that when 1-hr-starved *Tetrahymena* are exposed to cationized ferritin, the ferritin can be found in autophagic vacuoles within 5 to 10 min and occasionally in preformed food vacuoles as well. After 15 min the ferritin is also found in small dense bodies of ~0.3 μm in diameter. By 30 to 60 min most of the ferritin is found in autolysosomes and old preformed food vacuoles. Essentially the same results are obtained if the starved cells are induced to form food vacuoles by supplying the cells with nonnutritive carmine particles. As autolysosomes are induced by starvation and are not normally found in the log phase of growth, these structures can be discounted as the normal "sought-for" late endosome compartment. The small dense bodies, however, may represent profiles of such a compartment. In Fig. 14 of the paper by Nilsson and van Deurs (1983) such a dense body is illustrated and closely resembles the multivesicular body of mammalian cells. In macrophages multivesicular vesicles have been called endocytic carrier vesicles (Rabinowitz et al., 1992).

In *Paramecium* we are currently attempting to follow the fate of the small carrier vesicles to see where they are delivering their endocytic cargo. To do this we have attempted to use several markers to load the endocytic compartment. Horseradish peroxidase (HRP) is a useful marker, as the amount of electron opaque reaction product formed can be controlled by varying the duration of incubation in diaminobenzidine and H_2O_2. Other markers such as ferritin, colloidal gold, or thorotrast cannot be augmented in this way, and the amount of marker will depend on how much is taken up initially by the coated pits. To obtain a high enough concentration of these markers to be readily visible in the cell, the marker must either be selectively concentrated by the cell or be present in high enough concentration to be included in the fluid phase during coated vesicle formation.

We have found in *Paramecium* that HRP will sometimes enter compartments other than early endosomes, the acidosomes, very rapidly (Fig. 17). Acidosomes are organelles of about 1 μm in diameter that bind to and bring about digestive vacuole acidification (Allen and Fok, 1983). Following only 3 sec of exposure to HRP, some cells will already have marker in about half the acidosomes bordering the nascent vacuoles (R. D. Allen et al., 1993). Two possible routes of entry of exogenous marker into these acidosomes can be envisioned: either (1) the acidosomes are directly open to the external medium through membrane-lined connections, or (2) the endosomal system in the oral region can deliver cargo to acidosomes at a very rapid rate. We have eliminated the first possibility, that of direct contact to the outside, by using acridine orange to label these acidic organelles. A layer of acridine orange-containing organelles can be seen by fluorescence microscopy to line the surface of the nascent vacuole in living cells. As electron microscopy shows no other organelles bordering nascent vacuoles, we can be sure that these acridine orange-labeled organelles are acidosomes. These acidosomes must not be open to the outside or they would not appear red-orange, a color that indicates a high concentration of acridine orange.

Thus the possibility that the HRP in acidosomes derives from the endocytic system seems more likely. We do see some smaller vesicles in the vicinity of the cytopharynx that are labeled with HRP. Also we have often noted small vesicles next to the much larger acidosomes as well as small protrusions on the sectioned acidosomes. These small protrusions and what appear to be remnants of coats on the protrusions suggest that docking and fusion of vesicles with acidosomes are a commonplace event. Evidence for entry of an intramembrane particle-poor membrane into the more intramembrane particle-rich membrane of the acidosome has also been seen in deep-etch rotary-shadowed replicas (R. D. Allen et al., 1993). Thus we conclude that acidosomes, both those lying next to nascent vacuoles and those scattered around the cell, obtain marker via the endosome route. However, it is not yet clear whether the source of the HRP is through fusion of acidosomes with preendosomal vesicles directly, fusion of acidosomes with carrier vesicles derived from early endosomes, or is by some other route that is unknown at the present time.

The possibility that acidosomes are late endosomes needs to be considered in more detail. Late endosomes of mammalian cells are characterized by containing cargo derived from the endocytic pathway as well as the lysosomal hydrolase receptor known as the cation-independent mannose-6-phosphate receptor, which comes from the Golgi complex. In addition lysosome-associated membrane proteins and hydrolases are found in this compartment in various mammalian cells. Finally, rab 7, a small GTP-binding protein, is found only on late endosomes, on their cytosolic surface (Chavrier et al., 1990).

Applying these criteria to the acidosomes of *Paramecium*, we find that the acidosome membrane is similar to lysosome and Golgi complex membranes in their common ability to bind wheat germ agglutinin. This is also true of the trans-cisterna of the Golgi stack and the trans-Golgi network (Allen et al., 1989), as well as digestive vacuole membrane at all vacuole stages. In addition, as pointed out above, acidosomes are capable of receiving endocytic cargo. However, acidosomes do not appear to contain the common hydrolase indicator, acid phosphatase (Allen and Fok, 1983). Studies on the presence of hydrolase receptors or rab 7 have not yet been applied to the relevant systems of vesicles in any ciliate, so acidosomes have yet to be tested for these criteria. Two additional characteristics of acidosomes may be used to assist us in considering the question of their being late endosomes. First, we have obtained several monoclonal antibodies which are specific for acidosome membrane antigens and which also label a small group of putative pre-acidosome vesicles and cisternae in the cytosol. This compartment thus bears its own resident antigens which identify and separate its membrane from that of early endosomes, lysosomes, and the Golgi apparatus. Thus acidosome membrane cannot be considered to be identical to any of these three membrane pools, although they are related by their common ability to bind wheat germ agglutinin. The shape of acidosomes actually compares favorably with the large endosome "carrier" vesicles, as illustrated by Griffiths and Gruenberg (1991). Acidosomes in *Paramecium* are essentially spherical but frequently have indentations of their membrane extending into their lumens, particularly following fusion but before the merging of acidosomes with digestive vacuoles. They do not, however, contain free vesicles, as is usually implied of multivesicular vesicles.

Though we cannot be sure, we believe there is enough circumstantial evidence to suggest that acidosomes may fill the role either of a late endosome or of the endosome carrier vesicle as presented in the scheme of Griffiths and Gruenberg (1991) (see Section VII for further discussion).

VII. COMPARISON WITH MAMMALIAN ENDOSOMAL SYSTEMS

In mammalian cells it has been generally established that cargo entering the cell via coated vesicles is transferred to an early endosomal sorting compartment and that many plasma membrane-located receptors are sequestered and recycled back to the cell surface. The cargo and some membrane components are then passed on to a late endosomal sorting compartment, where they first come into contact with a battery of hydrolases arriving at this second compartment from the Golgi apparatus. Hydrolase receptors are presumably sequestered and removed from this compartment to be recycled back to the Golgi apparatus. The cargo is then passed to the lysosomes for digestion.

Variations on this general scheme have been noted, and whether these compartments are permanent preexisting structures or whether early endosomes mature into late endosomes is still under debate. In one study Hopkins et al. (1990) showed by confocal microscopy that the endosomal system in Hep-2 carcinoma cells can form a long continuous tubular reticulum in which the two sorting functions would presumably be carried out, one at the proximal end and the other at the distal end of the same reticulum. Cargo could simply be centripetally passed along the continuous lumen of this reticulum from its peripheral end to its perinuclear end. In discontinuous systems such as that found in the peritoneal macrophage (Rabinowitz et al., 1992), cargo apparently passes from the early to late endosomes via endosomal "carrier" vesicles which leave behind a persistent early endosome. The carrier vesicles then fuse with a late endosome "tubulo-reticular" compartment, which also receives the hydrolases as well as phagocytic cargo entering the cell by phagocytosis. Digestion will begin and may even be completed in this late endosome compartment in these cells before the nondigestible material accumulates in the heavier lysosomes (Rabinowitz et al., 1992).

The endosomal system of *Paramecium* may resemble more closely than does the macrophage endosomal system the earliest successful evolutionary attempt at developing an endosomal process. As summarized in Fig. 18, in *Paramecium* this system has a discontinuous early endosome compartment composed of many individual units. These units are more or less discoidal of about 0.2 to 0.7 μm in diameter and receive cargo by fusing with preendosomal vesicles arising from the cell surface. Early endosomes of the ciliates thus resemble those early endosomes described in BHK-21 cells (Marsh et al., 1986) in both size and general shape and differ only in the length of their extended arms. These arms are elongated tubules in BHK-21 cells but are reduced to mere buds in *Paramecium*.

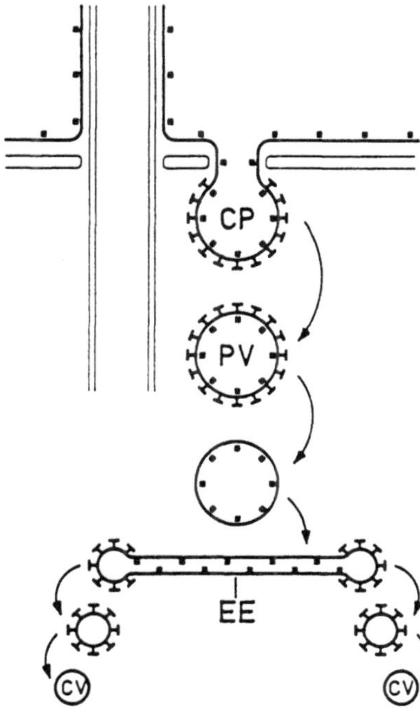

Figure 18. Drawing to illustrate the trafficking of endosomal membrane in ciliates. Coated pits (CP) give rise to coated preendosomal vesicles (PV) which lose their coats before docking at and fusing with the early endosome (EE). Coated evaginations form from the early endosome and are released as coated vesicles one-half to one-third the diameter of preendosomal vesicles. These smaller carrier vesicles (CV) transport cargo deeper into the cell. Dots on membranes indicate the sites of the C6 resident plasma membrane antigen in *Paramecium*, which is excluded from the smaller cargo-carrying vesicles.

The size and shape of the vesicles leaving early endosomes in *Paramecium*, on the other hand, differ from those described in one mammalian cell, the peritoneal macrophage. In the macrophage an endocytic marker moves from early endosomes into multivesicular bodies within 8 to 30 min, suggesting that these multivesicular bodies are endosomal carrier vesicles (Rabinowitz et al., 1992). In *Paramecium* the only vesicles arising from early endosomes appear to be the small 80-nm-diameter coated vesicles. As these vesicles lack the resident plasma membrane C6 antigen they do not appear to be recycling antigens to the cell surface. In macrophage the label then passes into the tubulo-reticular late endosomes by 15 min, but it can take up to 60 min to clear the label from early endosomes. This tubulo-reticular compartment contains lysosomal membrane antigens and in addition has rab 7 on its cytosolic side. The cation-independent mannose-6-phosphate receptor which binds lysosomal hydrolases was restricted to the regions of this compartment near the Golgi complex. This compartment also receives phagocytosed latex beads and bacteria. In *Paramecium*, though we have not yet determined with certainty which compartment plays the role of the late endosome, we can see a possible membrane flow scheme linking the endosomal system with the phagosomal system. In this scheme small 100-nm carrier vesicles arising from early endosomes may find their way to acidosomes and/or to early digestive vacuoles, where they will dock and fuse. Thus the endocytic cargo will enter the phagosomal compartment and will undergo denaturation and digestion together with the phagocytic cargo. In *Paramecium* the acidosomes, which are hydrolase negative, may be equivalent to the endocytic carrier vesicles of the macrophage, while the digestive vacuole itself serves the function of a late endosome. It is hoped that we will be able to determine with further work the true relationship of the acidosomes to the endosomal system and which compartment serves as a late endosome.

VIII. CONCLUSIONS

The endosomal system of ciliates, as exemplified by *Paramecium*, has many features in common with the endocytic system of mammalian cells, particularly the professional phagocytes. This implies that endosomal systems arose early in evolution and have been passed down to the present time with only minor modification. Coated pits with clathrin-like cages appear at all levels of cellular organization and these pits pinch off to form the preendosomal coated vesicles. In *Paramecium* these preendosomal vesicles can then fuse either with early endosomes or under some conditions they may fuse with acidosomes. Early endosomes function as a sorting compartment and give rise to smaller coated vesicles which carry cargo to a later compartment along the discontinuous endosomal pathway. Though not yet determined this later compartment seems to be either the acidosome or the digestive vacuole itself. In the end the endosomal cargo is transferred to the phagosomal compartment, and denaturation and digestion of cargo from both pathways are completed together.

The major questions remaining to be worked out are the following: (1) By what pathway do acidosomes receive HRP within 3 sec of exposing cells to this marker? (2) Where do the small carrier vesicles that arise from the early endosomes go? (3) What is the fate of C6 antigen once it enters the early endosomes? (4) Are acidosomes late endosomes or are they endosomal carrier vesicles? and, finally, (5) What compartment does play the role of a late endosome? One of the next phases of the study of the endosomal system will be to investigate the proteins needed to regulate this pathway of membrane traffic. A search for fusion proteins such as *N*-ethylmaleimide-sensitive fusion protein and soluble NSF attachment proteins as well as coat proteins such as β-COP (Duden et al., 1991) and small GTP-binding proteins homologous to the rab family of proteins must be made. Cell-free fusion systems will need to be developed to investigate the soluble factors required for membrane trafficking. Monoclonal antibodies have already been useful in showing us relationships between various membranes in the cytosol and will continue to provide additional information in the future. And for the ciliates it remains to be determined which compounds these cells acquire through this endocytic pathway and if receptor-mediated endocytosis does actually occur in these cells.

ABBREVIATIONS

β-COP:	beta coat protein
GTP:	guanosine triphosphate
H_2O_2:	hydrogen peroxide
HRP:	horseradish peroxidase
NSF:	N-ethylmaleimide sensitive fusion protein
SNAPs:	soluble NSF attachment proteins

ACKNOWLEDGMENTS

We thank Marilynn S. Ueno for technical support. Our current work is supported in part by National Science Foundation grants DCB 88 19182 and DCB 90 17455. The electron microscope facility used in our work has received support from a National Institutes of Health Research Centers in Minority Institutes grant GM 07684 and NSF instrumentation grants PCM 82 12316, PCM 84 00418, and BBS 87 21751.

REFERENCES

Aescht, E., Foissner, W., and Mulisch, M. (1991). Ultrastructure of the mycophagous ciliate *Grossglockneria acuta* (Ciliophora, Colpodea) and phylogenetic affinities of colpodid ciliates. *Eur. J. Protistol.* 26: 350–364.

Allen, R. D. (1967). Fine structure, reconstruction and possible functions of components of the cortex of *Tetrahymena pyriformis*. *J. Protozool.* 14: 553–565.

Allen, R. D. and Eckert, R. (1969). A morphological system in ciliates comparable to the sarcoplasmic reticulum-transverse tubular system in striated muscles. *J. Cell Biol.* 43: 4a–5a.

Allen, R. D. and Fok, A. K. (1980). Membrane recycling and endocytosis in *Paramecium* confirmed by horseradish peroxidase pulse-chase studies. *J. Cell Sci.* 45: 131–145.

Allen, R. D. and Fok, A. K. (1983). Nonlysosomal vesicles (acidosomes) are involved in phagosome acidification in *Paramecium*. *J. Cell Biol.* 97: 566–570.

Allen, R. D., Ma, L., and Fok, A. K. (1993). Acidosomes: recipients of multiple sources of membrane and cargo during development and maturation. *J. Cell Sci.* 106: in press.

Allen, R. D., Schroeder, C. C., and Fok, A. K. (1989). Intracellular binding of wheat germ agglutinin by Golgi complexes, phagosomes, and lysosomes of *Paramecium multimicronucleatum*. *J. Histochem. Cytochem.* 37: 195–202.

Allen, R. D., Schroeder, C. C., and Fok, A. K. (1992). Endosomal system of *Paramecium*: coated pits to early endosomes. *J. Cell Sci.* 101: 449–461.

Allen, R. D. and Wolf, R. W. (1974). The cytoproct of *Paramecium caudatum*: structure and function, microtubules, and fate of food vacuole membranes. *J. Cell Sci.* 14: 611–631.

Allen, R. D. and Wolf, R. W. (1979). Membrane recycling at the cytoproct of *Tetrahymena*. *J. Cell Sci.* 35: 217–227.

Antipa, G. A. (1971). Structural differentiation in the somatic cortex of a ciliated protozoan, *Conchophthirus curtus* Engelmann 1862. *Protistologica* 7: 471–501.

Balch, W. E. (1990). Small GTP-binding proteins in vesicular transport. *Trends Biochem. Sci.* 15, 473–477.

Bardele, C. F. (1983). Mapping of highly ordered membrane domains in the plasma membrane of the ciliate *Cyclidium glaucoma*. *J. Cell Sci.* 61: 1–30.

Berger, J. and Lynn, D. H. (1984). Somatic and oral cortical ultrastructure of the plagiopylid ciliates *Lechriopyla mystax* Lynch, 1930 and *Plagiopyla minuta* Powers, 1933. *J. Protozool.* 31: 433–443.

Bohatier, J. (1979). Morphogénese de régénération chez le cilié *Condylostoma magnum* (Spiegel): étude ultrastructurale. *J. Protozool.* 26: 404–414.

Bohatier, J., Iftode, F., Didier, P., and Fryd-Versavel, G. (1978). Sur l'ultrastructure des genres *Spathidium* et *Bryophyllum*, ciliés Kinetophragmophora (de Puytorac et al., 1974). *Protistologica* 14: 189–200.

Bradbury, P. C. (1966). The fine structure of the mature tomite of *Hyalophyse chattoni*. *J. Protozool.* 13: 591–607.

Bradbury, P. C. (1989). Evidence for hymenostome affinities in an apostome ciliate. *J. Protozool.* 36: 95–103.

Bradbury, P. C. and Goyal, V. (1976). The fine structure of a parasitic ciliate *Terebrospira* during ingestion of the exoskeleton of a shrimp *Palaemonetes*. *Tissue & Cell* 8: 573–582.

Chapman, G. B. and Kern, R. C. (1983). Ultrastructural aspects of the somatic cortex and contractile vacuole of the ciliate, *Ichthyophthirius multifiliis* Fouquet. *J. Protozool.* 30: 481–490.

Chapman-Andresen, C. (1962). Studies on pinocytosis in *Amoebae*. *C. R. Trav. Lab. Carlsberg* 33: 73–264.

Chavrier, P., Parton, R. G., Hauri, H. P., Simons, K., and Zerial, M. (1990). Localization of low molecular weight GTP binding proteins to exocytic and endocytic compartments. *Cell* 62, 317–329.

Corliss, J. O. (1979). *The Ciliated Protozoa*. Pergamon Press, New York.

Corliss, J. O. and Lom, J. (1985). An annotated glossary of protozoological terms. In: Illustrated Guide to the Protozoa (Lee, J. J., Hutner, S. H., and Bovee, E. C., eds.), pp. 576–602, Society of Protozoologists, Lawrence, Kansas.

Courtoy, P. J. (1989). Dissection of endosomes. In: *Intracellular Trafficking of Proteins* (Steer, C. and Hanover, J., eds.), pp. 103–156. Cambridge University Press, New York.

Daniel, W. A. and Mattern, C. F. T. (1965). Some observations on the structure of the peristomial membranelle of *Spirostomum ambiguum*. *J. Protozool.* 12, 14–27.

Didier, P., de Puytorac, P., Wilbert, N., and Detcheva, R. (1980). A propos d'observations sur l'ultrastructure du cilié *Cyrtolophosis mucicola* Stokes, 1885. *J. Protozool.* 27: 72–79.

Duden, R., Allan, V., and Kreis, T. (1991). Involvement of β-COP in membrane traffic through the Golgi complex. *Trends Cell Biol.* 1: 14–19.

Dunn, K. W. and Maxfield, F. R. (1992). Delivery of ligands from sorting endosomes to late endosomes occurs by maturation of sorting endosomes. *J. Cell Biol.* 117: 301–310.

Eperon, S. and Grain, J. (1983). Sur les fibres postciliaires de la polycinétie de *Thuricola folliculata* (O. F. Müller, 1786) (Ciliophora, Peritrichida) en bipartition. *Protistologica* 19: 123–129.

Estève, J.-C. (1970). Distribution of acid phosphatase in *Paramecium caudatum*: its relations with the process of digestion. *J. Protozool.* 17: 24–35.

Fahrni, J. F. (1982). Morphologie et ultrastructure de *Spirochona gemmipara* Stein, 1852 (Ciliophora, Chonotrichida). I. Structures corticales et buccales de l'adulte. *J. Protozool.* 29: 170–184.

Fauré-Fremiet, E. and André, J. (1968). Structure fine de l'*Euplotes eurystomus* (WRZ). *Archives d'Anatomie Microscopique* 57: 53–78.

Finley, H. E., Brown, C. A., and Daniel, W. A. (1964). Electron microscopy of the ectoplasm and infaciliature of *Spirostomum ambiguum*. *J. Protozool.* 11: 264–280.

Foissner, W. and Foissner, I. (1985). Oral monokinetids in the free-living haptorid ciliate *Enchelydium polynucleatum* (Ciliophora, Enchelyidae): ultrastructural evidence and phylogenetic implications. *J. Protozool.* 32: 712–722.

Franke, W. W. (1971). Membrane-microtubule-microfilament-relationships in the ciliate pellicle. *Cytobiologie* 4: 307–316.

Furness, D. N. and Butler, R. D. (1983). The cytology of sheep rumen ciliates. I. Ultrastructure of *Epidinium caudatum* Crawley. *J. Protozool.* 30: 676–687.

Glick, B. S. and Rothman, J. E. (1987). A possible role for fatty acyl-coenzyme A in intracellular protein transport. *Nature* 326: 309–312.

Grain, J. and Batisse, A. (1974). Étude ultrastructurale du cilié chonotriche *Chilodochona quennerstedti* Wallengren, 1985. I. Cortex et structures buccales. *J. Protozool.* 21: 95–111.

Grain, J., Iftode, F., and Fryd-Versavel, G. (1979). Étude des infraciliatures somatique et buccale de *Bryophrya bavariensis* et considérations systématiques. *Protistologica* 15: 581–595.

Grębecki, A. (1991). Participation of the contractile system in endocytosis demonstrated in vivo by video-enhancement in heat-pretreated amoebae. *Protoplasma* 160: 144–158.

Griffiths, G. and Gruenberg, J. (1991). The arguments for pre-existing early and late endosomes. *Trends Cell Biol.* 1: 5–9.

Grim, J. N. (1972). Fine structure of the surface and infraciliature of *Gastrostyla steinii*. *J. Protozool.* 19: 113–126.

Grimes, G. W. (1972). Cortical structure in nondividing and cortical morphogenesis in dividing *Oxytricha fallax*. *J. Protozool.* 19: 428–445.

Gruenberg, J. and Howell, K. E. (1989). Membrane traffic in endocytosis: insights from cell-free assays. *Annu. Rev. Cell Biol.* 5: 453–481.

Hansen, S. H., Sandvig, K., and van Deurs, B. (1991). The preendosomal compartment comprises distinct coated and noncoated endocytic vesicle populations. *J. Cell. Biol.* 113: 731–741.

Hausmann, K. (1979). Characteristics of the membranes of the pellicle of the ciliate *Pseudomicrothorax dubius*. *Protoplasma* 100: 199–213.

Hausmann, K. and Allen, R. D. (1976). Membrane behavior of exocytic vesicles. II. Fate of the trichocyst membranes in *Paramecium* after induced trichocyst discharge. *J. Cell. Biol.* 69: 313–326.

Hiller, S. A. (1991). *Bursellopsis spaniopogon* n. spec. (Ciliophora: Prostomatida). I. Infraciliature and ultrastructure of the somatic and oral cortex. *Eur. J. Protistol.* 27: 1–16.

Hiller, S. and Bardele, C. F. (1988). *Prorodon aklitolophon* n. spec. and the "dorsal brush" as a character to identify certain subgroups in the genus *Prorodon*. *Arch. Protistenkunde* 136: 213–236.

Hofmann, A. H. (1987). Stomatogenesis in cyrtophorid ciliates. II. *Chilodonella cyprini* (Moroff, 1902): the kinetofragment as an anlagen-complex. *Eur. J. Protistol.* 23: 165–184.

Hofmann, A. H. and Bardele, C. F. (1987). Stomatogenesis in cyrtophorid ciliates. I. *Trithigmostoma steini* (Blochmann, 1895): from somatic kineties to oral kineties. *Eur. J. Protistol.* 23: 2–17.

Hopkins, C. R., Gibson, A., Shipman, M., and Miller, K. (1990). Movement of internalized ligand-receptor complexes along a continuous endosomal reticulum. *Nature* 346: 335–339.

Huttenlauch, I. and Bardele, C. F. (1987). Light and electron microscopical observations on the stomatogenesis of the ciliate *Coleps amphacanthus* Ehrenberg, 1833. *J. Protozool.* 34: 183–192.

Ishida, M., Suzaki, T., and Shigenaka, Y. (1991). Studies on the mechanism of cell elongation in *Blepharisma japonicum*. 4. Three dimensional construction of the kinetosomal complex and its functional role on cell elongation. *Eur. J. Protistol.* 27: 46–54.

Kaneshiro, E. S. and Holz, G. G. (1976). Observations on the ultrastructure of *Uronema* spp., marine scuticociliates. *J. Protozool.* 23: 503–517.

Keen, J. H. (1990). Clathrin and associated assembly and disassembly proteins. Annu. Rev. Biochem. 59: 415–438.

Kennedy, J. R. (1965). The morphology of *Blepharisma undulans* Stein. *J. Protozool.* 12: 542–561.

Lipscomb, D. L. and Riordan, G. P. (1992). A reexamination of the ultrastructure of *Didinium nasutum* and a reanalysis of the phylogeny of the haptorid ciliates. *J. Protozool.* 39: 110–121.

Lobo-da-Cunha, A. and Azevedo, C. (1990a). Enzyme cytochemistry of the alveolar sacs and golgian-like cisternae in the ciliate *Ichthyophthirius multifiliis*. *J. Protozool.* 37: 206–211.

Lobo-da-Cunha, A. and Azevedo, C. (1990b). Ultrastructural localization of phosphatases with cerium in a ciliated protozoan. *Acta Histochem. Cytochem.* 23: 467–473.

Lom, J. and Corliss, J. O. (1971). Morphogenesis and cortical ultrastructure of *Brooklynella hostilis*, a dysteriid ciliate ectoparasitic on marine fishes. *J. Protozool.* 18: 261–281.

Lom, J. and Kozloff, E. N. (1969). Ultrastructure of the cortical regions of ancistrocomid ciliates. Protistologica 5: 173–192.

Lynn, D. H. (1976a). Comparative ultrastructure and systematics of the Colpodida: structural conservatism hypothesis and a description of *Colpoda steinii* Maupas. *J. Protozool.* 23: 302–314.

Lynn, D. H. (1976b). Comparative ultrastructure and systematics of the Colpodida: an ultrastructural description of *Colpoda maupasi* Enriquez, 1908. *Can. J. Zool.* 54: 405–420.

Lynn, D. H. (1976c). Comparative ultrastructure and systematics of the Colpodida (Ciliophora): structural differentiation in the cortex of *Colpoda simulans*. *Trans. Amer. Microsc. Soc.* 95: 581–599.

Lynn, D. H. (1976d). Comparative ultrastructure and systematics of the Colpodida: fine structural specializations associated with large body size in *Tillina magna* Gruber, 1880. *Protistologica* 12: 629–648.

Lynn, D. H. and Nicholls, K. H. (1985). Cortical microtubular structures of the ciliate *Lepidotrachelophyllum fornicis* Nicholls and Lynn, 1984 and phylogeny of the litostomate ciliates. *Can. J. Zool.* 63: 1835–1845.

Marsh, M., Griffiths, G., Dean, G. E., Mellman, I., and Helenius, A. (1986). Three-dimensional structure of endosomes in BHK-21 cells. *Proc. Natl. Acad. Sci. U.S.A.* 83: 2899–2903.

Mellman, I., Howe, C., and Helenius, A. (1987). The control of membrane traffic on the endocytic pathway. *Curr. Top. Membr. Transp.* 29: 255–288.

Mulisch, M., Barthlott, W., and Hausmann, K. (1981). Struktur und Ultrastruktur von *Eufolliculina* spec. Schwärmer und sessiles Stadium. *Protistologica* 17: 285–312.

Mulisch, M. and Hausmann, K. (1983). Lorica construction in *Eufolliculina* sp. (Ciliophora, Heterotrichida). *J. Protozool.* 30: 97–104.

Murphy, R. F. (1991). Maturation models for endosome and lysosome biogenesis. *Trends Cell Biol.* 1: 77–82.

Nilsson, J. R. and van Deurs, B. (1983). Coated pits and pinocytosis in *Tetrahymena*. *J. Cell Sci.* 63: 209–222.

Njine, T. and Didier, P. (1980). Étude ultrastructurale des ciliés du genre *Leptopharynx* Mermod, 1914. *Protistologica* 16: 155–166.

Noirot-Timothée, C. (1968). Les sacs parasomaux sont des sites de pinocytose: étude expérimentale à l'aide de thorotrast chez *Trichodinopsis paradoxa* (Ciliata Peritricha). *C.R. Acad. Sci. Paris* 267: 2334–2336.

Patterson, D. J. (1978). Membranous sacs associated with cilia of *Paramecium*. *Eur. J. Cell. Biol.* 17: 107–113.

Paulin, J. J. and Bussey, J. (1971). Oral regeneration in the ciliate *Stentor coeruleus*: a scanning and transmission electron optical study. *J. Protozool.* 18: 201–213.

Peck, R. K. (1977). Cortical ultrastructure of the scuticociliates *Dexiotricha media* and *Dexiotricha colpidiopsis* (Hymenostomata). *J. Protozool.* 24: 122–134.

Peck, R. K. (1978). Ultrastructure of the somatic and buccal cortex of the tetrahymenine hymenostome *Glaucoma chattoni*. *J. Protozool.* 25: 186–198.

Peck, R., Pelvat, B., Bolivar, I., and de Haller, G. (1975). Light and electron microscopic observations on the heterotrich ciliate *Climacostomum virens*. *J. Protozool.* 22: 368–385.

Peterson, O. W. and van Deurs, B. (1983). Serial section analysis of coated pits and vesicles involved in adsorptive pinocytosis in cultured fibroblasts. *J. Cell. Biol.* 96: 277–281.

Puytorac, P. de and Njine, T. (1970). Sur l'ultrastructure des *Loxodes* (ciliés holotriches). *Protistologica* 6: 427–444.

Rabinowitz, S., Horstmann, H., Gordon, S. and Griffiths, G. (1992). Immunocytochemical characterization of the endocytic and phagolysosomal compartments in peritoneal macrophages. *J. Cell. Biol.* 116: 95–112.

Rasmussen, L. and Orias, E. (1975). *Tetrahymena*: growth without phagocytosis. *Science* 190: 464–465.

Rudzinska, M. A. (1977). Uptake of ferritin from the medium by *Tokophrya infusionum*. *Experientia* 33: 1595–1598.

Rudzinska, M. A. (1980). Internalization of macromolecules from the medium in suctoria. *J. Cell. Biol.* 84: 172–183.

Ruiz, S. and Anadon, R. (1986). The oral structures of the peritrich ciliate *Opercularia coarctata* (Clap. and Lachm., 1858): a light microscope and ultrastructural study. *Protistologica* 22: 291–299.

Schroeder, C. C., Fok, A. K., and Allen, R. D. (1990). Vesicle transport along microtubular ribbons and isolation of cytoplasmic dynein from *Paramecium*. *J. Cell. Biol.* 111: 2553–2562.

Sénaud, J. and Grain, J. (1972). Étude ultrastructurale préliminaire de *Cochliatoxum periachtum* Gassovsky, 1919, cilié entodiniomorphe endocommensal du Cheval. *Protistologica* 8: 65–82.

Small, E. B. and Lynn, D. H. (1985). Phylum Ciliophora Doflein, 1901. In: *An Illustrated Guide to the Protozoa* (Lee, J. J., Hutner, S. H., and Bovee, E. C., eds.), pp. 393–575, Society of Protozoologists, Lawrence, KS.

Stelly, N., Mauger, J. P., Claret, M., and Adoutte, A. (1991). Cortical alveoli of *Paramecium*: a vast submembranous calcium storage compartment. *J. Cell. Biol.* 113: 103–112.

Stockem, W. (1977). Endocytosis. In: *Mammalian Cell Membranes* (Jamieson, G. A. and Robinson, D. M., eds.), vol. 5, pp. 151–195. Butterworths, London.

Stockem, W. and Wohlfarth-Bottermann, K. E. (1969). Pinocytosis (endocytosis). In: Handbook of Molecular Cytology (Lima de Faria, A., ed.), pp. 1373–1400, North Holland, Amsterdam.

Suchard, S. J. and Goode, D. (1982). Microtubule-dependent transport of secretory granules during stalk secretion in a peritrich ciliate. *Cell Motil.* 2: 47–71.

Tucker, J. B. (1971). Development and deployment of cilia, basal bodies, and other microtubular organelles in the cortex of the ciliate *Nassula*. *J. Cell Sci.* 9: 539–567.

Van Deurs, B., Petersen, O. W., Olsnes S., and Sandvig, K. (1989). The ways of endocytosis. *Int. Rev. Cytol.* 117: 131–177.

Weidman, P. J., Melançon, P., Block, M. R., and Rothman, J. E. (1989). Binding of an *N*-ethylmaleimide-sensitive fusion protein to Golgi membranes requires both a soluble protein(s) and an integral membrane receptor. *J. Cell. Biol.* 108: 1589–1596.

Westcot, D. M., Fok, A. K., and Allen, R. D. (1985). Flow of membrane markers from the coated pits in *Paramecium caudatum*. *J. Cell. Biol.* 101: 52a (abstract).

Wicklow, B. J. (1983). Ultrastructure and cortical morphogenesis in the euplotine hypotrich *Certesia quadrinucleata* Fabre-Domergue, 1885 (Ciliophora, Protozoa). *J. Protozool.* 30: 256–266.

Williams, D. B., Williams, B. D., and Hogan, B. K. (1981). Ultrastructure of the somatic cortex of the gymnostome ciliate *Spathidium spathula* (O.F.M.). *J. Protozool.* 28: 90–99.

Wilson, D. W., Whiteheart, S. W., Orci, L., and Rothman, J. E. (1991). Intracelluar membrane fusion. *Trends Biochem. Sci.* 16: 334–337.

Wilson, D. W., Whiteheart, S. W., Wiedmann, M., Brunner, M., and Rothman, J. E. (1992). A multisubunit particle implicated in membrane fusion. *J. Cell. Biol.* 117: 531–538.

MEMBRANE FLOW IN THE DIGESTIVE CYCLE OF *PARAMECIUM*

Agnes K. Fok and Richard D. Allen

I. INTRODUCTION

The digestive system of *Paramecium* is possibly one of the most complex membrane flow systems currently known in any organism, plant, animal, or protist. This

Advances in Cell and Molecular Biology of Membranes, Volume 2B
Membrane Traffic in Protozoa, pages 311–337

ISBN: 1-55938-628-2

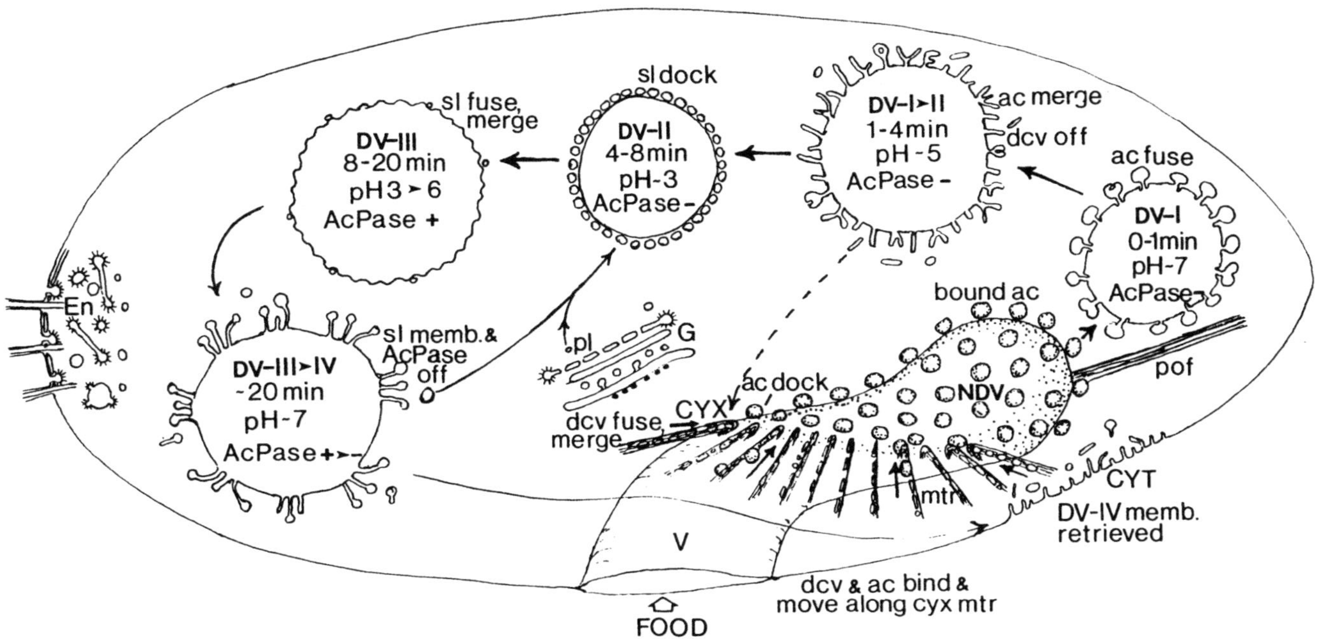

Figure 1. Annotated map of the flow of membrane through the phagosome–lysosome system in *Paramecium*. Consult the text for a more detailed explanation of the events occurring at the various compartments. ac = acidosome; AcPase = acid phosphatase; CYT = cytoproct; CYX = cytopharynx; dcv = discoidal vesicle; DV-I to DV-IV = digestive vacuole stages; En = endosomes (with clathrin coats indicated by stipples); G = Golgi complex; mtr = microtubular ribbons; NDV = nascent digestive vacuole; pl = primary lysosome; pof = postoral fiber; sl = secondary lysosome; V = vestibulum.

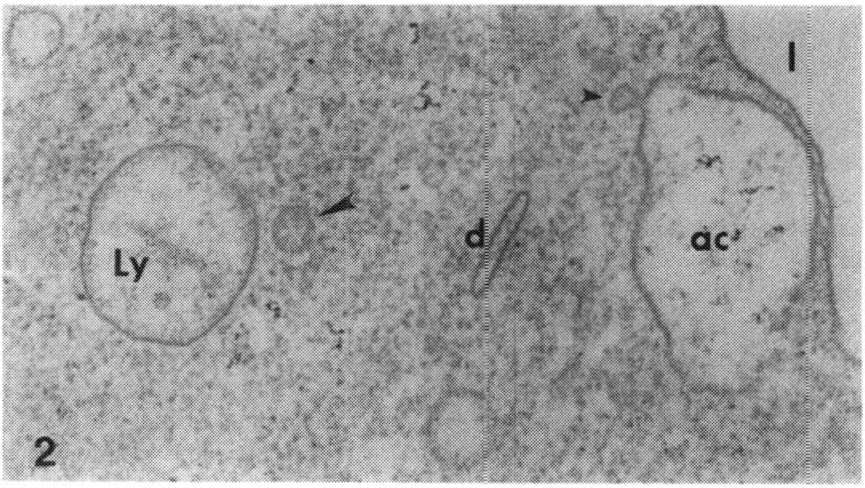

Figure 2. Profiles of the vesicles taking part in the digestive vacuole cycle. These include the acidosome (ac), which is bound here to the DV-I membrane (I), the lysosome (Ly), the discoidal vesicle (d), and a 100-nm vesicle (large arrowhead). The small arrowhead indicates a vesicle either going on or coming off the acidosome. (Unpublished micrograph). ×50,000.

system begins with the formation of a large digestive vacuole at the cytopharynx and ends with the retrieval of spent vacuole membrane at the cytoproct. (Figure 1 should be consulted throughout this chapter for a summary of the structures and membrane flow patterns discussed.) Several types of vesicles each bearing their own antigenic determinants enter and leave the digestive vacuole membrane through the course of the vacuole's progression through the cell. These vesicles include discoidal vesicles, acidosomes, and lysosomes (Fig. 2). In addition recent evidence suggests an intimate association of small "shuttle" vesicles with at least some stages of digestive vacuoles, indicating another possible branch of membrane flow.

In this chapter we will describe this elaborate membrane flow system, and whenever possible we will comment on the source of the membrane for the various vesicle populations, vesicle recruitment, sequence of their entry into the vacuole, the consequences of their entry on the processes of vacuole function, the fate of vesicle membranes or membrane macromolecules after incorporation into the vacuole membrane, and what is known, if anything, about the regulation of the steps of membrane flow. Problems for future consideration and study will be given at the end of this chapter. Other aspects of this membrane flow system in *Paramecium* have been included in two previous reviews (Fok and Allen, 1988, 1990).

II. SOURCE AND RECRUITMENT OF DIGESTIVE VACUOLE (PHAGOSOME) MEMBRANE

Digestive vacuoles or phagosomes form from the cytopharynx of *Paramecium.* The cytopharynx (Fig. 3) is delineated by a single limiting membrane along the dorsal posterior half of the oral region and defines the posterior border of the buccal cavity proper (Allen, 1974). On its left side the cytopharynx is in close contact with about 40 ribbons of cytopharyngeal microtubules which arise from the filamentous reticulum and fan out into the posterior region of the cell, reaching as far as the cytoproct (Allen and Wolf, 1974).

The membrane of the cytopharynx is continuous with the plasma membrane, which covers the whole cell, but is notably different in its composition, both in its freeze-fracture characteristics and its lectin-binding and antigenic properties (Figs. 4, 5). Following freeze-fracture the cytopharyngeal membrane is seen to bear a very high number of intramembrane particles (IMPs) on its E-fracture face, so many that it appears to be entirely covered with IMPs (Allen and Staehelin, 1981). In contrast the E-fracture face of the plasma membrane has far fewer IMPs (Allen, 1976). Within the membrane there is a distinct band of packed IMPs separating the plasma membrane from the cytopharynx (Fig. 3) (Allen, 1978). This band of IMPs runs parallel to the terminal margins of the underlying alveoli and may help bind the cell's outer membrane to the alveolar membrane system (Allen, 1978). This band of IMPs may also prevent intermixing of the integral membrane proteins between the plasma membrane and the cytopharyngeal membrane.

That intermixing of IMPs between the plasma membrane and cytopharynx membrane does not occur is borne out by the differences in antigens found in these two membrane regions and in their ability to bind lectin. Wheat germ agglutinin (WGA) binds to the luminal surfaces of all phagosome-lysosome membranes of *Paramecium multimicronucleatum* (Allen et al., 1989) including the external

---

Figure 3. Freeze-fracture view of the left side of the oral region exposing a portion of the cytopharynx (CYX), which, although continuous with the plasma membrane (pm) along one side, is separated from it by a band of intramembrane particles (between large black arrowheads) that follows the terminus of the underlying alveolar sac (as). The cytopharynx merges into the nascent digestive vacuole (ndv) on its other side. Discoidal vesicles (d) accumulate next to the cytopharynx and fuse with the cytopharyngeal membrane (small black arrowheads) between the ridges where the cytopharyngeal microtubular ribbons abut upon the membrane. Acidosomes dock at the growing vacuole membrane in a line (curved large white arrow) bordering these ridges. Imprints of a row of acidosomes can be seen (small white arrows) against the nascent vacuole, and fractured acidosomes (ac) are found in the adjacent cytoplasm. C = cytostomal cord; T = trichocyst. Reproduced from Allen (1978) by permission of Academic Press, San Diego. ×15,000.

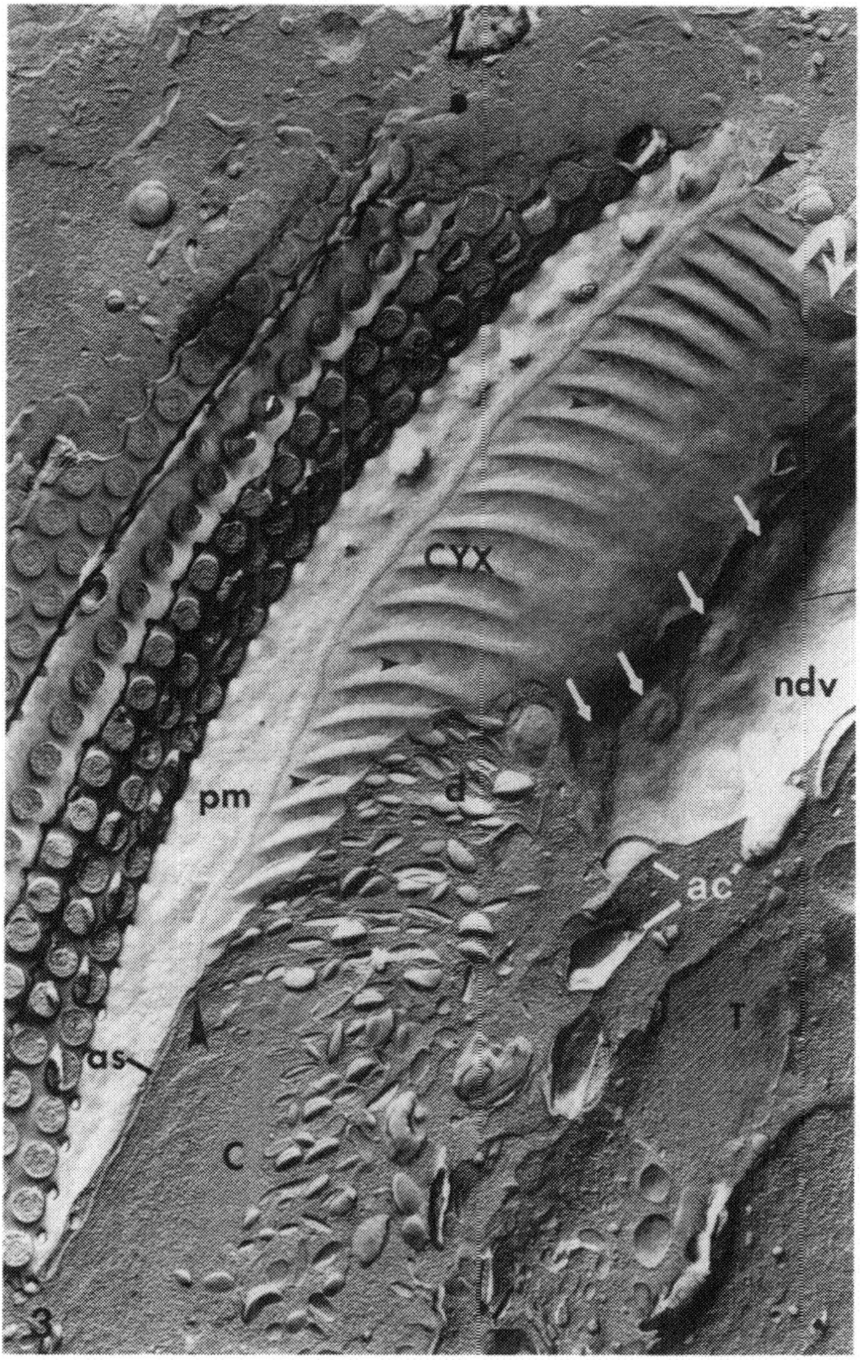
CYX
ndv
pm
d
ac
T
as
C
3

surface of the cytopharynx. Similar results were also reported for *P. tetraurelia* (Lüthe and Plattner, 1986). However, WGA does not bind to the plasma membrane's outer surface in cryosections. In contrast, the resident plasma membrane antigen, a 190 kD protein called C6, is found over the entire external surface of the plasma membrane including the ciliary membranes, but is not present on the cytopharynx membrane (Fig. 4) (Allen et al., 1992). The reverse is true of the B2 antigen, which is found on the cytopharynx but not on the plasma membrane (Fig. 5). The PC-1 monoclonal antibody reported in Fok et al. (1986) reacts with the B2 antigen. In all these cases there is a sharp line of demarcation between these labeled and unlabeled zones at the line where the alveolar system ends and where the band of packed IMPs presumably binds the cell's outer membrane to the underlying intracellular alveolar membrane.

Whereas the phagosomal membrane in phagocytic cells, such as amoeba and polymorphonuclear leukocytes, is derived from the plasma membrane (Moore et al., 1978; Bowers, 1980; reviewed in Steinman et al., 1983), this is not true of *Paramecium*. In *Paramecium*, as in other ciliates, the phagosome membrane is derived from a pool of disk-shaped vesicles which are transported to the cytopharynx, where they will fuse with the cytopharynx membrane to provide for the growth in surface area of the developing vacuole (Allen, 1974). Discoidal vesicles, like the cytopharynx, also exhibit the B2 antigen (Fig. 6). These discoidal vesicles measure around 0.3 μm in diameter and are usually about 60 nm wide. Their membrane contains a very regular luminal lining (Allen and Fok, 1984a), probably composed of glycosylated proteins, as this lining is reactive with WGA and the reaction is completely inhibited by chitotriose (Allen et al., 1989). WGA binding indicates the presence of *N*-acetylglu-

---

Figures 4–6. Cytopharynx region. Fig. **4.** Cryosection through the cytopharynx (CYX) immunogold-labeled with a monoclonal antibody specific for the C6 resident plasma membrane antigen. Label is found on the plasma and ciliary membranes but not on the cytopharynx membrane, which begins where the alveoli end (arrowhead). (Unpublished micrograph) ×25,000.

Fig. **5.** Epon-embedded section of cell exposed to the immunogold-labeled procedure prior to embedment. A monoclonal antibody to the B2 antigen found to occur only on discoidal vesicles (d), the cytopharynx (CYX), and DV-I was used. The cell was fixed in a paraformaldehyde–lysine–periodic acid mixture and is not well preserved, but the monoclonal antibody to B2 and secondary antibody–gold complexes were still unable to penetrate through the cellular membranes. Thus only the externally exposed cytopharynx is labeled. The plasma membrane (beginning at the arrowhead) and membranes of cilia (ci) are not labeled. (Unpublished micrograph) ×25,000.

Fig. **6.** Discoidal vesicles lying next to the cytopharynx (CYX) contain the B2 antigen. The small black arrowhead points to the compact 15-nm-thick luminal fringe found in these vesicles. (Unpublished micrograph) ×50,000.

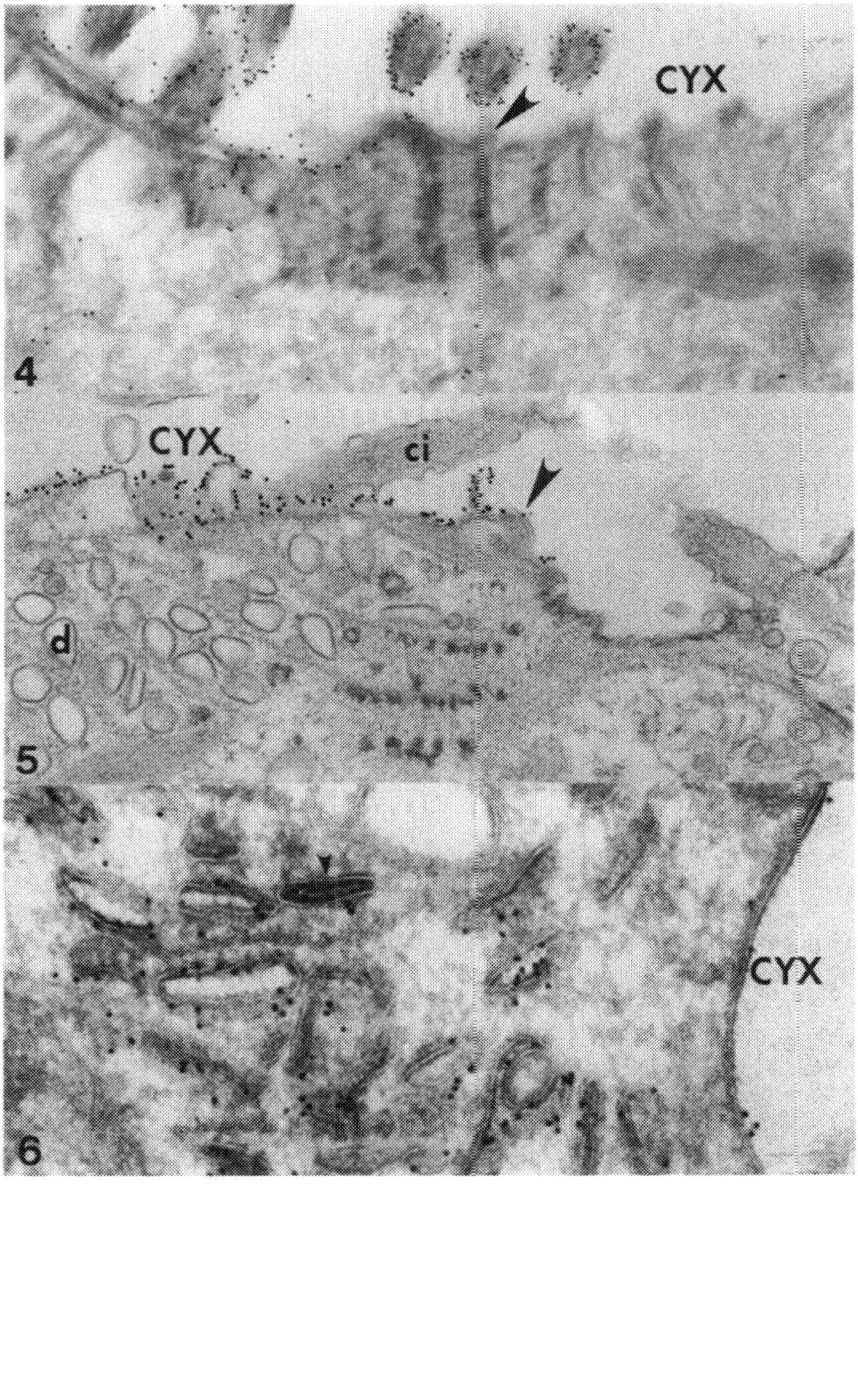
CYX
4
CYX
ci
d
5
CYX
6

cosamine probably linked together as β-(1–4) oligomers. This luminal lining appears to completely fill the limited space inside the flattened discoidal vesicles, suggesting that these vesicles have no other content. The only known role of discoidal vesicles in *Paramecium* is to provide membrane for phagosome growth. In other ciliates such as *Euplotes* the discoidal vesicles contain electron-opaque filamentous material which is released into the nascent vacuole when these vesicles fuse with the cytopharynx membrane (Kloetzel, 1974). This suggests additional functions for these vesicles in other ciliates besides providing membrane for vacuole growth.

The origin of discoidal vesicles has not been entirely worked out. At least some of these vesicles are recycled from spent vacuole membrane returning from the cytoproct following vacuole defecation (Fig. 7) (Allen and Fok, 1980). *Paramecium* is uniquely designed to efficiently bring about this recycling. Tubular vesicles pinching off from the spent vacuole come into contact with and bind to those cytopharyngeal ribbons of microtubules which extend toward the cytoproct (Fig. 8). These vesicles then become flat and disk-shaped and move along one side only of these ribbons, where they may be powered by cytoplasmic dynein, a minus-end-directed microtubule-based motor which we have recently isolated from *Paramecium* (Schroeder et al., 1990).

Another likely source of discoidal vesicles is the recycled membrane from early phagosomes, which are condensing and undergoing membrane replacement as they fuse with acidosomes and become acidified (Fig. 9) (also discussed in Section IV). However, as yet we have not been able to unambiguously distin-

---

Figures 7–10. Retrieval and transport of membrane. Fig. **7.** Spent vacuole membrane is reengulfed at the open cytoproct (CYT) by the rapid tubularization and fission of tubules. A microfilamentous system promotes this process. One lip of the cytoproct (arrowhead) from which one set of microtubules arises is shown. Reproduced from Allen and Fok (1985) by permission of Wissenschaftliche Verlagsgesellschaft mbH, Stuttgart. ×40,000.

Fig. **8.** Retrieved discoidal vesicles, some containing HRP obtained at the open cytoproct, align in single file along only one side of a cytopharyngeal microtubular ribbon and are transported in this manner back to the cytopharynx. (Unpublished micrograph) ×20,000.

Fig. **9.** Membrane replacement occurs in transition vacuoles between stages I and II. Acidosomes (ac) fuse at this time while the discoidal vesicle-derived membrane is removed (small arrowheads) as tubules form and pinch off. (Unpublished micrograph) ×30,000.

Fig. **10.** Quick-freeze deep-etch rotary-shadow replica of chemically unfixed cell. Small 100-nm vesicles bind to ribbons of microtubules and are presumably transported along the ribbons. Reproduced from Fok and Allen (1990) by permission of Academic Press, San Diego. ×80,000.

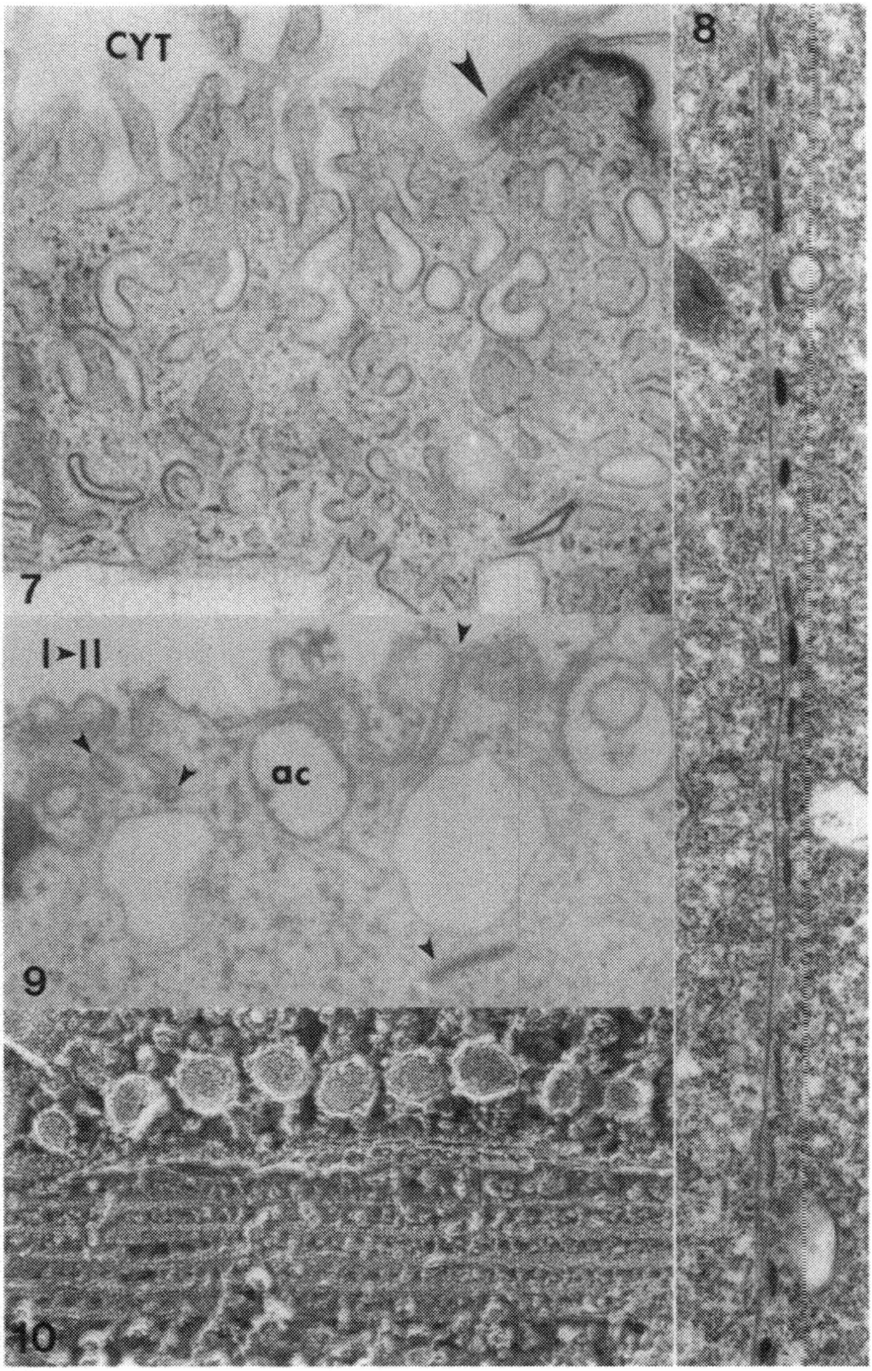
CYT
7
I➤II
ac
9
10
8

guish the HRP (horseradish peroxidase)-labeled vesicles, which arise from these young vacuoles, from those that arise from the spent vacuoles at the cytoproct.

This still leaves unsolved the question of the source of new (vs. recycled) discoidal vesicle membrane. As these discoidal vesicles share the B2 antigen with the cytopharynx and the DV-I (see Fig. 1 for the classification of digestive vacuole stages) and is not found elsewhere in the cell, it appears that this is a separate membrane pool and the resident antigens in this pool must derive from the endoplasmic reticulum and Golgi complex pathway. Additional work needs to be done to show where the B2 antigen is produced and how it becomes part of a discoidal vesicle.

In addition to the discoidal vesicles, a recent study has shown that a pool of 100-nm vesicles binds to and may also be transported along the cytopharygeal microtubular ribbons (Fig. 10) (Schroeder et al., 1990). Whether these small vesicles fuse with the nascent DV and/or acidosomes has not been determined. If they do fuse with the nascent DVs, these 100-nm vesicles would contribute membrane components to the nascent DVs that may perform some phagosomal functions not yet discovered.

III. SOURCE AND RECRUITMENT OF ACIDOSOMES

A second population of larger vesicles (Allen and Fok, 1983a, 1983b), called acidosomes (Allen and Fok, 1983c), accumulate in the region of the growing phagosome. Acidosomes are roughly 1 μm in diameter and, although often irregular in shape, are more rounded than the smaller discoidal vesicles. This is particularly evident following quick freezing (Fig. 11). For the most part, acidosomes appear

Figures 11–14. Acidosomes and the membrane of DV-I and DV-II.

Fig. **11.** Quick-freeze deep-etch rotary-shadow replica of a collapsed region of a DV-I containing an engulfed bacterium (Ba). This DV-I membrane is lined with acidosomes. Both P-fracture (P) and E-fracture (E) faces of acidosomes are exposed as well as the luminal labyrinth-like content in a cross-fractured acidosome (arrowhead). (Unpublished micrograph) ×30,000.

Fig. **12.** The DV-I (I) membrane is positive for the B2 antigen (labeled with gold) but not the membranes of the acidosomes which are fusing (arrowheads) with this DV-I. Gold is also found in retrieved vesicles. (Unpublished micrograph) ×30,000.

Fig. **13.** The DV-I (I) membrane is negative for the L1 antigen but the antigen is found in the docked acidosomes. Polystyrene latex beads of 1.1 μm diameter located in the vacuole indicate that this vacuole is 0–3 min old. (Unpublished micrograph) ×30,000.

Fig. **14.** The DV-II (II) membrane is positive for the L1 antigen but not the lysosomes (Ly) docked at its cytosolic surface. This DV-II contains 0.8 μm diameter latex beads, so it is 5–8 min old. (Unpublished micrograph) ×30,000.

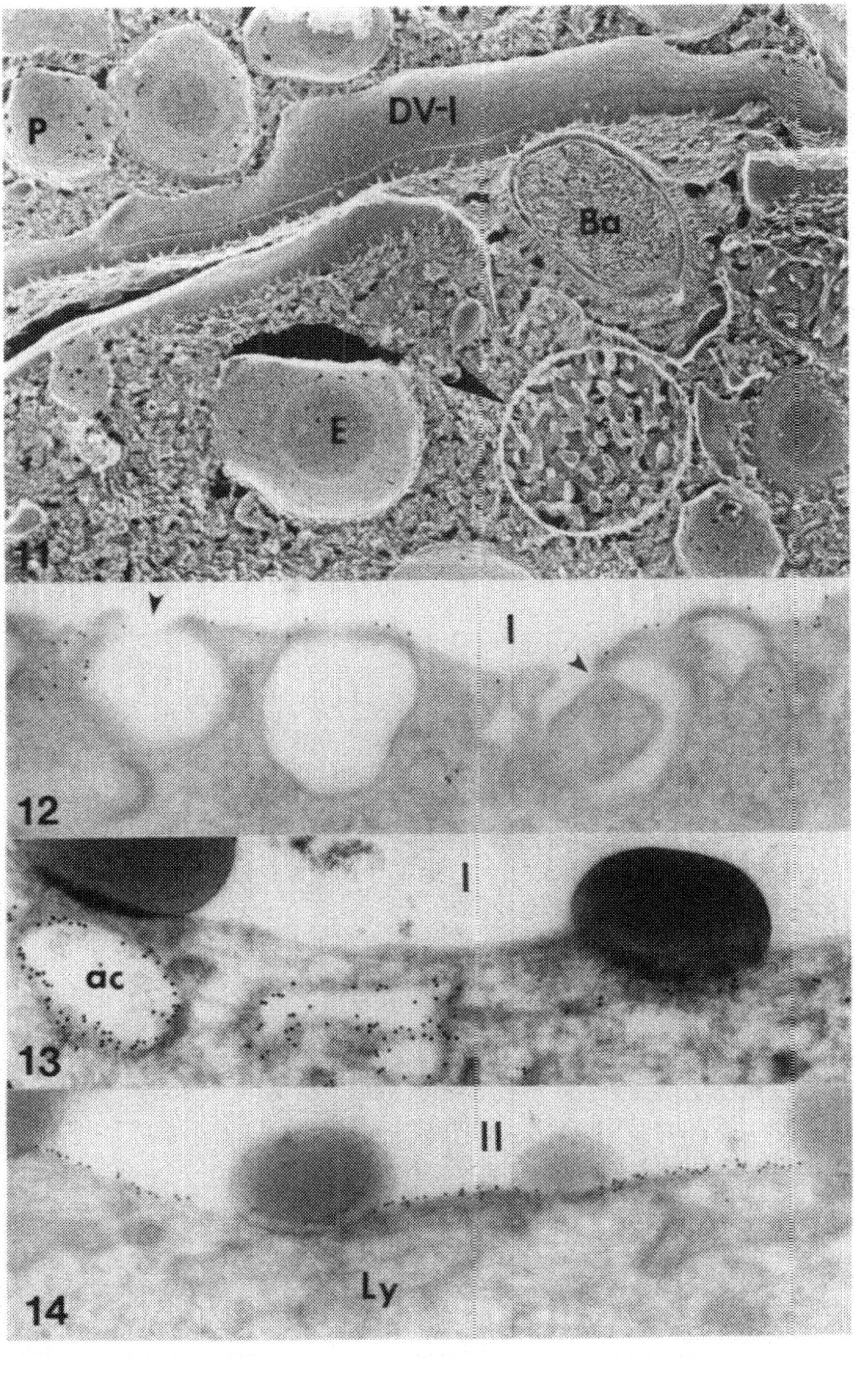

P
DV-I
Ba
E
11
12
I
13
ac
I
14
II
Ly

to be devoid of electron-opaque material and so appear empty in thin-section electron-microscopic studies. Acidosomes also move along the cytopharyngeal ribbons toward the cytopharynx (Schroeder et al., 1990). However, as they arrive at the cytopharynx they do not fuse with but dock at the nascent vacuole membrane (Fig. 3), where they remain as the nascent vacuole grows in size. Additional acidosomes subsequently dock with the newly incorporated nascent vacuole membrane, resulting in a layer of docked acidosomes surrounding the nascent phagosome as it grows and is released from the cytopharynx.

Acidosomes have a low pH. This is indicated by their ability to accumulate weak bases such as acridine orange and neutral red in the living cell (Allen and Fok, 1983c). Acidosomes fluoresce red-orange when the cell has been in the presence of acridine orange for a few minutes, indicating the accumulation of a high concentration of this dye.

The membrane of acidosomes differs antigenically from that of the B2-positive discoidal vesicles, cytopharynx and DV-I (Fig. 12), in its freeze-fracture appearance and in the morphological characteristics of its luminal lining. Antigenically this membrane contains another antigen, L1, which, apart from the acidosomes (Fig. 13), is found only in the acidified vacuoles (Fig. 14) and in some flattened cisternae and smaller vesicles in the cytosol of the cell (Allen et al., 1993). In freeze-fractured as well as quick-freeze deep-etch replicas (Fig. 11), the fractured acidosome membrane has an IMP-free E-fracture face and a P-fracture face with prominent IMPs that leave behind distinct pits on the E-fracture face (Allen and Fok, 1983b; Fok and Allen, 1990). The luminal membrane lining of these acidosomes is thinner and more corrugated in profile than the smooth compact lining of the discoidal vesicles (Allen and Fok, 1984a). However, like discoidal vesicles, the acidosome membrane also binds WGA (Allen et al., 1989).

Acidosomes appear to arise in the cytoplasm as smaller vesicles or flat cisternae (preacidosomes) which bear the L1 antigen (Allen et al., 1993). These vesicles and cisternae appear to actively coalesce with a second population of small vesicles of about 100 nm in diameter. These small vesicles do not contain the L1 antigen and they may or may not be different from the pool of 100-nm vesicles discussed in Section II (Fig. 10). Acidosomes at all stages of formation, up to the time of docking at the nascent vacuole membrane, show extensive signs of fusion with small vesicles. Small vesicles are often found docked at the acidosomes and small projections frequently arise from acidosomes. Regions of heterogeneity of IMPs in the deep-etch rotary-shadowed replicas of the acidosomes also suggest active fusion of these acidosomes with incoming vesicles. Thus acidosomes may be a focal point for the flow of many small vesicles into the phagosome system. Small vesicles may also leave acidosomes but we have no markers to follow such activity.

Thus, acidosome membrane very probably has a heterogeneous origin within the cell consisting of preacidosome vesicles bearing the L1 antigen as well as an input of L1-negative membrane from small vesicles. The pool of acidosome membrane, because of the resident L1 antigen, can be distinguished as a unique pool apart from discoidal vesicles, lysosomes, and the early endosome compartments (for the latter see chapter by Allen and Fok in this volume).

The fate of the L1 antigen once incorporated into the vacuole membrane has not been determined. Two possibilities are (1) that all or part of the acidosome membrane along with the L1 antigen is retrieved at the time that vacuoles fuse with lysosomes, or (2) that the antigen is inactivated or sloughed off the vacuole membrane into the vacuole and degraded as lysosomes fuse. In cryosections, only vacuoles in stage II, acidified vacuoles, label with the monoclonal antibody to L1 antigen. (Using cryosectioning DVs either have abundant L1 epitopes or none at all, as a gradual loss of the L1 epitope in DVs probed by this technique is not usually observed.) Based on the tendency of cells to reuse their structural molecules and their membranes, we suspect that recycling of acidosome membrane probably occurs and will eventually be observed.

IV. MEMBRANE REPLACEMENT IN EARLY DIGESTIVE VACUOLES

As stated in Section II the digestive vacuole membrane is derived, for the most part, from discoidal vesicle membrane with some possible input from at least one population of 100-nm vesicles. At some point the vacuole pinches off from the cytopharynx and is quickly propelled to the posterior end of the cell. Cohen et al. (1984a, 1984b) have shown that this occurs concomitantly with the polymerization of actin around the DV-I and along the bundles of postoral microtubular fibers, which arise along the right side of the buccal cavity and which extend toward the cell's posterior pole. That actin is involved in vacuole release has been confirmed by microinjecting phalloidin, an F-actin binding drug, into *Paramecium* (Kersken et al., 1986). At 20 μg/ml phalloidin strongly inhibits vacuole detachment.

Vacuole movement to the posterior end of the cell is rapid and requires only seconds to complete. Once the vacuole reaches the end of the microtubular bundles it is released to the cytosol. Shortly after this release a profound change takes place in the vacuole membrane, the acidosomes fuse with the vacuole concomitant with the retrieval of the original membrane of the vacuole. The latter is achieved by the formation of tubular extensions which protrude into the cytosol and which pinch off (Fig. 9). The net result is the almost complete replacement of the vacuole membrane around the vacuole's content (Allen and Fok, 1983b) at the start of vacuole acidification. Membrane replacement is accompanied by a complete change of the freeze-fracture appearance from that of a DV-I (Fig. 15), which is identical to discoidal vesicles (Fig. 16), to a DV-II (Fig. 17), which is identical to acidosome membranes (Allen and Staehelin, 1981; Allen and Fok, 1983b). The E-fracture face loses its many IMPs and becomes almost IMP-free while the P-fracture face receives many prominent particles like those on the acidosome P-fracture face. In addition, in cryosections the DV-II membrane no longer contains the B2 antigen but now is positive for the L1 antigen (Figs. 12, 14) (Allen et al.,

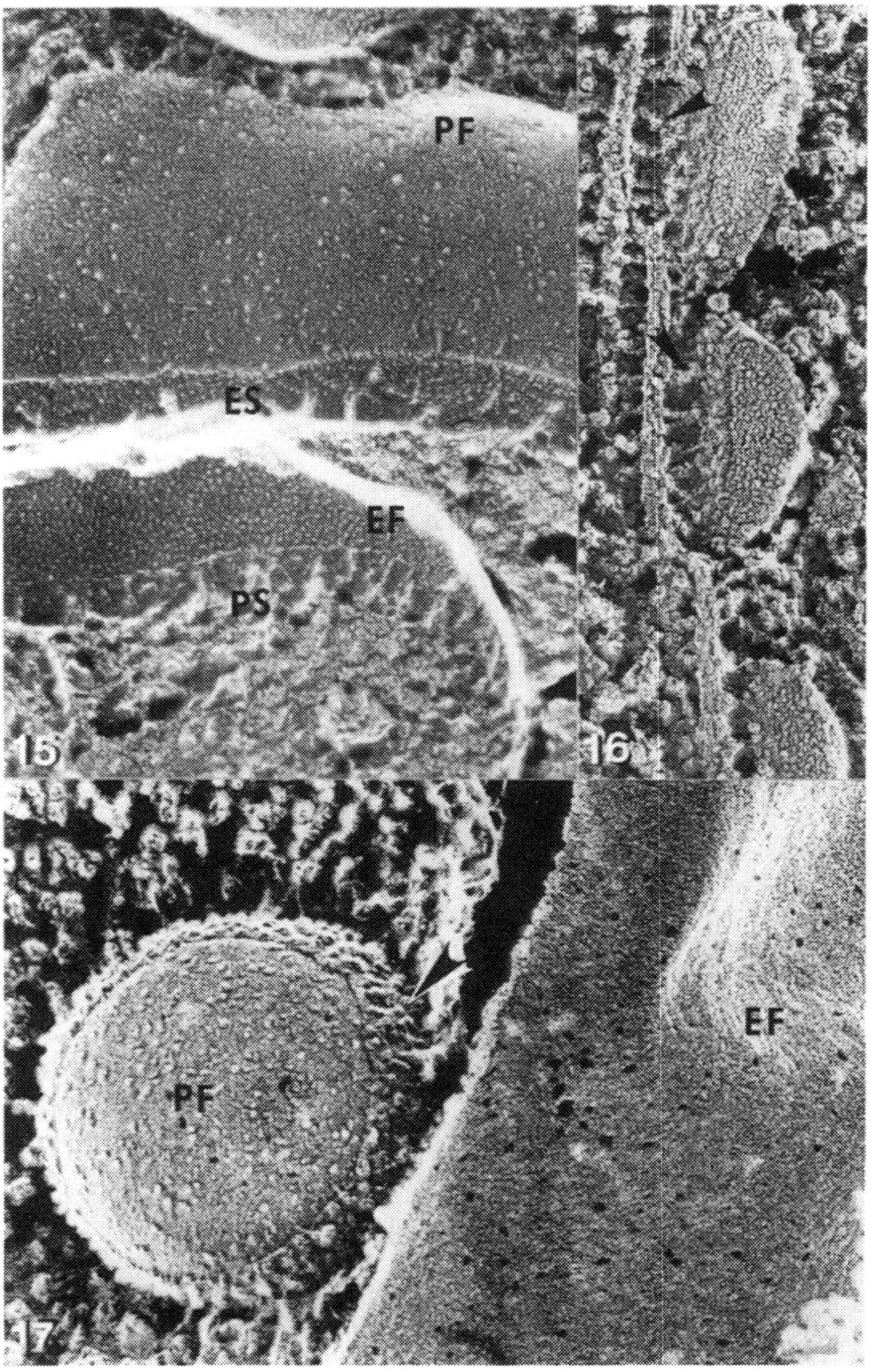
PF
ES
EF
PS
15
16
PF
EF
17

1993). This is also observed in immunofluorescence microscopy (Fok et al., 1986). In this way the membrane of a DV-I changes from that similar to discoidal vesicles into a DV-II containing mostly acidosome-like membrane. Taken together the freeze-fracture and antigenic data indicate that membrane replacement is complete, or nearly so.

What is the purpose for this membrane replacement? Two events occur while the nascent vacuole is surrounded by discoidal vesicle-type membrane. First, acidosomes bind to the nascent vacuole. Second, the nascent vacuole pinches off from the cytopharynx and moves to the cell's posterior end. These events probably require docking sites on the nascent vacuole membrane's cytosolic surface for acidosomes and sites on the DV-I for interaction with the motors which propel the vacuole along the microtubular bundles of the postoral fiber, be they kinesin-like motors (producing movement along a microtubule in a plus-end direction [Vale et al., 1985]) or an actin-based motor system, as is suggested by the work of Cohen et al. (1984a, 1984b). Also, sites for the attachment of putative cytoskeletal proteins presumably required for membrane retrieval via tubule formation and fission must also be present on DV-I membrane. Discoidal vesicle-type and/or 100-nm vesicle-type membrane apparently contribute the required docking sites for molecules involved in all three of these events. Once the vacuole has moved posteriorly and the acidosomes have fused with the vacuole these sites may no longer be needed and, indeed, to retain some of these sites may be detrimental to subsequent steps in the vacuole's normal processing. In addition the vacuole may now need to acquire the mechanism to bring about the acidification of the vacuole as well as

---

Figures 15–17. Appearance of membrane faces of DV-I, DV-II, and discoidal vesicles in deep-etch rotary-shadow replicas.

Fig. **15.** Enlargement of a portion of the DV-I illustrated in Fig. 11. All four surfaces that are exposed following the quick-freeze deep-etch rotary-shadow technique are present. The highly particulate E-fracture (EF) face and the more moderately particulate P-fracture (PF) face are shown as well as the true external membrane surface (ES), which bears subunits in a whorled pattern and the almost hidden true protoplasmic surface (PS). (Unpublished micrograph) ×100,000.

Fig. **16.** The E-fracture face of discoidal vesicles is like the E-fracture face of the DV-I in Fig. 15. Also exposed by the etching, and indicated by arrowheads, are bridges (motors?) linking the vesicles to the microtubules and limited views of the true protoplasmic surface of discoidal vesicles. (Unpublished micrograph) ×100,000.

Fig. **17.** Quick-freeze deep-etch rotary-shadow replica of a DV-II, which (because it bore several blebs protruding from its membrane into the vacuole lumen) has both its particle-free E-fracture (EF) face and particle-rich P-fracture (PF) face exposed. The EF-face has several holes or pits where IMPs were removed during fracturing. The rough or bumpy luminal lining of the DV-II is also exposed on the surface of the bleb (arrowhead). (Unpublished micrograph) ×100,000.

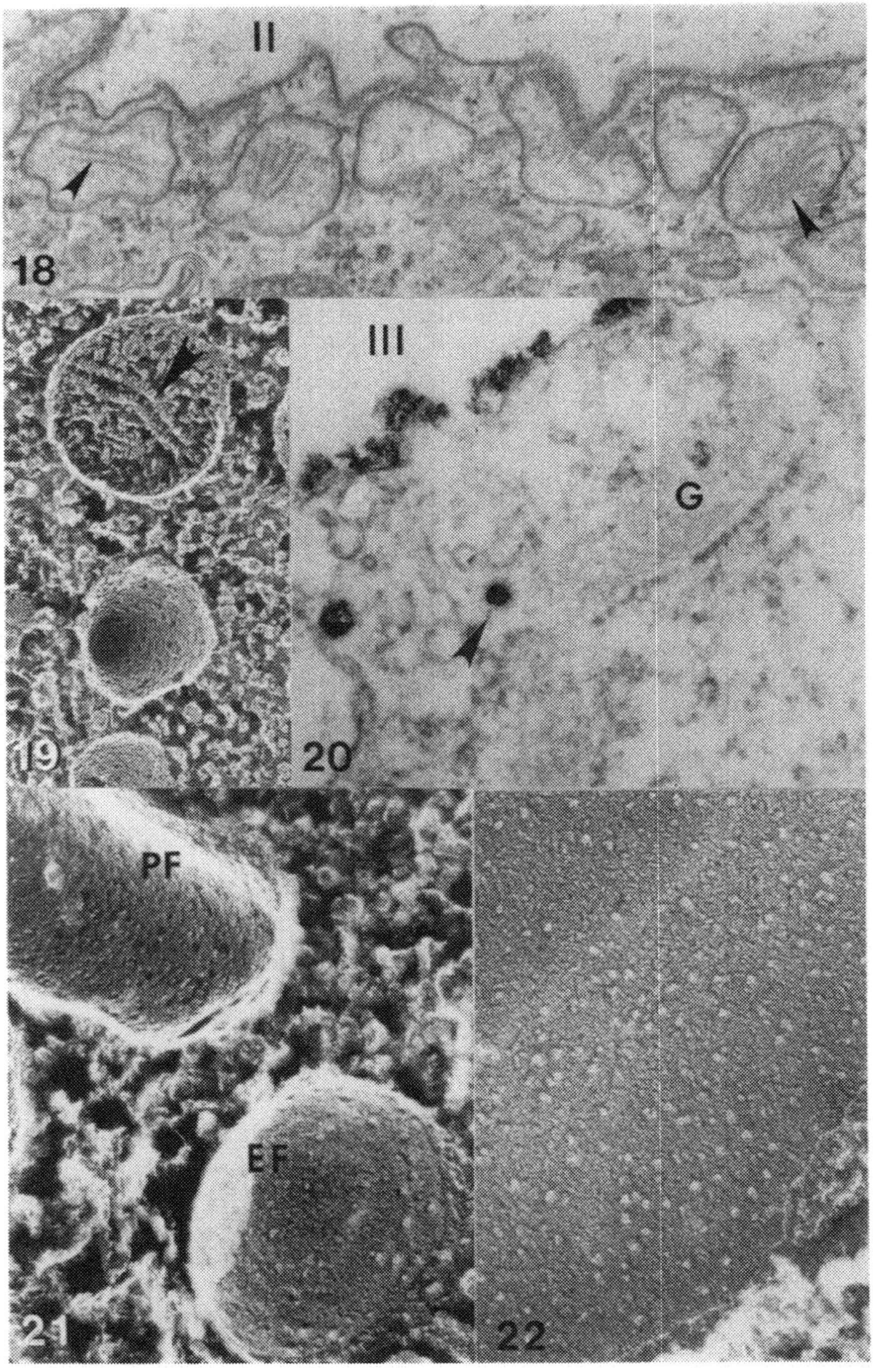

II
18
III
G
19
20
PF
EF
21
22

the docking sites for the lysosomes. These components would presumably be provided by the acidosome membrane. However, it is interesting to note that prior to acidosome binding and incorporation into the vacuole, acidosomes do not noticeably bind to free discoidal vesicles nor do lysosomes bind to the free acidosomes.

In the case of acidification, we do know that when acidosome fusion with the vacuole is inhibited by cytochalasin B, the vacuole will not become acidic (Allen and Fok, 1983c). We also know that few lysosomes will dock and fuse with the vacuole unless the vacuole first becomes acidic (Fok et al., 1987). Apparently membrane replacement not only removes docking sites for acidosomes and for the motors that power the vacuoles along the postoral fiber but also provides the mechanism for vacuole acidification and the docking sites for lysosomes. These changes are necessary to transform a DV-I into a DV-II, whose membrane morphology is now identical to that of the acidosome membrane.

V. SOURCE AND RECRUITMENT OF LYSOSOMES

Concomitant with vacuole acidification are the approach and docking of lysosomes with the vacuole membrane. Lysosomes are smaller than acidosomes and have a different morphology. Whereas acidosomes are irregular in shape and appear to be empty, lysosomes contain a more electron-opaque content and are spherical or prolate in shape (Fig. 18). These lysosomes contain a battery of acid hydrolases which includes acid phosphatase (Fok and Paeste, 1982). They also display a 30-nm-wide glycocalyx on their luminal surface (Fok and Allen, 1979; Allen and Fok, 1984a) and usually one or more sheets of paracrystalline material in their lumens (Fig. 18) (Fok and Allen, 1979, 1981). In deep-etch rotary-shadowed

Figures 18–22. Lysosomes and DV-III membrane.

Fig. **18.** Secondary lysosomes docked at a DV-II (II). These lysosomes contain a prominent luminal glycoprotein lining and paracrystalline sheets (arrowheads). (Unpublished micrograph) ×50,000.

Fig. **19.** Quick-freeze deep-etch rotary-shadow replica of lysosomes showing paracrystalline sheets (arrowhead), a globular content, and P-fracture face with pits. (Unpublished micrograph) ×50,000.

Fig. **20.** Primary lysosome (arrowhead) filled with acid phosphatase reaction product near a Golgi complex (G) and a DV-III (III), which also contains this reaction product. (Unpublished micrograph) ×50,000.

Fig. **21.** P-fracture face (PF) and E-fracture face (EF) of lysosomes prepared by quick-freeze deep-etch and rotary shadowing. A few large IMPs are found on the EF-face. (Unpublished micrograph) ×100,000.

Fig. **22.** Quick-freeze deep-etch rotary-shadow replica of DV-III EF-face. A moderate number of prominent IMPs are exposed. (Unpublished micrograph) ×100,000.

images they contain a loosely packed nonetchable globular content (Fig. 19) rather than the labyrinthine pattern seen in cross-fractured acidosomes (Fig. 11). Under normal conditions lysosomes have never been found docked at the nascent vacuole nor are they found along the cytopharyngeal microtubular ribbons. Lysosomes dock with acidified vacuoles but not with free acidosomes from which the acidified vacuole membrane is derived. Lysosomes first begin to appear next to the DV-II, which are around 3 min old and continue to accumulate until, in many cases, a closely packed layer of lysosomes has formed around the vacuole. At some signal that occurs for vacuoles that are around 7 to 8 min old, the lysosomes fuse with the DV-II seemingly all at one time, thus transforming the vacuole into a phagolysosome or DV-III.

Digestion by the lysosomal hydrolases occurs only at this vacuole stage. These hydrolases are synthesized by the ribosomal protein synthetic machinery on the membrane of the endoplasmic reticulum. They are then transferred from the endoplasmic reticulum to the Golgi stack and move through the stack to the trans-cisterna, where they are released from the Golgi stack in small 75-nm membrane-limited vesicles (Fok et al., 1984). These hydrolase-containing vesicles represent the primary lysosomes in *Paramecium* (Fig. 20). Although their fate has not yet been determined, these vesicles may either pass to the trans-Golgi network, dock and fuse with the larger secondary lysosomes, or dock and fuse directly with the membrane of the DV-III.

Paramecium does seem to have a trans-Golgi network which, like the trans-cisterna of the Golgi stack and the lysosomes, are labeled with WGA (Allen et al., 1989). The trans-Golgi network does not appear to be closely associated with the Golgi stack. No studies have been made to determine if the trans-Golgi network contains acid hydrolases. Recent high resolution scanning electron microscopic studies of time-defined isolated vacuoles show that a DV-III, like the DV-I and DV-II, has a population of small vesicles docked at its cytosolic surface (L. Ma, R. D. Allen, and A. K. Fok, unpublished observations). However, we do not know if any of these vesicles contain hydrolases. On the other hand, small coated vesicles containing WGA-binding sites are released from the trans-Golgi network and are found next to somewhat larger WGA-positive vesicles as though several of these vesicles are coalescing to form the larger lysosome-like bodies. We have concluded that at least some of the larger lysosomes that bind to the DV-II arise in this fashion (Allen et al., 1989).

VI. MEMBRANE MODIFICATION GIVING RISE TO PHAGOLYSOSOMES

Another profound change occurs in the vacuole membrane when lysosomes fuse. Again this is borne out both in freeze-fracture images and in the membrane's antigenicity. The E-fracture face of DV-III membranes (Fig. 22) is identical to the split lysosome membrane, which exhibits a few prominent IMPs (Fig. 21) (Allen and Staehelin, 1981). The P-fracture face has a moderate number of IMPs (Fig. 21), but like the lysosomes these are not as distinctive as those on the P-fracture faces of the acidosomes. A third antigen, B4, which occurs on the lysosome membrane, is present in these hydrolase-active vacuoles. This antigen is not found in acidosomes, discoidal vesicles, or in the DV-I and DV-II.

Thus a second membrane replacement, possibly total, may precede digestion. This replacement may potentially remove docking sites for lysosomes as well as the mechanism for acidification of vacuoles, and may also provide the putative active-transport mechanisms that allow the passage of the products of digestion from the vacuole lumen through the vacuole membrane into the cytosol. This putative replacement would occur concomitant with the rise in pH that the vacuole undergoes at this time (Fok et al., 1982) and with the appearance of acid phosphatase activity in the vacuole (Fok et al., 1984). The luminal surface of the vacuole, like the lysosomes, bears a 30-nm-wide moderately electron-opaque fringe (Fok and Allen, 1979; Allen and Fok, 1984a) which binds WGA (Allen et al., 1989) and is reactive for the monoclonal antibody to the B4 antigen. In thin sections it is also clear that at this time the paracrystalline material of the lysosomes is transferred to the vacuole (Fok and Allen, 1979).

VII. LYSOSOME MEMBRANE RECYCLING

At about 20 min old, a vacuole becomes defecation competent and is called a DV-IV (Fok et al., 1982). During the transition of a DV-III to the DV-IV another round of membrane modification occurs (Allen and Fok, 1984b). Long narrow, 45-nm-wide tubules are formed from the vacuole's membrane and project into the cytosol (Figs. 23, 24). These tubules expand at their ends into spherical bodies which resemble lysosomes both morphologically and in their acid phosphatase activity (Fig. 23) (Allen and Fok, 1984b). These tubules and their bulbous ends also bear the B4 antigen (Fig. 24) characteristic of the free lysosomes surrounding the acidic DV-II (R. D. Allen and A. K. Fok, unpublished observations).

These observations indicate that the lysosome membrane and acid phosphatase are being retrieved from the DV-III. By removing lysosome membrane it seems likely that the mechanisms required for transport of digested products into the cytosol may also be removed. The retrieved organelles can then reenter the pool from which lysosomes are continuously recruited for docking and fusing with acidified vacuoles. We suspect that these retrieved lysosomes may need to be recharged with hydrolases, which could occur by their fusion with the smaller primary lysosomes arising from the Golgi complex.

The process of lysosome membrane retrieval probably prepares a vacuole for defecation. Vacuoles that have not passed through this series of modifications may either not be defecated at all or defecated at a much slower rate (Fok and Allen, 1990). At this time one more event needs to occur. Defecation-competent vacuoles are capable of binding to microtubules which guide them to the cell's cytoproct

(Fig. 25). Receptors for a motor, which in this case may be cytoplasmic dynein, need to be introduced into the vacuole's membrane if they have been previously removed (see Section IV). How might this occur? Another population of vesicles, not yet recognized, may provide receptors for these motors, if the motors are in solution. If the motors are integral membrane complexes, they could be added to the DV-IV by fusion of a population of motor-carrying vesicles at about the time that lysosomal membrane is being retrieved. Recent unpublished high resolution scanning electron microscopic studies indicate the presence of a possible third population of small vesicles of about 100 nm diameter, which are docked at the vacuole's cytosolic surface from DV-I up to the DV-III stage but not in DV-IV. Whether these are identical to the vesicles of this size that are capable of binding to microtubules by arms, which we strongly suspect are cytoplasmic dynein motors (Fig. 10) (Schroeder et al., 1990), is not known. If such motor-associated vesicles fuse with the vacuole, they could provide the motors or motor receptors that would be needed to link the vacuoles to the microtubules that extend into the cytosol from the cytoproct. These motors would then be instrumental in moving the vacuole to the cytoproct.

The membrane of these DV-IV vacuoles is somewhat modified from the DV-III. In thin sections these vacuoles bear a less prominent luminal fringe than DV-III. Small tubules can be found continuous with the DV-IV but they are not as elaborate as those found during active lysosome membrane retrieval. These vacuoles are now acid phosphatase negative and are the only vacuoles that are defecation competent.

Figures 23–26. Membrane retrieval from DV-III and defecation.

Fig. **23.** Acid phosphatase reaction product in a DV-III (III) and in the rounded distal ends of some of the retrieved tubules (small arrowheads) that arise from the DV-III membrane. (Unpublished micrograph) ×35,000.

Fig. **24.** B4 antigen is localized by immunogold labeling on the vacuole membrane (vm) and in retrieved tubules (arrowheads) of digestive vacuoles in transition between stages III and IV. (Unpublished micrograph) ×50,000.

Fig. **25.** A bundle of microtubules (arrowhead) extending from a basal body that is located next to the cytoproct (not shown) lies against the DV-IV (IV) membrane and thereby the vacuole is guided to the cytoproct. (Unpublished micrograph) ×35,000.

Fig. **26.** The spent vacuole membrane at the open cytoproct (CYT) bears an HRP-positive luminal lining when exposed to this marker during defecation. This lining is thick enough to account for either retrieved vesicles returning as discoidal vesicles or as lysosomes. (Unpublished micrograph) ×40,000.

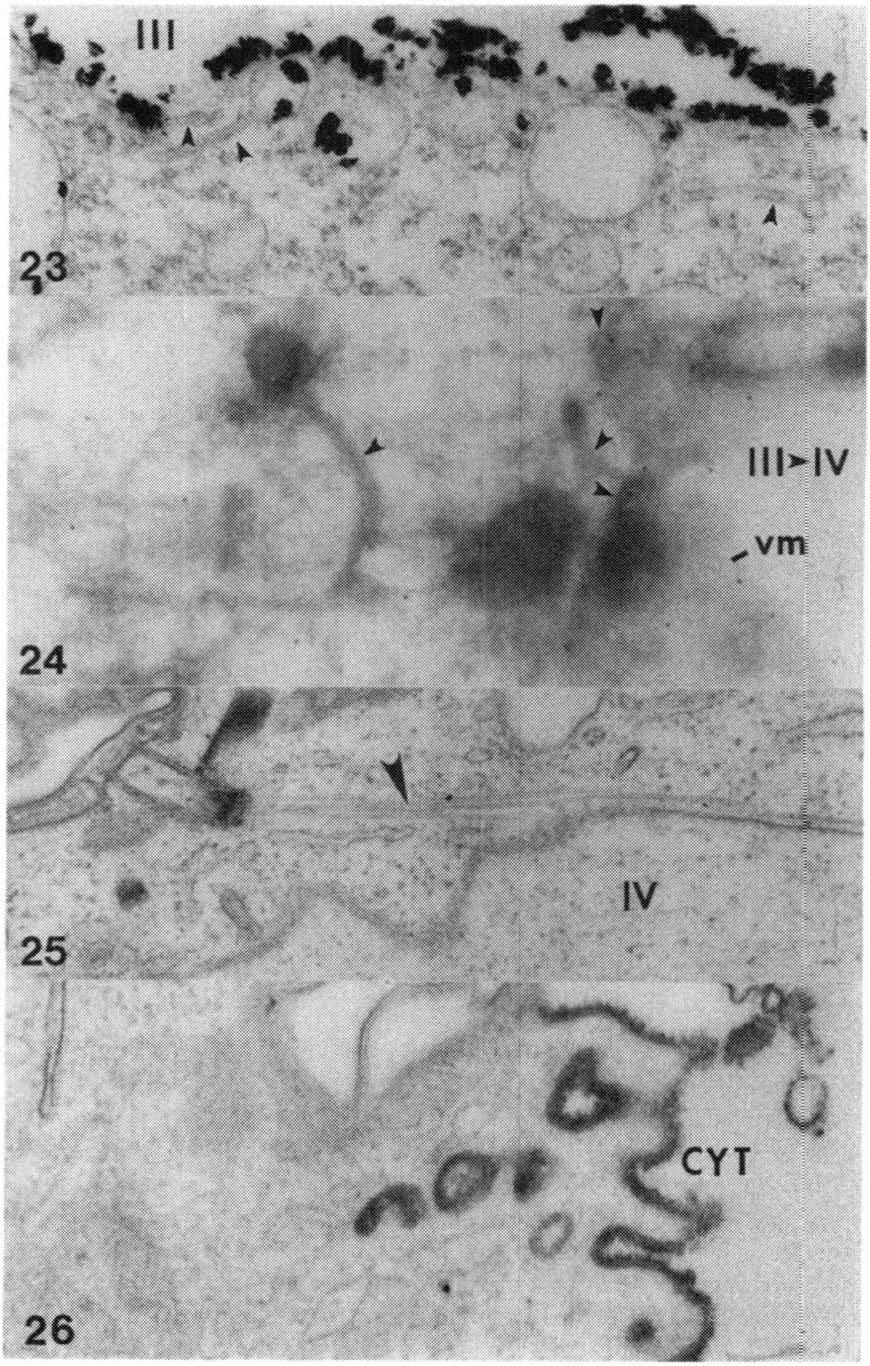
III
23
III➤IV
vm
24
IV
25
CYT
26

VIII. DIGESTIVE VACUOLE MEMBRANE RECYCLING FROM THE CYTOPROCT

The spent vacuole is eventually guided to the cytoproct along two sets of microtubules, one that extends from the cytoproct ridge and a second that extends from the proximal ends of the basal bodies that border the cytoproct (Fig. 25) (Allen and Wolf, 1974). The cytoproct ridge is pulled inward and expands to accommodate the approaching vacuole until the plasma membrane at the cytoproct ridge and the vacuole membrane come close enough to fuse. Fusion permits the cytoproct to zip open and allows the vacuole contents to stream out of the cell, but not the vacuole membrane. In thin sections a distinct glycosylated layer coats the spent vacuole membrane's externally exposed luminal surface at the open cytoproct (Fig. 26). This spent vacuole membrane is retrieved by a process of membrane tubularization and fission, resulting in tubule release into the cell. When all the membrane is thereby reengulfed, the cytoproct lips will be brought together to reform the resting cytoproct ridge.

These retrieved membrane-limited tubules will contain HRP (Fig. 26) if a cell defecates the vacuole in the presence of HRP (Allen and Fok, 1980). Many of these tubules will bind to microtubular ribbons, reshape into discoidal vesicles, and return to the cytopharynx along these ribbons (Fig. 8), where they will be used in a new round of vacuole formation. Other retrieved tubules will round up into lysosome-like vesicles and will eventually be found next to DV-II. Presumably the macromolecules destined for lysosomes are sorted from those destined for discoidal vesicles, and the appropriate "address" molecules that will determine with what membrane-bound compartments these different vesicles will dock and fuse will be located on their cytosolic surfaces. Thus, in order for some of the recycled vacuole membrane to reenter the discoidal vesicle pool, this membrane may need to regain the B2 antigenicity, which is not found in DV-IV. How this modification is brought about is currently unclear, as there is no evidence at all that these retrieved vesicles return to the Golgi stacks, or even to the trans-Golgi network.

Membrane tubularization at the cytoproct appears to require the polymerization of a layer of microfilaments on the cytosolic side of the spent vacuole membrane. Several lines of evidence suggest that these microfilaments are actin-like and that these are needed for membrane retrieval. Treating cells with cytochalasin B, a drug that prevents actin polymerization, results in cells having open cytoprocts with the spent vacuole membrane still attached. The microfilamentous network, normally well developed around the membrane at open cytoprocts, is reduced or absent (Allen and Fok, 1985). In such cases it is obvious that the tubularization and membrane retrieval process has been inhibited so that the cytoproct cannot close. Another drug that has a similar effect is trifluoperazine, a calmodulin antagonist. (In *Tetrahymena* the lectin concanavalin A has been reported to produce a similar effect, resulting in open cytoprocts in this ciliate [Scott and Hufnagel, 1983]). Thus it seems that membrane retrieval at the cytoproct requires an actin-based system that may be under the regulation of calcium. Any substance that can disrupt this actin system or the regulation of its function apparently has the potential of preventing membrane retrieval and, in turn, preventing cytoproct closure.

We can also make the broader generalization that, whenever membrane of digestive vacuoles is to be retrieved, the membrane must first be transformed from a nearly flat sheet into a tubule of around 45 nm in diameter. This process of tubularization may always require the polymerization of actin. Such tubules form from DV-I, from DV-III, and from spent vacuole membrane at the cytoproct, that is, in all cases where membrane retrieval has been observed. Though membrane retrieval from the late DV-II has not been reported, we are currently looking for evidence of such retrieval using the monoclonal antibody specific for the L1 antigen. Preliminary work suggests that retrieval via tubules does occur here as well (R. D. Allen, L. Ma, and A. K. Fok, unpublished observations). Around all these tubules is found a zone containing hints of microfilament-like networks (Allen and Fok, 1983a, 1984a, b, 1985).

Acidosomes are also typically associated with microfilaments (Allen and Fok, 1983a) which may be involved in producing the frequently observed tubular extensions as well as the indentations of these organelles. When cells were treated with cytochalasin B, large clumps of filamentous material accumulated next to the DV-I (Allen and Fok, 1985). Acidosomes remained bound to the DV-I membrane under this treatment but they did not fuse with the vacuole as long as cytochalasin B was present (Allen and Fok, 1983c). These filamentous clumps were lined with discoidal vesicles (Allen and Fok, 1985), suggesting that these vesicles that provide NDV membrane must be capable of binding both to microtubule-based motors as well as a microfilamentous-based cytoskeletal system. Presumably this is also true of NDV and DV-I membranes which are derived from discoidal vesicles.

Thus digestive vacuole membranes may need to contain receptors for actin-binding molecules in order to modulate their shape. Because actin may bind to vacuoles of all stages, such receptors may be a continuous part of the vacuole. Actin binding and unbinding with these receptors may be tightly regulated during the vacuole cycle and so play a crucial role in membrane tubularization, fusion, and fission.

IX. INPUT FROM THE ENDOSOMAL SYSTEM

One of the mysteries which has yet to be explained in *Paramecium* is how and why acidosomes sometimes obtain HRP within a few seconds of the cell's exposure to this marker. The most probable explanation is that the marker enters the cell by endocytosis and that the preendosomal vesicles pass directly to the acidosomes, where they fuse, thereby bypassing the early endosomes. Thus the endosome pathway appears to empty into the phagosome–lysosome pathway at one or more points along its route. Acidosomes that acquire HRP will follow their normal route of docking at the nascent vacuole. These acidosomes will then ride the vacuole to the cell's posterior end, where they will fuse with and empty their contents into the vacuole. Endocytically derived cargo (e.g., HRP) will thereby enter the phagosome pathway prior to vacuole acidification, and this cargo will be subjected to the full range of acidification and digestion in the same way as the contents acquired by phagocytosis. As discussed in the chapter by Allen and Fok in this book, we also have circumstantial evidence that shuttle vesicles arise from early endosomes, move along cytopharyngeal ribbons, and bind to and possibly fuse with acidosomes. Further studies on the interconnections between digestive vacuole membrane flow and endocytic membrane flow are needed to give a complete picture of the total possibilities of endocytic and phagocytic membrane interrelatedness and flow in this ciliate.

X. CONCLUSIONS AND UNSOLVED PROBLEMS

Many but probably not all of the pathways of membrane flow in the phagosome–lysosome system of *Paramecium* have been determined (Fig. 1). The major loop in this recycling process is the formation of vacuoles from a pool of discoidal vesicles and the passage of these vacuoles through the cell to the cytoproct, where the membrane of spent vacuoles is retrieved and returned, at least in part, to the discoidal vesicle pool. Superimposed on this loop are at least two, and possibly three, shorter loops: (1) retrieval of membrane from DV-I and presumably the return of this membrane to the discoidal vesicle pool, (2) retrieval of acidosome membrane from DV-II and the return of this membrane to the pool of acidosomes, and (3) retrieval of lysosome membrane from the DV-III and the return of this to the pool of lysosomes. This leads to the obvious question: If all membrane replacement at each step is total, from what source are the membranes of DV-IV derived? We can suggest two possibilities in answer to this question. Membrane retrieval at each step may not be total, even though the freeze-fracture images make it appear they are or are nearly so, or there are additional sources of membrane input yet to be characterized. Preliminary data from high-resolution scanning electron microscopy suggest that there is another pool of membrane not yet factored into our equation. Small shuttle-size vesicles are tightly bound to vacuoles at various stages, which could provide another potential source of membrane that must be accounted for in our digestive vacuole membrane balance sheet.

The major tools and methodology used to follow these pathways have been to use latex bead-marked vacuoles whose ages can be determined *in situ* at both the light and electron microscopic levels and to use tracer studies with HRP as the marker to follow the flow of retrieved vesicles from the cytoproct. More recently the use of monoclonal antibodies has allowed us to identify subsets of membrane-bound compartments bearing specific epitopes. This has greatly aided our attempts to identify the shorter loops of membrane flow. High-resolution scanning electron microscopy of magnetic-bead-containing isolated vacuoles promises to provide new information about this membrane flow system in the near future.

Another major area of research waiting to be investigated is that of the membrane components which are required to regulate this complex flow system. We need to look for the possible presence of small GTP-binding proteins which are similar to those of the rab family of proteins known to be involved in the regulation of membrane flow within the mammalian Golgi complex and endocytic pathway (Zahraoui et al., 1989). Homologues of many of these proteins have also been identified and studied in yeast (reviewed in Balch, 1990). Also, coat proteins similar to β-COP (reviewed in Duden et al., 1991) and fusion proteins similar to the *N*-ethylmaleimide sensitive fusion protein and its SNAPs (reviewed in Wilson et al., 1991, and Rothman and Orci, 1992) are probably needed for membrane recognition, docking, and fusion in the Golgi complex of *Paramecium* as they are in mammals and yeast. Similar proteins may be needed in the phagosome–lysosome system. So far almost nothing is known about any of these molecules in any ciliate. Only one report, on small GTP-binding proteins (Peterson, 1991), has so far appeared. Ultimately these problems of membrane recognition, docking, and fusion in the phagosome–lysosome membrane flow system of *Paramecium* will require the tools of molecular biology as well as cell-free reconstitution systems to determine the range of molecules required to bring about and regulate these processes.

ABBREVIATIONS

DV-I to IV:	four stages of digestive vacuoles
HRP:	horseradish peroxidase
IMP:	intramembrane particle
NDV:	nascent digestive vacuole
WGA:	*Triticum vulgaris* (wheat germ) agglutinin

ACKNOWLEDGMENTS

We thank Marilynn S. Ueno for technical support. Our current work is supported in part by National Science Foundation grants DCB 88 19182 and DCB 90 17455. The electron microscope facility used in our work has received support from a National Institutes of Health Research Centers in Minority Institutes grant GM 07684 and NSF instrumentation grants PCM 82 12316, PCM 84 00418, and BBS 87 21751.

REFERENCES

Allen, R. D. (1974). Food vacuole membrane growth with microtubule-associated membrane transport in *Paramecium. J. Cell Biol.* 63: 904–922.

Allen, R. D. (1976). Freeze-fracture evidence for intramembrane changes accompanying membrane recycling in *Paramecium. Cytobiologie* 12: 254–273.

Allen, R. D. (1978). Particle arrays in the surface membrane of *Paramecium*: junctional and possible sensory sites. *J. Ultrastruct. Res.* 63: 64–78.

Allen, R. D. and Fok, A. K. (1980). Membrane recycling and endocytosis in *Paramecium* confirmed by horseradish peroxidase pulse-chase studies. *J. Cell Sci.* 45: 131–145.

Allen, R. D. and Fok, A. K. (1983a). Phagosome fusion vesicles of *Paramecium*. I. Thin-section morphology. *Eur. J. Cell Biol.* 29: 150–158.

Allen, R. D. and Fok, A. K. (1983b). Phagosome fusion vesicles of *Paramecium*. II. Freeze-fracture evidence for membrane replacement. *Eur. J. Cell Biol.* 29: 159–165.

Allen, R. D. and Fok, A. K. (1983c). Nonlysosomal vesicles (acidosomes) are involved in phagosome acidification in *Paramecium*. *J. Cell Biol.* 97: 566–570.

Allen, R. D. and Fok, A. K. (1984a). Stages of digestive vacuoles in *Paramecium*: membrane surface differences and location. *Eur. J. Cell Biol.* 35: 149–155.

Allen, R. D. and Fok, A. K. (1984b). Retrieval of lysosomal membrane and acid phosphatase from phagolysosomes of *Paramecium caudatum*. *J. Cell Biol.* 99: 1955–1959.

Allen, R. D. and Fok, A. K. (1985). Modulation of the digestive lysosomal system in *Paramecium caudatum*. III. Morphological effects of cytochalasin B. *Eur. J. Cell Biol.* 37: 35–43.

Allen, R. D., Ma, L., and Fok, A. K. (1993). Acidosomes: recipients of multiple sources of membrane and cargo during development and maturation *J. Cell Sci.* 106, in press.

Allen, R. D., Schroeder, C. C., and Fok, A. K. (1989). Intracellular binding of wheat germ agglutinin by Golgi complexes, phagosomes, and lysosomes of *Paramecium multimicronucleatum*. *J. Histochem. Cytochem.* 37: 195–202.

Allen, R. D., Schroeder, C. C., and Fok, A. K. (1992). Endosomal system of *Paramecium*: coated pits to early endosomes. *J. Cell Sci.* 101: 449–461.

Allen, R. D. and Staehelin, L. A. (1981). Digestive system membranes: freeze-fracture evidence for differentiation and flow in *Paramecium*. *J. Cell Biol.* 89: 9–20.

Allen, R. D. and Wolf, R. W. (1974). The cytoproct of *Paramecium caudatum*: structure and function, microtubules, and fate of food vacuole membranes. *J. Cell Sci.* 14: 611–631.

Balch, W. E. (1990). Small GTP-binding proteins in vesicular transport. *Trends Biochem. Sci.* 15: 473–477.

Bowers, B. (1980). A morphological study of plasma and phagosome membranes during endocytosis in *Acanthamoeba*. *J. Cell Biol.* 84: 246–260.

Cohen, J., Garreau de Loubresse, N., and Beisson, J. (1984a). Actin mobilization during phagocytosis in *Paramecium*. *J. Submicrosc. Cytol.* 16: 103–104.

Cohen, J., Garreau de Loubresse, N., and Beisson, J. (1984b). Actin microfilaments in *Paramecium*: localization and role in intracellular movements. *Cell Motil.* 4: 443–468.

Duden, R., Allan, V., and Kreis, T. (1991). Involvement of β-COP in membrane traffic through the Golgi complex. *Trends Cell Biol.* 1: 14–19.

Fok, A. K. and Allen, R. D. (1979). Axenic *Paramecium caudatum*. I. Mass culture and structure. *J. Protozool.* 26: 463–470.

Fok, A. K. and Allen, R. D. (1981). Axenic *Paramecium caudatum*. II. Changes in fine structure with culture age. *Eur. J. Cell Biol.* 25: 182–192.

Fok, A. K. and Allen, R. D. (1988). The lysosome system. In: *Paramecium*. (Görtz, H.-D., ed.), pp. 301–324, Springer-Verlag, Berlin.

Fok, A. K. and Allen, R. D. (1990). The phagosome-lysosome membrane system and its regulation in *Paramecium*. *Int. Rev. Cytol.* 123: 61–94.

Fok, A. K., Lee, Y., and Allen, R. D. (1982). The correlation of digestive vacuole pH and size with the digestive cycle in *Paramecium caudatum*. *J. Protozool.* 29: 409–414.

Fok, A. K., Muraoka, J. H., and Allen, R. D. (1984). Acid phosphatase in the digestive vacuoles and lysosomes of *Paramecium caudatum*: a timed study. *J. Protozool.* 31: 216–220.

Fok, A. K. and Paeste, R. M. (1982). Lysosomal enzymes of *Paramecium caudatum* and *Paramecium tetraurelia*. *Expl. Cell Res.* 139: 159–169.

Fok, A. K., Ueno, M. S., and Allen, R. D. (1986). Differentiation of *Paramecium* phagosome membrane and stages using monoclonal antibodies. *Eur. J. Cell Biol.* 40: 1–8.

Fok, A. K., Ueno, M. S., Azada, E. A., and Allen, R. D. (1987). Phagosomal acidification in *Paramecium*: effects on lysosomal fusion. *Eur. J. Cell Biol.* 43: 412–420.

Kersken, H., Momayezi, M., Braun, C., and Plattner, H. (1986). Filamentous actin in *Paramecium* cells: functional and structural changes correlated with phalloidin affinity labeling in vivo. *J. Histochem. Cytochem.* 34: 455–465.

Kloetzel, J. A. (1974). Feeding in ciliated protozoa. I. Pharyngeal disks in *Euplotes*: a source of membrane for food vacuole formation? *J. Cell Sci.* 15: 379–401.

Lüthe, N. and Plattner, H. (1986). Lectin binding sites in *Paramecium tetraurelia* cells. II. Labeling analysis predominantly of non-secretory components. *Histochemistry* 85: 377–388.

Moore, P. L., Bank, H. L., Brissie, N. T., and Spicer, S. S. (1978). Phagocytosis of bacteria by polymorphonuclear leukocytes: a freeze-fracture, scanning electron microscope, and thin-section investigation of membrane structure. *J. Cell Biol.* 76: 158–174.

Peterson, J. B. (1991). Small GTP-binding proteins associated with secretory vesicles of *Paramecium*. *J. Protozool.* 38: 495–501.

Rothman, J. E. and Orci, L. (1992). Molecular dissection of the secretory pathway. *Nature* 355: 409–415.

Schroeder, C. C., Fok, A. K., and Allen, R. D. (1990). Vesicle transport along microtubular ribbons and isolation of cytoplasmic dynein from *Paramecium*. *J. Cell Biol.* 111: 2553–2562.

Scott, S. M. and Hufnagel, L. A. (1983). The effect of concanavalin A on egestion of food vacuoles in *Tetrahymena*. *Exp. Cell Res.* 144: 429–441.

Steinman, R. M., Mellman, I. S., Muller, W. A., and Cohn, Z. A. (1983). Endocytosis and recycling of plasma membrane. *J. Cell Biol.* 96: 1–27.

Vale, R. D., Schnapp, B. J., Mitchison, T., Steuer, E., Reese, T. S., and Sheetz, M.P. (1985). Different axoplasmic proteins generate movement in opposite directions along microtubules in vitro. *Cell* 43: 623–632.

Wilson, D. W., Whiteheart, S. W., Orci, L., and Rothman, J. E. (1991). Intracellular membrane fusion. *Trends Biochem. Sci.* 16: 334–337.

Zahraoui, A., Touchot, N., Chardin, P., and Tavitian, A. (1989). The human *Rab* genes encode a family of GTP-binding proteins related to yeast YPT1 and SEC4 products involved in secretion. *J. Biol. Chem.* 264: 12394–12401.

SIGNAL COUPLING DURING ENDOCYTOSIS IN *AMOEBA PROTEUS*

Robert D. Prusch

Advances in Cell and Molecular Biology of Membranes, Volume 2B
Membrane Traffic in Protozoa, pages 339–354

ISBN: 1-55938-628-2

I. INTRODUCTION

All cells are capable to some extent of accumulating material from the external environment by a variety of transport and controlled diffusion mechanisms. Just important, most cells are able to take up external material by surface membrane infolding and vesiculation, or endocytosis. This includes both pinocytosis and phagocytosis, generally referring to solute and particulate matter, respectively, being taken up by the cell. Amoeboid cells such as *Amoeba proteus* have served as model systems to investigate endocytosis for some time. For example, *Amoeba proteus* feeds primarily by phagocytosis and is also capable of internalizing a wide variety of solutes through pinocytosis. Endocytosis is initiated by the interaction of specific inducers or signal molecules with external cell surface in the amoeba, triggering a complex sequence of events ultimately leading to the activation of cytoskeletal elements eliciting the appropriate response on the part of the cell. These processes are easily observed at a superficial level in the amoeba, but the mechanism by which the amoeba accomplishes these feats remains for the most part unknown. Somehow, the amoeba like other cells has a mechanism(s) by which information from the external environment is transferred across the cell membrane to the appropriate effector or signal coupling. The purpose of this discussion is to explore some of the more recent observations concerning the onset of endocytosis in the amoeba. Signal coupling mechanisms in other cell systems will first be briefly reviewed to provide a basis of comparison for what may be happening in the amoeba and to perhaps suggest additional experimental approaches in examining this complex process in the amoeba. For some other aspects on endocytosis in amoebae, see Chapters by Hausmann and Radek and by Jeon in this volume.

II. *DICTYOSTELIUM DISCOIDEUM*

The cellular slime mold *Dictyostelium discoideum* has been a model cell system examining signal coupling during cell aggregation (chemotaxis) and differentiation, and much of the mechanistic basis of these processes has been experimentally established. These organisms begin their life cycle as free-living amoebae (DDA = *Dictyostelium discoideum* amoebae), which feed by phagocytosis on bacteria in the soil. During this phase of their life cycle the cells reproduce asexually by binary fission. When environmental conditions deteriorate such as a decrease in the availability of food or desiccation, cells send out signals which in turn attract other cells toward the initial signal source. The activated DDA in turn send out signals in an oscillatory fashion recruiting additional cells to the initial focus, forming a multicellular "slug" or reproductive structure. The slug, which is made up of a large number of individual DDA, migrates for a time along the substratum and then differentiates into a spore-forming structure at the tip of a stalk of now dead cells. The spore-forming structure gives rise to individual haploid spores which are capable of withstanding harsh environmental conditions. When the environment again become favorable for DDA, the spores hatch out into amoeboid cells to repeat the cycle.

The primary signal or messenger used by DDA is cAMP (cyclic adenosine monophosphate), which is released by the initial amoeba or founder cell. The cAMP then binds to receptors on the surface or other amoebae, which attracts these cells to the original signal and causes them in turn to produce and release cAMP, attracting an increasing number of cells to the point of initial cAMP release. In this particular system, cAMP is being primarily used as an intercellular signal, and in fact this same signal can be used for a number of different functional responses (Fontana et al., 1991).

A. Surface Receptors

The response of DDA to signals in the environment is initiated by the interaction of the signal with a receptor on the external surface of the cell membrane. DDA respond to a number of signals in the environment, implying the existence of a number of different surface receptors. In addition to cAMP, folic acid also serves as a chemotactic substance in DDA (Pan et al., 1972), which stimulates production, elongation, and branching of filopodia (Rifkin and Ikek, 1984). Also involved in phagocytic recognition are lectin-type receptors (specific for glucose residues) and apparent nonspecific receptors (Vogel et al., 1980).

The cAMP receptor on the external surface of DDA is responsible for the initiation of the aggregation response of these cells. The receptor itself may be a transmembrane protein with an extracellular cAMP binding site, a hydrophobic transmembrane sequence, and a cytoplasmic domain subject to covalent modification (Klein et al., 1987). Activation of the receptor requires a stereospecific interaction between the receptor active site and the phosphate moiety of cAMP (Van Haaster and Klein, 1983). As is the case for many other cellular signaling mechanisms, the affinity of the cAMP receptor for cAMP can be modified or down-regulated. Initial evidence suggested that down regulation of the DDA cAMP receptor involved a transition to a higher molecular weight form (Juliani and Klein, 1981), as opposed to physical removal of surface cAMP receptors. There appears to be some question as to the actual number of different cAMP receptors on the surface of DDA (Klein et al., 1986), but it is apparent that binding of cAMP to its receptor(s) brings about a phosphorylation of the receptor (Meier and Klein, 1988). That is, DDA exposed to cAMP respond by decreasing receptor binding activity when the receptor is phosphorylated by an unusual protein kinase which preferentially phosphorylates the ligand–receptor complex (Meier and Klein, 1988; Anschütz et al., 1991). Phosphorylation of the receptors uncouples the stimulated receptor from subsequent signal-coupling components of the signal cascade (Klein et al., 1987).

B. Adenylate Cyclase

Activation of the cAMP surface receptor in DDA in turn activates adenylate cyclase, which is the primary event in the generation of intercellular signals in this cell system (Roos et al., 1977). In general, activation of adenylate cyclase is short-lived (Roos and Gerisch, 1976), which may be associated with the previously described modification of the occupied cAMP receptor. Adenylate cyclase is a membrane-bound protein localized on the inner surface of the plasma membrane (Farnham, 1975). In stimulated cells, adenylate cyclase activity is maximal after approximately 1 to 2 min (Dinauer et al., 1980a) and then rapidly declines to basal levels (Dinauer et al., 1980b). The cAMP which is produced is intended primarily for intercellular signaling as opposed to intracellular signaling.

C. G-Proteins

Stimulation of surface receptors in DDA by cAMP leads to intracellular activation through a G-protein(s) (Van Haastert, 1984; Theibert and Devreotes, 1986). Evidence in support of the involvement of G-proteins in signal transduction includes the binding of GTP (guanosine triphosphate) by isolated DDA membranes which is not observed in intact cells (DeWit and Snaar-Jagalska, 1985). In this case, GTP binding by isolated membranes is increased in the presence of cAMP or folate and inhibited in the presence of guanylyl imidodiphosphate. Furthermore, cAMP stimulates GTPase in isolated DDA membranes (Snaar-Jagalska et al., 1988). Sodium dodecyl sulfate-polyacrylamide gel electrophoresis (SDS-PAGE) has revealed the presence of a major 39 kDa protein band in DDA, which is ADP (adenosine diphosphate) ribosylated by both cholera and pertussis toxins (Khachatrian et al., 1987). Additional evidence suggesting a role for a G-protein in DDA activation is the phosphorylation of a 36 kDa membrane protein in the presence of GDP which varies during development (Anschütz et al., 1989).

Up to this point, the processes described in stimulated DDA are associated with intercellular signaling, that is, the production and release of cAMP to in turn stimulate other cells. Another obvious component of the signaling process is intracellular signaling, which is the process by which an individual cell itself responds to the external cAMP signal by directed cell movement or chemotaxis, which ultimately is a result of cytoskeletal activation. As mentioned previously, intracellular signaling in this system does not apparently involve adenylate cyclase and cAMP but may be associated with elements of the phosphatidyl 4,5-inositol (PIP_2) cascade.

D. Phosphatidylinositol

PIP_2 breakdown by activated phospholipase C (PLC) yields two separate messengers: diacylglycerol (DG) and inositol trisphosphate (IP_3). Stimulation of permeabilized DDA with cAMP results in peak levels of IP_3 at 5, 15, and 30 sec

(Europe-Finner and Newell, 1987). Exposure of these cells to GDP resulted in only a slight production of IP_3, while exposure to GTP induced a transient elevation of IP_3, suggesting the involvement of a G-protein in this signal pathway. Further evidence implicating IP_3 in intracellular signal coupling in DDA was obtained by exposing intact cells to ^{3}H-inositol and then stimulating the cells with cAMP. It was found that labeled IP_3 accumulated transiently after cAMP stimulation (Europe-Finner and Newell, 1985).

As is the case in other cell systems, IP_3 induces a rapid release of Ca^{2+} from a nonmitochondrial pool in DDA (Europe-Finner and Newell, 1986). An additional source of Ca^{2+} during activation of DDA may be from the extracellular medium. DDA stimulated with either folate or cAMP took up Ca^{2+} at the same rate as resting cells for the first 6 to 9 sec (Milne and Coukell, 1991) and then exhibited a large increase in Ca^{2+} influx which continued for 20 to 25 sec. In this case, although folate and cAMP bind to different receptors, they appear to be activating the same Ca^{2+} uptake systems. The cAMP-induced Ca^{2+} uptake was insensitive to Ca^{2+}-channel blockers such as verapamil and was not inhibited extensively by La^{3+}, which is known to interfere with Ca^{2+} movements in mammalian systems. In addition, Ca^{2+} entry in cAMP-stimulated DDA is apparently not influenced by changes in intracellular cyclic nucleotides (Milne and Coukell, 1991).

DDA stimulated by external cAMP responds by increasing intracellular cAMP through the activation of membrane-bound adenylate cyclase as described above. In addition, stimulation of DDA with cAMP also increases cytoplasmic levels of cGMP, which precedes pseudopod formation (Mato et al., 1977), with cGMP increasing ten-fold 25 to 30 sec after cAMP stimulation. There appears to be a variation in the cGMP response associated with the developmental stage of the cells (Wurster et al., 1977). Stimulation of preaggregation DDA results in a rapid monophasic increase in cGMP while a biphasic change in cytoplasmic cGMP is observed when late preaggregation cells are excited. Both cAMP and folic acid stimulate cGMP, but the stimuli are not additive (Van Haastert, 1983), suggesting that the two different stimuli are using common elements of the signal transduction pathway leading to the activation of guanylate cyclase. Activation of guanylate cyclase may be through the PIP_2 cascade since both IP_3 and Ca^{2+} mimic cAMP in stimulating cGMP production (Europe-Finner and Newell, 1985).

E. Cytoskeleton

Alteration in cGMP levels in DDA are in turn associated with activation of elements of the cytoskeleton. One mutant strain of DDA, streamer F mutants, forms long pseudopods when stimulated with cAMP and forms smaller than normal fruiting bodies. It has been established that this mutant strain has a defect in cGMP metabolism (Ross and Newell, 1981; Liu and Newell, 1988). In normal cells, an intracellular cGMP-specific phosphodiesterase is responsible for the degradation of cGMP (DeWit et al., 1987). Mutant cells apparently have a defective cGMP-phosphodiesterase and when stimulated with cAMP guanylate cyclase action elevate cGMP to a persistent high level. Correlated with this extended peak in cytoplasmic cGMP in mutant cells is a broad peak in myosin heavy chain activation (Liu and Newell, 1988).

Actin polymerization is associated with chemotactic stimulation of DDA, as indicated by the increased association of actin with the Triton-insoluble cytoskeleton 5 and 60 sec after stimulation (McRobbie and Newell, 1983). Pseudopod extension in response to chemotactic stimulation involves an increase in the level of F-actin (Hall et al., 1988). F-actin in DDA pseudopods is organized into a network of filaments, and presumably motile activity in these structures is driven in part by the assembly of this network (Wolosewick and Condeelis, 1986). Participating in the formation of this actin network and active pseudopod extension are a number of actin-binding proteins, including ABP-120, which becomes selectively concentrated at the site of new pseudopod formation (Wolosewick and Condeelis, 1986).

Also involved in cytoskeletal activation in DDA is a 30 kDA Ca^{2+}-binding protein that is involved in reorganization of filopodia and pseudopodia during directed cell movement. This protein has structural sequences similar to α-actinin and fimbrin (Fechheimer et al., 1991). Severin is another Ca^{2+}-activated actin binding protein found in DDA that severs actin filaments, nucleates actin assembly, and caps fast-growing ends of actin filaments (Eichinger et al., 1991). The protein contains two active actin-binding sites located next to each other responsible for the breaking and nucleating functions, and a single binding site near the N-terminal end used for capping of actin filaments.

In order for actin polymerization and pseudopod formation to be localized responses in amoeboid cells, it has been suggested that part of the signal-coupling mechanism may be nondiffusible (Condeelis et al., 1988). Extraction of DDA with nonionic detergents produces a cytoskeleton that retains cAMP-binding sites (Galvin et al., 1984). The fact that a receptor which transmits information to the cytoskeleton is in turn physically linked to the cytoskeleton suggests a functional interaction as well. In this context, it is interesting to note that concanavalin A (Con A) receptors from DDA also bind to F-actin (Shiozawa et al., 1989). The actual connection of cytoskeletal elements in pseudopods to the surface membrane may be through ABP-240 (Condeelis et al., 1988).

III. *AMOEBA PROTEUS*

Calcium is intimately involved at a number of different levels of endocytosis in *Amoeba proteus*. This involvement of Ca^{2+} in endocytosis ranges from influencing the association of endocytic inducers with the external cell surface to transient elevations in cytoplasmic Ca^{2+} levels which may be involved in the signal transduction sequence leading to pinocytotic channel or phagosome production. The onset of endocytosis comes about through the interaction of various signal molecules with different aspects of the external cell surface. An examination of signal binding with the amoeba surface is the point at which to begin an elucidation of the coupling mechanism by which endocytosis is initiated in the amoeba.

A. Interaction of Signals with the Cell Surface

Pinocytosis in the amoeba can be initiated with a wide variety of inducers (Chapman-Andresen, 1962) ranging from inorganic cations to high molecular weight proteins. The only factor in common to these inducers is that they must possess a net positive charge, suggesting in part that the interactions of pinocytotic inducers are with negative sites on the external cell surface. The association of two inducers of pinocytosis, Alcian blue and Na^+, has been extensively examined (Prusch, 1986). Alcian blue is a copper-phthalate dye that binds irreversibly to the amoeba surface while Na^+ associates rather loosely with the external cell surface or glycocalyx. The association of Alcian blue with the cell surface increases as the level of the dye in the external medium is increased up to approximately 400 μg/ml. Increase in the dye level above this point fails to bring about further increase in the amount of dye associated with the cell surface, or apparent saturation. The maximal level of dye associated with the amoeba surface is on the order of 4.2 μg/mg cells.

The association of Na^+ with the amoeba is quite different from that observed with Alcian blue. In this case, Na^+ associated with the amoeba surface is a linear function of the level of Na^+ in the external medium ranging from 1 to 10 mM, with no indication of saturation. The association of these two inducers with the amoeba surface then is quite different: Alcian blue binds tightly in an irreversible fashion to the cell surface, while Na^+ appears to associate nonspecifically and loosely with the cell surface. The association of both of these pinocytotic inducers with the cell surface though is influenced by the level of Ca^{2+} in the external medium.

At low external Ca^{2+} levels, the association of both Alcian blue and Na^+ is minimal, but increases as the external Ca^{2+} is increased. Inducer association with the cell surface reaches a maximum level when the external Ca^{2+} level approaches 0.15 mM. Increases in external Ca^{2+} above this level brings about a decrease in the amount of both inducers associated with the cell surface. The intensity of pinocytosis in *Amoeba proteus* measured in a variety of ways is also a function of external Ca^{2+}, increasing as Ca^{2+} is increased up to an optimum and then decreasing with further increases in external Ca^{2+}. The level of pinocytotic intensity then may be associated with the amount of inducer bound to the cell surface which in turn is controlled by external Ca^{2+}. Previous studies had shown that inducers of pinocytosis clearly displace surface associated Ca^{2+} in the amoeba, suggesting a simple competition between pinocytotic inducers and Ca^{2+} for similar sites on the cell surface (Prusch and Hannafin, 1979). The observation that Ca^{2+} controls both the association of inducers with the cell surface and correspondingly pinocytotic intensity at an optimal level suggests that the interaction between Ca^{2+} and pinocytotic inducers is quite complex and may involve Ca^{2+} influencing the structure of the glycocalyx or cell membrane.

Different categories of pinocytotic binding sites having variations in distribution, as well as sensitivity to pH and Ca^{2+} have been described in *A. proteus* (Kukulies et al., 1986). These include anionic binding sites, which are in general sensitive to changes in both external pH and Ca^{2+}, and Con A receptors, which apparently function independently of changes in either external pH or Ca^{2+}. The function of these Con A receptors in the amoeba is presently unknown (Taylor et al., 1980b; Kukulies et al., 1986), although they probably may also play a role in the initiation of pinocytosis in this system. In addition, receptors for so-called pinocytosis regulating factor have been described in *A. proteus* (Arvidson and Josefsson, 1982).

The association of inducers of phagocytosis with the surface of *Amoeba proteus* has been examined using *n*-formyl-methionyl-leucyl-phenylalanine (NFMLP) (Prusch and Wells, 1990), which had previously been shown to be highly chemotactic and to induce phagosome formation (Prusch and Britton, 1987). NFMLP binds to the cell surface in a saturable ($K_d = 7.96 \times 10^{-9}$) and reversible fashion. A series of competition studies demonstrated that substances structurally related to NFMLP were effective competitors, but inducers of pinocytosis did not displace surface-associated NFMLP. This effectively demonstrates that inducers of pinocytosis and phagocytosis are interacting at different sites on the amoeba surface. Given that *Amoeba proteus* feeds on a number of different target cells, there may be a number of different types of phagocytic receptors on the membrane, allowing the cell to respond to a variety of environmental stimuli.

The amount of labeled NFMLP associated with cells increases linearly with time at room temperature but remains at a constant level at reduced temperature. This would imply that phagocytic receptors in the amoeba are being internalized at room temperature and replaced with unoccupied receptors. The ability of *Amoeba proteus* to influence the distribution and cycling of receptors is also observed in cells exposed to constant, relatively high signal levels in the external environment or down regulation. In this case, amoebae were exposed at room temperature to high levels of NFMLP in the external environment and washed with fresh medium, and their ability to bind labeled NFMLP was monitored with time. Under these conditions, the amount of NFMLP associated with the cell surface decreased to a minimum within 20 min after exposure to an elevated signal level. The cells were able to recover from down regulation by apparently increasing the number of receptors associated with the cell surface. This recovery of signal binding in down-regulated amoebae occurred in the presence of inhibitors of protein synthesis, indicating that the cells are capable of cycling receptors from the cytoplasm to the surface membrane, as has been observed in other cell systems.

B. Calcium

Inducers of endocytosis bind to receptors on the external cell surface and then elicit either pinocytosis or phagocytosis. The mechanism by which activation of surface receptors in the amoeba elicits the proper endocytic response is for the most part unknown although Ca^{2+} clearly is involved in some aspects of the signal-coupling process (Gawlitta et al., 1980; Taylor et al., 1980a). Calcium influences the initial association of the signal with the cell surface, which in turn displaces surface-associated Ca^{2+}. Changes in surface-associated Ca^{2+} may in turn lead to the modification of membrane permeability in stimulated cells. Measurements of Ca^{2+} fluxes during the onset of phagocytosis suggest that changes in intracellular Ca^{2+} ion activity may be part of the signal-coupling process itself in the amoeba (Prusch and Vedovatti, 1989).

The movement of Ca^{2+} across the plasmalemma of *Amoeba proteus* from the external medium into the cytoplasm is essentially a diffusional process influenced by the Ca^{2+}-electrochemical gradient. Initiation of pinocytosis in this system was previously shown to be accompanied by Ca^{2+} movements into the cell (Prusch and Hannafin, 1979), and pinocytosis can be influenced by the Ca^{2+}-ionophore A23187 (Prusch, 1980). The onset of phagocytosis in the amoeba is also associated with a Ca^{2+} influx. Induction of phagocytosis in the amoeba with NFMLP brings about an increase in Ca^{2+} movement into the cell which can be inhibited with verapamil, flunarazine, and La^{3+} (Prusch and Vedovatti, 1989). This suggests the possibility of Ca^{2+} channels in the amoeba plasmalemma which may be activated by inducers of endocytosis, allowing an increase in Ca^{2+} influx elevating at least transiently cytoplasmic Ca^{2+} levels.

There are also suggestions that *Amoeba proteus* possesses intracellular Ca^{2+}-binding sites or compartments which can release Ca^{2+} into the cytoplasm, increasing the Ca^{2+} ion activity. Gollnick et al. (1991), utilizing the fura-2 technique for measuring intracellular Ca^{2+} ion activity, have shown that stimulation of *Amoeba proteus* with 30 mM KCl produces a large transient increase in intracellular Ca^{2+}. This Ca^{2+} transient was independent of extracellular Ca^{2+}, indicating that the source of this Ca^{2+} was intracellular compartments. The possibility exists, therefore, that increases in intracellular Ca^{2+} associated with signaling processes in endocytosis may come from both intra- and extracellular compartments, as has been observed in other cellular systems (Milne and Coukell, 1991).

C. Membrane Lipids

Another possible component of the signal-coupling process associated with the onset of endocytosis in *Amoeba proteus* may be derivatives of fatty acid metabolism. For example, it has been previously shown that prostaglandins ($PGF_{2\alpha}$, 1–5 mg/ml) are capable of eliciting food vacuole formation in the amoeba (Stockem et al., 1983). Exposure of *A. proteus* to PGE_2 (10^{-6} M) brings about increased cytoplasmic streaming, pseudopod formation, and after 10 to 15 min exposure to PG, the formation of food vacuoles (Prusch et al., 1989). Vacuoles formed in the presence of PGE_2 (Fig. 1) are transient in nature, initial vacuoles disappearing as new ones are formed. This response of *Amoeba proteus* to PGE_2 begins when the PG level in the medium reaches approximately 10^{-8} M and increases with increasing PG levels up to a maximum response at 10^{-6} M. Paralleling the time course of vacuole formation in the amoeba elicited with PG is the uptake of labeled sucrose

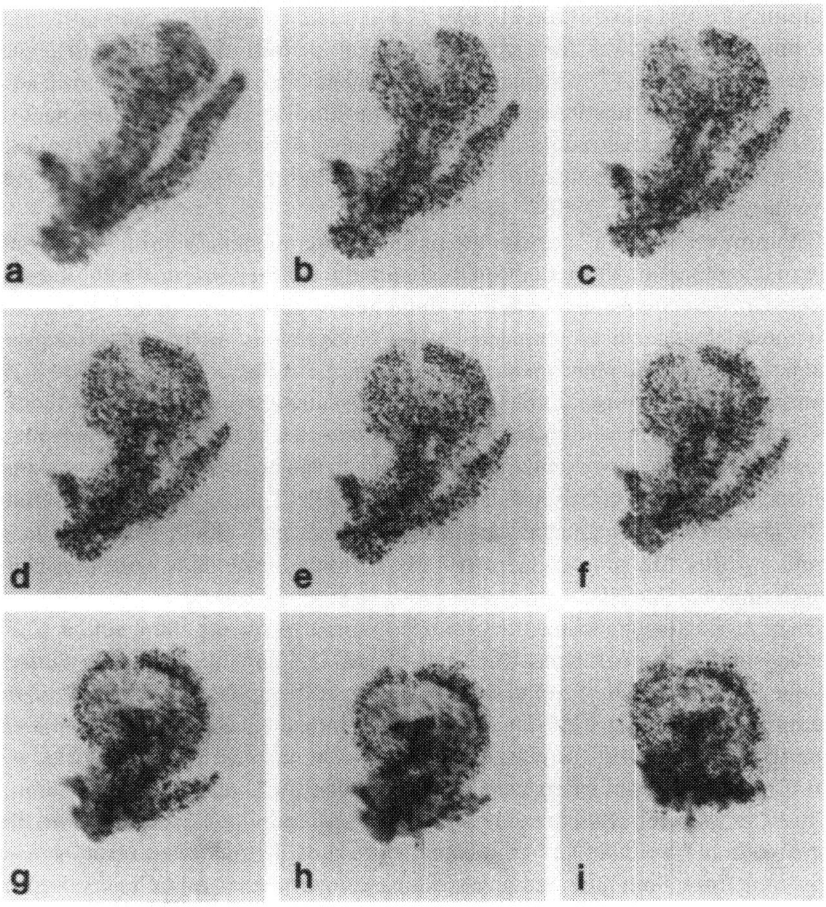

Figure 1. Timed sequence of a single *Amoeba proteus* exposed to PGE_2 (10^{-7} M) for approximately 15 min, forming a large, single food vacuole. Photographs were taken 10 sec apart. 140×. Reproduced from Prusch et al. (1989) by permission of Cell and Tissue Research (Churchill Livingstone, Edinburgh).

from the external medium, demonstrating that vacuole formation is not just a cytoplasmic event but results from pseudopod fusion with the entrapment and internalization of the external medium. Vacuole formation induced with PGE_2 is also Ca^{2+} dependent. At low levels of external Ca^{2+} PG-stimulated phagocytosis is minimal and increases with increasing external Ca^{2+} levels to a maximum response at 10^{-5} M. Increases in external Ca^{2+} above this level decreases the level of vacuole formation.

Arachidonic acid, the fatty acid precursor of PGE_2, is also capable of inducing vacuole formation in *Amoeba proteus* at levels ranging from 0.005 to 0.5 mg/ml.

The response of the cells to arachidonate is generally much slower than for PGE_2, requiring almost 30 min for maximum vacuole formation as opposed to 10 to 15 min for PGE_2. This effect of arachidonate on vacuole formation in the amoeba can be partially inhibited (50 to 70%) by indomethacin (0.16 μM). Indomethacin is a cyclo-oxygenase inhibitor which has been shown in other studies to at least partially inhibit the conversion of arachidonic acid to PG (Flower, 1974). The cyclo-oxygenase inhibitor acetylsalicylic acid was toxic to the amoeba at levels below that generally used to inhibit arachidonic metabolism. Indomethacin was also shown to decrease phagocytic activity in the amoeba induced experimentally with peptide or in the uptake of *Tetrahymena*.

Pinocytosis in *A. proteus* can be at least partially inhibited by external application of inhibitors of phospholipase A (PLA) activity (Josefsson et al., 1988). Pinocytotic activity in this case could be restored by the addition of low levels of lysolecithin to the external medium. In addition, low levels of various lysophospholipids and lecithins were shown to increase pinocytotic intensity in the amoeba. The stimulatory effect of lecithins but not the corresponding lysolecithins could be blocked by inhibitors of PLA. These observations, along with the previously described effects of PG on phagocytosis in the amoeba, strongly suggest that membrane lipids may play a pivotal role in the signal-coupling process during the initiation of endocytosis.

D. Cytoskeleton

Whatever the mechanism of signal coupling during the initiation of endocytosis in the amoeba, activated components of the cytoskeleton mediate the specific response on the part of the stimulated cell. The cytoskeleton of *A. proteus* consists in part of a cortical filamentous layer containing actin and myosin in considerable amounts and in a rather high degree of structural order (Stockem et al., 1983; Sonobe et al., 1986). At least four different types of filaments have been found to be associated with the amoeba cytoskeleton (Christiani et al., 1986). These include a series of 5 to 7 nm filaments organized as a more or less cortical network at the internal face of the plasma membrane and probably representing F-actin; 10 to 12 nm filaments running separately or slightly aggregated through the cytoplasm and probably representing intermediate filaments; 24 to 26 nm filaments forming a lose network and probably representing microtubules; and 2 to 4 nm filaments as connecting elements between the other cytoskeleton constituents. Numerous thick filaments are localized in the uroid region of the cell and probably represent polymerized myosin (Brix and Stockem, 1990). In addition, *A. proteus* contains α-actinin in a thin layer at the internal face of the plasma membrane. In amoebae undergoing pinocytosis, there is an increase in the thickness of the actomyosin layer immediately beneath the plasma membrane, showing an equal diameter along the entire cell surface (Stockem et al., 1983). Furthermore, the microfilament system is responsible for the generation of motive force necessary to promote endocytosis, endosome coalescence, fusion, sequestration, and exocytosis (Christofidou-Solomidou et al., 1989).

IV. CONCLUSION

It was mentioned previously that phospholipase A_2 may be associated with the onset of pinocytosis (Joseffson et al., 1988), and both PGF_2 (Prusch et al., 1989) and diacylglycerol stimulate phagosome formation in *Amoeba proteus* (unpublished results). These observations suggest that membrane lipids may be involved in the signal-coupling process associated with the onset of endocytosis in the amoeba. In this context, it is interesting to note that DGs may act as second messengers in mammalian lymphocytes, eliciting pinocytosis, changes in cell shape, and actin polymerization (Keller et al., 1989). In human neutrophils, DG produces biphasic changes in cell shape and results in the formation of large intracellular vacuoles (Zimmermann et al., 1988). In this instance, exposure of the cells to PMA induces phagocytosis. Chemotactic stimulation of neutrophils brings about a biphasic increase in intracellular DG levels (Reibman et al., 1988). More recently, it has been demonstrated that membrane permeant DGs trigger actin polymerization in DDA (Shariff and Luna, 1992). In this case, one or more functional membrane-bound proteins are required for the DG-mediated increase in actin polymerization. It is conceivable that the direct linkage of endocytic membrane receptors through the plasma membrane to elements of the cytoskeleton may in part be responsible for the generation of DG, which in turn mediates a local increase in actin nucleation activity, bringing about a localized response on the part of the cell, for example, pinosome or phagosome formation.

Amoeba proteus represents an ideal model system in which to investigate the mechanisms by which the transmembrane signal-coupling mechanism is activated during the initiation of endocytosis. Both forms of endocytosis, namely, pinocytosis and phagocytosis, are classically exhibited in this cell system and can be initiated independently by a variety of inducers. For example, Na^+ and albumin can be used to initiate pinocytosis, while certain peptides and prostaglandins can be used to stimulate phagocytosis. In regard to phagocytosis, it is clear that potentially a wide variety of environmental signals can probably initiate phagocytosis, as is the case in other cell systems. There is some question as to whether pinocytosis and phagocytosis are separate processes in the amoeba or represent modifications of the same fundamental cellular process (Stockem and Klopocka, 1988). Clearly, amoeba respond for the most part to a specific signal by either pinocytosis or phagocytosis, although there is some evidence that a mixed signal stimulus can modulate the response on the part of the cell (Klein et al., 1988). Surface characteristics of the amoeba differ markedly during pinocytosis and phagocytosis (Opas, 1981), which may reflect differences in cytoskeletal arrangement and cell surface properties during different forms of endocytosis. This in turn suggests that the two forms of endocytosis in the amoeba are fundamentally different. This question can best be resolved by examining the signal-coupling processes associated with each form of endocytosis in the amoeba.

In order for the mechanism by which information is transmitted across the cell surface in the amoeba is to be established, considerably more work needs to be done with this cell system. Some information regarding the initial interaction of signals with the cell surface in respect to both pinocytosis (Prusch, 1986) and phagocytosis (Prusch and Wells, 1989) have been done, but clearly much more needs to be done in establishing more definitely which types of signals induce endocytosis in the amoeba and the possible interaction between different types of signals. Another area that needs particular attention in the amoeba and is a focus of this laboratory is an examination of changes in cytoplasmic levels of substances involved in signal coupling in other cell systems, for example, cAMP, cGMP, IP_3, and DG. Finally, how is this signal process coupled to elements of the cytoskeleton bringing about either pinocytosis or phagocytosis?

ABBREVIATIONS

ADP:	adenosine diphosphate
cAMP:	cyclic adenosine monophosphate
cGMP:	cyclic guanosine monophosphate
Con A:	concanavalin A
DAG:	diacylglycerol
DDA:	*Dictyostelium discoideum* amoebae
GDP:	guanosine diphosphate
GTP:	guanosine triphosphate
NFMLP:	*n*-formyl-methionyl-leucyl-phenylalanine
PG:	prostaglandin
PIP_2:	phosphatidylinositol 4,5-bisphosphate
PLA:	phospholipase A
PLC:	phospholipase C
SDS-PAGE:	sodium dodecyl sulfate-polyacrylamide gel electrophoresis

ACKNOWLEDGMENT

This work was supported in part by grants from the National Science Foundation (DCB-8818901 and MCB-9118411).

REFERENCES

Anschütz, A., Howlet, A., and Klein, C. (1989). GDP stimulates the phosphorylation of a 36-kDa membrane protein in *Dictyostelium discoideum*. *Proc. Natl. Acad. Sci. U.S.A.* 86: 3665–3668.

Anschütz, A., Um, H. D., Tao, Y. P., and Klein, D. (1991). Resolution of protein phosphorylation in *Dictyostelium discoideum*. *Dev. Genet.* 12: 14–18.

Arvidson, G. and Josefsson, J. O. (1982). Lysolecithin as a modifier of induced pinocytosis in *Amoeba proteus*. *Eur. J. Cell Biol.* 28: 34–38.

Brix, K. and Stockem, W. (1990). Dynamics of the cytoskeleton in *Amoeba proteus*. III. Influence of microinjected antibodies on the organization and function of the microfilament system. *Eur. J. Cell Biol.* 51: 279–284.

Chapman-Andresen, C. (1962). Studies on pinocytosis in amoebae. *C.R. Trav. Lab. Carlsberg* 33: 73–264.

Christiani, A., Hugelmeyer, P., and Stockem, W. (1986). Morphological evidence for the existence of a more complex cytoskeleton in *Amoeba proteus*. *Cell Tissue Res.* 246: 163–168.

Christofidou-Solomidou, M., Brix, K., and Stockem, W. (1989). Induced pinocytosis and cytoskeletal organization in *Amoeba proteus*: a combined fluorescence and electron microscopic study. *Eur. J. Protistol.* 24: 336–345.

Condeelis, J., Hall, A., Bresnick, A., Warren, V., Hock, R., Bennett, H., and Ogihara, S. (1988). Actin polymerization and pseudopod extension during amoeboid chemotaxis. *Cell Motil. Cytoskel.* 10: 77–90.

DeWit, R. J. W., Bulgakov, R., Dominaar, T. A., and DeWit, T. F. R. (1987). Differential effects of stimulus termination of excitation and desensitization of folic acid receptors and guanylate cyclase in *Dictyostelium discoideum*. *Biochim. Biophys. Acta* 930: 1–9.

DeWit, R. J. W. and Snaar-Jagalska, B. E. (1985). Folate and cAMP modulate GTP binding to isolated membranes of *Dictyostelium discoideum*: functional coupling between cell surface receptors and G-proteins. *Biochem. Biophys. Res. Commun.* 129: 11–17.

Dinauer, M. C., Mackay, S. A., and Devreotes, P. N. (1980a). Cyclic 3,5-AMP relay in *Dictyostelium discoideum* III. The regulation of cAMP synthesis and secretion during the cAMP signaling response. *J. Cell Biol.* 86: 537–545.

Dinauer, M. C., Steck, T. L., and Devreotes, P. N. (1980b). Cyclic 3,5-AMP relay in *Dictyostelium discoideum* IV. Recovery of the cAMP signaling after adaptation to cAMP. *J. Cell Biol.* 86: 545–553.

Eichinger, L., Noegel, A. A., and Schleicher, M. (1991). Domain structure in actin-binding proteins: expression and functional characterization of truncated severin. *J. Cell Biol.* 112: 665–676.

Europe-Finner, G. N. and Newell, P. C. (1985). Inositol 1,4,5-trisphosphate induced cyclic GMP formation in *Dictyostelium discoideum*. *Biochem. Biophys. Res. Commun.* 130: 1115–1122.

Europe-Finner, G. N. and Newell, P. C. (1986). Inositol 1,4,5-trisphosphate induces calcium release from a non-mitochondrial pool in amoebae of *Dictyostelium*. *Biochim. Biophys. Acta* 887: 335–340.

Europe-Finner, G. N. and Newell, P. C. (1987). GTP analogues stimulate inositol trisphosphate formation in *Dictyostelium*. *J. Cell Sci.* 87: 513–518.

Farnham, C. J. M. (1975). Cytochemical localization of adenylate cyclase and 3′,5′-nucleotide phosphodiesterase in *Dictyostelium*. *Exp. Cell Res.* 91: 36–46.

Fechheimer, M., Murdock, D., Carney, M. and Glover, C. V. C. (1991). Isolation and sequencing of complementary DNA clones encoding the *Dictyostelium discoideum* 30,000 dalton actin-binding protein. *J. Biol. Chem.* 266: 2883–2889.

Flower, R. J. (1974). Drugs which inhibit prostaglandin biosynthesis. *Pharmacol. Rev.* 26: 33–67.

Fontana, D. R., Price, P. L., and Phillips, J. C. (1991). Cell-cell contact mediates cyclic AMP secretion in *Dictyostelium discoideum*. *Dev. Genet.* 112: 54–62.

Galvin, N. J., Stockhausen, D., Meyers-Hutchins, B. L., and Frazier, W. A. (1984). Association of the cyclic AMP chemotaxis receptor with the detergent-insoluble cytoskeleton of *Dictyostelium discoideum*. *J. Cell Biol.* 98: 584–595.

Gawlitta, W., Stockem, W., Wehland, J., and Weber, K. (1980). Pinocytosis and locomotion of amoebae. XV. Visualization of Ca^{++}-dynamics by chlorotetracycline (CTC) fluorescence during induced pinocytosis in living *Amoeba proteus*. *Cell Tissue Res.* 213: 9–20.

Gollnick, R., Meyer, R., and Stockem, W. (1991). Visualization and measurement of calcium transients in *Amoeba proteus* by fura-2 fluorescence. *Eur. J. Cell Biol.* 55: 262–271.

Hall, A., Schlein, A., and Condeelis, J. (1988). Relationship of pseudopod extension to chemotactic hormone-induced polymerization in amoeboid cells. *J. Cell. Biochem.* 37: 285–299.

Josefsson, J. O., Arvidson, G., and Cobbold, P. (1988). Possible regulation of cation-induced pinocytosis in *Amoeba proteus* by phospholipase A. *Eur. J. Cell Biol.* 46: 200–206.

Juliani, M. H. and Klein, C. (1981). Photoaffinity labeling of the cell surface adenosine 3′,5′-monophosphate receptor in *Dictyostelium discoideum* and its modification in down-regulated cells. *J. Biol. Chem.* 256: 613–619.

Keller, H. U., Niggli, V., and Zimmermann, A. (1989). Diacylglycerols and PMA induce actin polymerization and distinct shape changes in lymphocytes: relation to fluid pinocytosis and locomotion. *J. Cell Sci.* 93: 457–465.

Khachatrian, L., Klein, C., and Howlett, A. (1987). Pertussis and cholera toxin ADP-ribosylation in *Dictyostelium discoideum* membranes. *Biochem. Biophys. Res. Commun.* 149: 975–981.

Klein, C., Sadeghi, H., and Simons, S. (1986). Immunological analysis of chemotactic receptor of *Dictyostelium discoideum*: identification of cDNA clones. *J. Biol. Chem.* 261: 15192–15196.

Klein, H. P., Koster, B., and Stockem, W. (1988). Pinocytosis and locomotion of Amoebae. XVIII. Different morphodynamic forms of endocytosis and microfilament organization in *Amoeba proteus. Protoplasma Suppl.* 2: 76–87.

Klein, P., Vaughan, R., Borleis, J., and Devreotes, P. N. (1987). The surface cAMP receptor in *Dictyostelium discoideum*: levels of ligand-induced phosphorylation, solubilization, identification of primary transcript, and developmental regulation of expression. *J. Biol. Chem.* 262: 358–364.

Kukulies, J., Ackermann, G., and Stockem, W. (1986). Pinocytosis and locomotion in amoebae XIV. Demonstration of two different receptor sites on the cell surface of *Amoeba proteus. Protoplasma* 131: 233–243.

Liu, G. and Newell, P. C. (1988). Evidence that cyclic GMP regulates myosin interaction with the cytoskeleton during chemotaxis of *Dictyostelium. J. Cell Sci.* 90: 123–129.

McRobbie, S. J. and Newell, P. C. (1983). Changes in actin associated with the cytoskeleton following chemotactic stimulation of *Dictyostelium discoideum. J. Cell Sci.* 68: 139–151.

Mato, J. M., Van Haastert, P. J. M., Krens, F. A., Rhijnsburger, E. H., Dobbe, F. C., and Konijn, T. M. (1977). Cyclic AMP and folic acid mediated cyclic GMP accumulation in *Dictyostelium discoideum. FEBS Lett.* 79: 331–336.

Meier, K. and Klein, C. (1988). An unusual protein kinase phosphorylates the chemotactic receptor of *Dictyostelium discoideum. Proc. Natl. Acad. Sci. U.S.A.* 85: 2181–2185.

Milne, J. L. and Coukell, M. B. (1991). A Ca^{++} transport system associated with the plasma membrane of *Dictyostelium discoideum* is activated by different chemoattractant receptors. *J. Cell Biol.* 112: 103–110.

Opas, M. (1981). Effects of induction of endocytosis on adhesiveness of *Amoeba proteus. Protoplasma* 107: 161–169.

Pan, P., Hall, E. M., and Bonner, J. T. (1972). Folic acid as a second chemotactic substance in the cellular slime molds. *Nature New Biol.* 237: 181–182.

Prusch, R. D. (1980). Endocytotic sucrose uptake in *Amoeba proteus* induced with the calcium ionophore A23187. *Science* 209: 691–692.

Prusch, R. D. (1986). Calcium and the initial surface binding phase of pinocytosis in *Amoeba proteus. Am. J. Physiol.* 251: C153–C158.

Prusch, R. D. and Britton, J. C. (1987). Peptide stimulation of phagocytosis in *Amoeba proteus. Cell Tissue Res.* 250: 589–593.

Prusch, R. D., Goette, S. M., and Haberman, P. L. (1989). Prostaglandins may play a signal-coupling role during phagocytosis in *Amoeba proteus. Cell Tissue Res.* 255: 553–557.

Prusch, R. D. and Hannafin, J. (1979). Sucrose uptake by pinocytosis in *Amoeba proteus* and the influence of external calcium. *J. Gen. Physiol.* 74: 523–535.

Prusch, R. D. and Vedovatti, P. A. (1989). Calcium movements associated with peptide induced phagocytosis in *Amoeba proteus. J. Comp. Physiol.* B159: 287–292.

Prusch, R. D. and Wells, S. M. (1990). Binding characteristics of peptide phagocytic receptors on the surface of *Amoeba proteus. Cell Tissue Res.* 261: 445–450.

Reibman, J., Korchak, H. M., Vosshall, L. B., Haines, K. A., Rich, A. M., and Weismann, G. (1988). Changes in diacylglycerol labeling, cell shape, and protein phosphorylation distinguish "triggering" from "activation" of human neutrophils. *J. Biol. Chem.* 263: 6322–6328.

Rifkin, J. L. and Isik, R. (1984). Effects of folic acid upon filopodia of *Dictyostelium discoideum* vegetative amoebae. *Cell Motil. Cytoskel.* 4: 129–135.

Roos, W. and Gerisch, G. (1976). Receptor-mediated adenylate cyclase activation in *Dictyostelium discoideum. FEBS Lett.* 68: 170–172.

Roos, W., Scheidegger, C., and Gerisch, G. (1977). Adenylate cyclase activity oscillations as signals for cell aggregation in *Dictyostelium discoideum. Nature* 266: 259–261.

Ross, F. M. and Newell, P. C. (1981). Streamers: chemotactic mutants of *Dictyostelium discoideum* with altered cyclic GMP metabolism. *J. Gen. Microbiol.* 127: 339–350.

Shariff, A. and Luna, E. J. (1992). Diacylglycerol-stimulated formation of actin nucleation sites at plasma membranes. *Science* 256: 245–247.

Shiozawa, J. A., Brandts, J. F., and Jacobson, B. S. (1989). Binding of plasma membrane glycoproteins to the cytoskeleton during patching and capping is consistent with an entropy-enhancement model. *Biochim. Biophys. Acta* 980: 361–366.

Snaar-Jagalska, B. E., Kesbeke, F., Pupillo, M., and Van Haastert, P. J. M. (1988). Immunological detection of G-protein α-subunits in *Dictyostelium discoideum. Biochem. Biophys. Res. Commun.* 156: 757–761.

Sonobe, S., Takahashi, S., Hatano, S., and Kuroda, K. (1986). Phosphorylation of *Amoeba* G-actin and its effect on actin polymerization. *J. Biol. Chem.* 261: 14837–14843.

Stockem, W., Hoffman, H. U., and Gruber, B. (1983). Dynamics of the cytoskeleton in *Amoeba proteus.* I. Redistribution of microinjected fluorescein-labeled actin during locomotion, immobilization and phagocytosis. *Cell Tissue Res.* 232: 79–96.

Stockem, W. and Klopocka, W. (1988). Ameboid movement and related phenomena. *Int. Rev. Cytol.* 112: 137–183.

Taylor, D. L., Blinks, J. R., and Reynolds, G. (1980a). Contractile basis of ameboid movement VII. Aequorin luminescence during ameboid movement, endocytosis, and capping. *J. Cell Biol.* 86: 599–607.

Taylor, D. L., Wang, Y. L., and Heiple, J. M. (1980b). Contractile basis of ameboid movement. VII. The distribution of fluorescently labelled actin in living amebas. *J. Cell Biol.* 86: 590–598.

Theibert, A. and Devreotes, P. N. (1986). Surface receptor-mediated activation of adenylate cyclase in *Dictyostelium*: regulation by guanine nucleotides in wild type cells and aggregation-deficient mutants. *J. Biol. Chem.* 261: 15121–15125.

Van Haastert, P. J. M. (1983). Relationship between adaptation of the folic acid and the cAMP mediated cGMP response in *Dictyostelium. Biochem. Biophys. Res. Commun.* 115: 130–136.

Van Haastert, P. J. M. (1984). Guanine nucleotides modulate cell surface cAMP-binding sites in membranes from *Dictyostelium discoideum. Biochem. Biophys. Res. Commun.* 124: 597–604.

Van Haastert, P. J. M. and Klein, E. (1983). Binding of cAMP derivatives to *Dictyostelium discoideum* cells: activation mechanism of the cell surface cAMP receptor. *J. Biol. Chem.* 258: 9636–9642.

Vogel, G., Thilo, L., Schwarz, H., and Steinhart, R. (1980). Mechanism of phagocytosis in *Dictyostelium discoideum*: phagocytosis is mediated by different recognition sites as disclosed by mutants with altered phagocytic properties. *J. Cell Biol.* 86: 456–465.

Wolosewick, J. J. and Condeelis, J. (1986). Fine structure of gels prepared from actin-binding proteins and actin: comparison to cytoplasmic extracts of cells of *Dictyostelium discoideum. J. Cell. Biochem.* 30: 227–243.

Wurster, B., Schubiger, K., Wick, U., and Gerisch, G. (1977). Cyclic GMP in *Dictyostelium discoideum*: oscillations and pulses in response to folic acid and cyclic AMP signals. *FEBS Lett.* 76: 141–144.

Zimmermann, A., Gehr, P., and Keller, H. U. (1988). Diacylglycerol-induced shape changes, movements and altered F-actin distribution in human neutrophils. *J. Cell Sci.* 90: 657–666.

MEMBRANE RECYCLING AND TURNOVER IN LARGE, FREE-LIVING AMOEBAE

Kwang W. Jeon

Advances in Cell and Molecular Biology of Membranes, Volume 2B
Membrane Traffic in Protozoa, pages 355–369

ISBN: 1-55938-628-2

I. INTRODUCTION

Since an amoeba was first observed under the microscope as "little Proteus" by Rösel von Rosenhof (1755), large, free-living amoebae (amoebae belonging to *Amoeba, Chaos and Pelomyxa*) have been widely used in studies of various aspects of cell biology, and amoebae were one of the first cell types in which cell membranes were studied. As motile and phagocytic cells, amoebae exhibit extensive membrane activities during both normal locomotion and capture of preys. However, it has been difficult to study the membrane traffic in these cells due to the lack of membrane-specific markers. Thus, not much is known about the membrane turnover and recycling in spite of their popularity as experimental cells for many years. For example, multiple pathways whereby components are known to flow between the plasma membrane and intracellular membranes in many other types of cells have not been firmly established in amoebae, except for occasional suggestions that such flow may exist. Since the activity, turnover, and recycling of amoeba membranes were reviewed previously (Chapman-Andresen, 1972; Stockem, 1972; Komnick et al., 1973), some new information has been obtained. In this chapter, I will summarize what is known about the membrane traffic in amoebae at present.

II. MEMBRANES OF AMOEBAE

The free-living amoebae have all the usual cellular membranes with known structure and properties (Daniels, 1973; Flickinger, 1973), but the plasma membrane has an unusually thick glycocalyx attached to it in most species. Only *Pelomyxa palustris* (Daniels and Breyer, 1967) and *Amoeba dubia* (Sanders and Bell, 1970) are thought to have no thick glycocalyx. The following is a brief survey of the general properties of cellular membranes of amoebae.

A. The Plasmalemma

The term *plasmalemma* has been used to denote the limiting surface layer of amoebae since before the structure of the cell surface became known (e.g., Mast, 1926). Electron microscopic examination has shown that the plasmalemma consists of the typical plasma membrane covered with glycocalyx, which has a 20 nm-thick amorphous layer and another fibrous layer extending up to 200 nm (Fig. 1) (Pappas, 1959). Such an ultrastructural morphology is maintained even after the plasmalemma has been isolated (O'Neill and Wolpert, 1961; Allen and Winzler, 1973; Ahn and Jeon, 1982).

The chemical composition of the plasmalemma of *A. proteus* was studied in some detail by O'Neill (1964) and by Allen and Winzler (1973; Allen et al., 1974) but no unusual components were found. When isolated plasmalemma is analyzed by sodium dodecyl sulfate-polyacrylamide gel electrophoresis, several bands of peri-

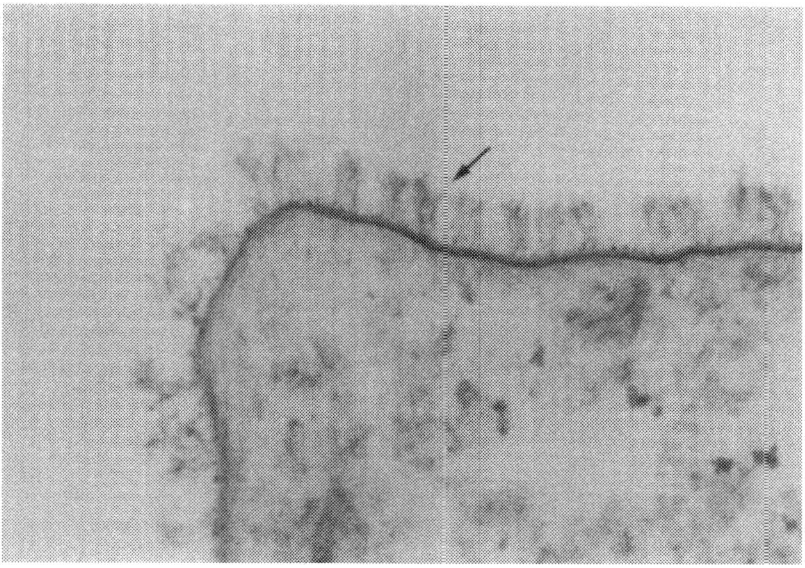

Figure 1. Electron micrograph of the plasmalemma of *Amoeba proteus* to show the glycocalyx (arrow). Note the thin amorphous zone and the fibrous extension. Jeon, unpublished. × 43,300.

odic acid-Schiff reagent-staining glycoproteins are detected in addition to numerous polypeptides of different sizes (Fig. 2) (Ahn and Jeon, 1982). One notable finding is that there is only one kind of polypeptide on the outside of the plasmalemma that can be iodinated by lactoperoxidase (Fig. 2) (Ahn and Jeon, 1982, 1985). This contrasts with plasma membranes of other cells such as macrophages (Muller et al., 1980) that have many such polypeptides. When amoebae are treated with neuraminidase or trypsin, more proteins are exposed and iodinated, indicating that the amoeba's glycocalyx prevents the exposure of some polypeptides on the external surface (Ahn and Jeon, 1985).

Recently, 225/220 kDa spectrin was found on the cytoplasmic side of the plasmalemma, as detected with monoclonal antibodies (mAb) as probes (Choi and Jeon, 1989). When spectrin was precipitated by the injection of anti-spectrin mAb, cell morphology was adversely affected (Fig. 3) as well as other movement-related cell activities (Choi and Jeon, 1992a), indicating that spectrin plays a structural role in amoebae.

B. Intracellular Membranes

In addition to membranes surrounding or forming regular organelles such as nuclei, mitochondria, lysosomes, Golgi complexes, endoplasmic reticulum, dense

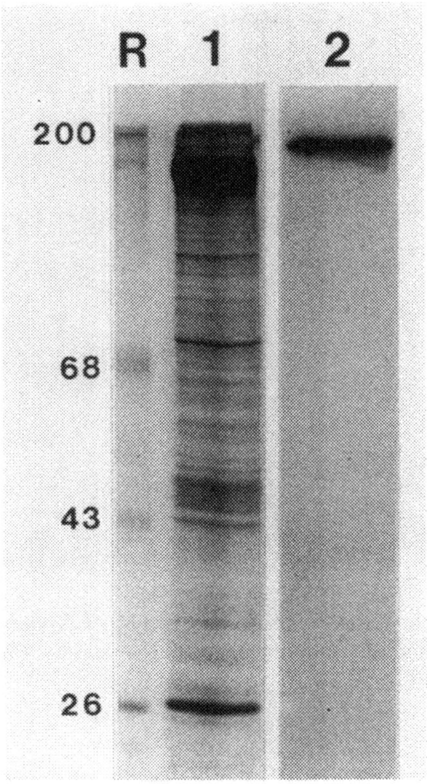

Figure 2. SDS-PAGE pattern of amoeba plasmalemma after staining with Comassie blue (lane 1) and radioautography (lane 2). Lane R, molecular size markers (kDa). The sample in lane 2 had been obtained from whole amoebae iodinated prior to the isolation of plasmalemma and subsequent electrophoresis. After Ahn and Jeon (1982,1985).

bodies, and a contractile vacuole, amoebae have an extensive vesicular membrane system whose function remains unclear. The ultrastructure of internal membranes was described in detail (Daniels, 1973; Flickinger, 1973), and the relation between Golgi complex, cytoplasmic vesicles, and the plasmalemma was studied by Flickinger (1975, 1984) in a series of experiments in which radioactively labeled mannose was injected and traced by electron-microscopic radioautography. Wise and Flickinger (1970) also used electron-dense markers and periodic acid-silver methenamine stain to study the relation between Golgi complexes and the glycocalyx. Their results indicated that Golgi membranes were not derived from the plasmalemma but that they may contribute to the formation of the glycocalyx. Wise (1972) found cisternae of the Golgi complex to be connected with rough endoplasmic reticulum. The cell nucleus plays an important role in the maintenance and

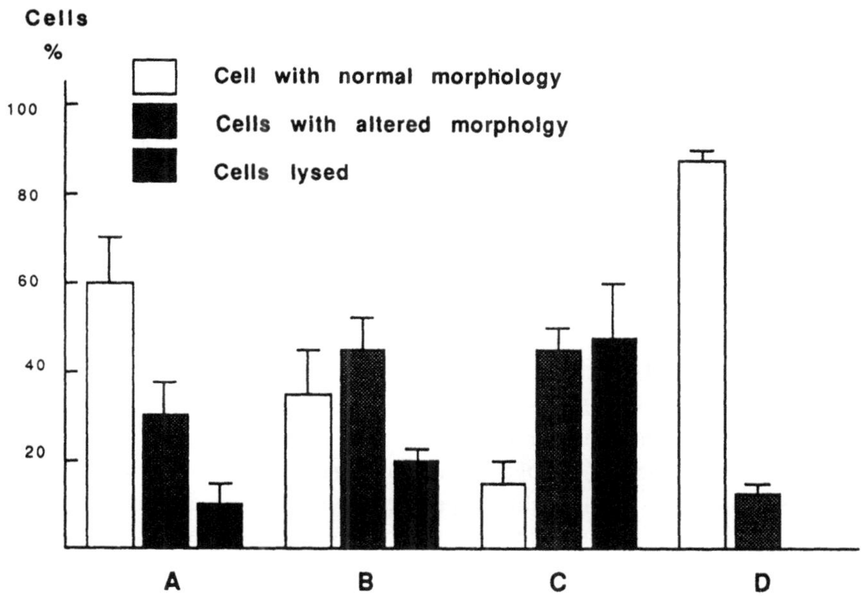

Figure 3. Graphs to show the adverse effect on cell morphology of precipitating amoeba spectrin by the injection of antispectrin monoclonal antibody. **A**, 1 mg/ml; **B**, 2.5 mg/ml; **C**, 5 mg/ml; **D**, amoebae injected with antilipopolysaccharide monoclonal antibody at 2.5 mg/ml for comparison. Reproduced from Choi and Jeon (1992) by permission of Academic Press.

regeneration of some membranous organelles (Flickinger, 1973), but the mechanism is not known.

In symbiont-bearing amoebae, a separate membrane system exists that forms the "symbiosomes" (Roth et al., 1988). Symbiosomes have unique components and do not fuse with lysosomes (Jeon and Jeon, 1976b; Ahn et al., 1990; Choi and Jeon, 1992b). See also chapter by Reisser in this volume.

III. MEMBRANES IN NORMAL CELL MOVEMENT

The large, free-living amoebae constantly change their shape and surface area/volume ratio, and there probably is a continuous change in the amount (surface area) of the plasmalemma of a moving cell. Since the early days of studies on amoeboid movement, there has been much controversy concerning the behavior of cell membrane associated with normal locomotion (cf. De Bruyn, 1947, for earlier refs.) In one view, the plasmalemma is fluid and is a passive participant in cell movement, flowing in the direction of locomotion simply to accommodate changes in cell shape and form (Mast, 1926; Griffin and Allen, 1960). In this case, no major membrane turnover would be required during normal locomotion. In another view, the plasmalemma is active and dynamic, and new cell surface is thought to be constantly formed as a cell moves (Goldacre and Lorch, 1950; Goldacre, 1961; Bell, 1961), requiring a rapid turnover of the cell surface. Goldacre and Lorch (1950) were the first to propose a possible mechanism for such membrane turnover by suggesting that the cell membrane dissolves into the cytoplasm in the rear (uroid) and forms de novo in the front. This suggestion was based on their observation that the plasmalemma did not move forward during locomotion, as had been believed by many earlier authors. However, some authors consider the view that the plasmalemma plays an active role and rapidly turns over during normal locomotion to be no longer tenable (Stockem, 1972; Komnick et al., 1973; Grębecki, 1987, 1990).

Wolpert and O'Neill (1962) used a plasmalemma-specific antibody in an attempt to label and follow the behavior of membrane and its turnover in amoebae. These authors calculated the half-life of a normally moving amoeba's surface to be about 5 hr, based on results from labeling experiments with a fluorescent antibody. The authors recognized that the antibody was probably directed against a component of the glycocalyx, and hence the suggested turnover rate may not apply to that of the plasmalemma as a whole. The probability that the behavior of amoeba's glycocalyx during movement did not represent that of the underlying plasma membrane proper was also suggested by similar experiments carried out by Jeon and Bell (1964). It appeared that the glycocalyx of *Amoeba proteus* was fluid and flowed forward over the newly formed pseudopods and that it could be stripped off an amoeba or could be prevented from flowing forward. However, some authors (Grębecki, 1987) still consider the movement of attached particles to represent the behavior of the plasmalemma as a whole.

There has been a suggestion that the cell membrane is slowly but continuously internalized at the uroid region of moving amoebae via "permanent pinocytosis" (Wohlfarth-Bottermann and Stockem, 1966). The phenomenon would be similar to that suggested earlier by Goldacre (1961) except that the rate would be much slower. However, such a scheme has been difficult to prove without a specific marker for the membrane components. Involvement of micropinocytosis in the membrane turnover among secreting cells has been well established (Wollman, 1989).

One normal cellular activity that is obviously accompanied by membrane recycling is the vesiculation of contractile vacuoles at systole (McKanna, 1973). During diastole, the vacuole enlarges by fusing with smaller vesicles (see also Couillard and Bourdeau, 1983), and the membrane of a mature vacuole fuses with the plasmalemma at systole, discharging its contents. But it appears that the entire contractile vacuole membrane does not return to the plasmalemma at systole but a portion of it recycles within the cytoplasm as small vesicles.

As an amoeba moves along the substratum, some glycocalyx is probably lost at the focal contact points (Bell and Jeon, 1963; Jeon and Jeon, 1976a; Opas, 1981), as is the case with *Naegleria* (Preston and King, 1978). However, no definitive studies have been made on the "footprints" of large amoebae and it is not known how much glycocalyx is lost or how such lost glycocalyx is replaced.

In summary, except for the partial recycling of contractile vacuole membranes, there is as yet no conclusive evidence to prove or disprove a regular membrane turnover in normally moving amoebae. One of the difficulties of studying amoeba membranes has been the unavailability of a convenient way to use chemical or metabolic markers to label membrane components. For example, radioactive precursors usually have to be given to amoebae indirectly via their food organism *Tetrahymena*, and such a procedure is not suitable for pulse labeling. Thus, radioactive mannose for the glycoproteins or choline for the lipid of membranes has had to be directly injected into individual amoebae (Flickinger, 1975; Flickinger and Reed, 1982). Also, amoebae lack membrane-specific enzymes or receptor molecules. It has been much easier to study the turnover of membrane components and to determine their half-lives in cells that secrete proteins or have specific receptors on the membrane (Hare, 1990).

IV. MEMBRANE DYNAMICS IN ENDOCYTOSIS

During endocytosis, a large amount of the plasmalemma is internalized and hence an equal amount of membrane would have to be returned somehow to the cell surface. Therefore, it is easier to consider membrane turnover and recycling during endocytosis than in normal locomotion. In amoebae, there appears to be no independent exocytic process, such as secretion of proteins. Thus, any exocytosis would be a reversal of the previous endocytosis, such as seen during defecation of undigested material.

A. Phagocytosis

Large, free-living amoebae rely exclusively on phagocytosis for the acquisition of their nutrients, and the phenomenon of phagocytosis has been extensively studied for morphology, physiology, biochemistry, and behavior. As much as 40–50% of the plasmalemma of an amoeba may be internalized by phagocytosis in 1 hr (Chapman-Andresen and Holter, 1960). According to some estimates, an amoeba may internalize more than 10 times the amount of its plasmalemma in 24 hr (Christiansen and Marshall, 1965), necessitating an extensive turnover of the plasmalemma.

Various labels, including basic dyes, antibodies, proteins, and electron-dense dyes, have been used to mark the plasmalemma and to follow its fate during phagocytosis and associated cellular activities (see Chapman-Andresen, 1972, for earlier refs.). However, most of the markers used have been directed to components of the glycocalyx, which is removed from the membranes of phagolysosomes within a few hours after ingestion of objects (Ahn and Jeon, 1985), and the fate of the underlying plasma membrane cannot be traced with such markers. It is desirable

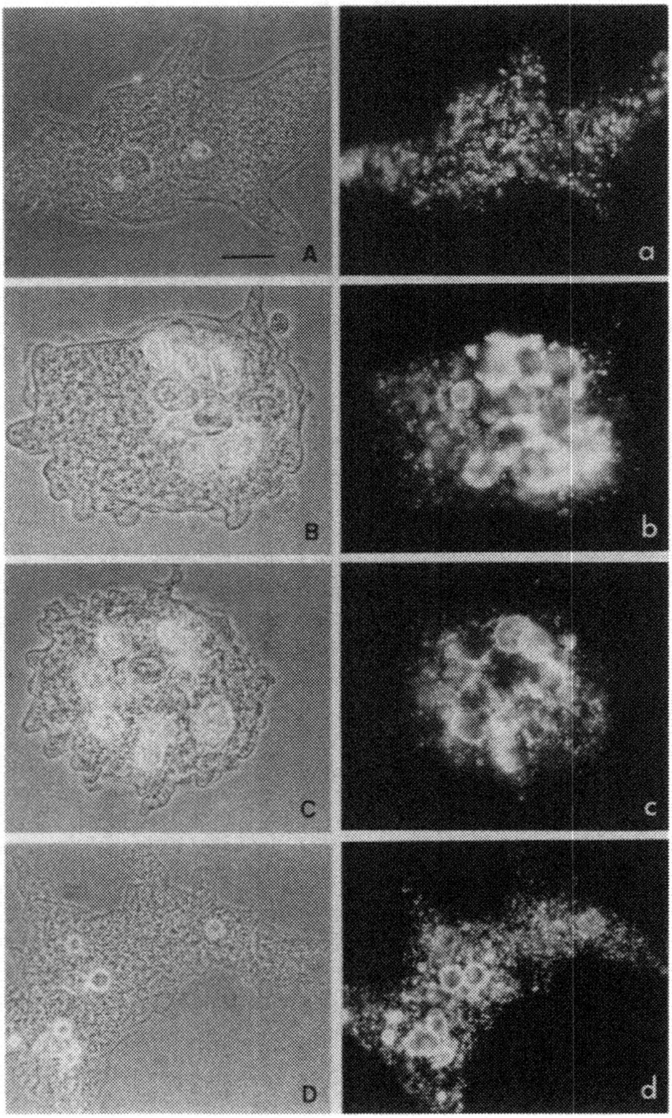

Figure 4. Phase-contrast (A–D) and immunofluorescence (a–d) micrographs of amoebae fixed at different intervals after phagocytosis. Amoebae had been allowed to ingest *Tetrahymena* for 5 min and were chased for 5 min (**B**, **b**), 15 min (**C**, **c**), and 2 hr (**D**, **d**). A and a represent an amoeba before phagocytosis. In a–d, fixed amoebae were treated for immunofluorescence with an mAb against a lysosomal membrane protein of amoebae as the probe. × 133. After Choi et al. (1992).

to have membrane-specific markers such as enzymes and recognizable receptor proteins on the amoeba plasmalemma for studies of membrane recycling as in other cells that are active in receptor-mediated endocytosis (Brown et al., 1983; Hare, 1990).

In one study (Ahn and Jeon, 1985), a membrane protein was found to be recycled and reused during endo- and exocytosis, as revealed by lactoperoxidase-mediated iodination. A 175 kDa protein exposed on the outside of the plasmalemma appeared to be broken down into two small polypeptides (70 and 35 kDa) found on phagolysosomal membranes within 2 hr after phagocytosis, and the smaller 35 kDa component was returned to the plasmalemma 22 hr after the initial ingestion. In the same study, a portion of the 175 kDa protein was found to be recycled intact without any breakdown.

If amoebae are fed regularly, undigested residues are defecated within 2–3 days after ingestion and the vacuole membranes fuse with the plasmalemma, completing the membrane recycling (Chapman-Andresen, 1973). Recently, Choi et al. (1991, 1992) obtained specific mAbs against lysosomal membrane proteins of *Amoeba proteus* and followed phagosome–lysosome fusion after ingestion of *Tetrahymena* (Fig. 4). Lysosomal membrane proteins recognized by the mAb started to appear on phagolysosomal membranes about 5 min after the initial ingestion. Such mAbs could also be used in studying the turnover and recycling of the lysosomal membrane system of amoebae in general, although antibody binding may artificially enhance the turnover rate of membrane proteins (Hare, 1990).

B. Pinocytosis

A massive pinocytosis does not appear to be a normal activity in amoebae. However, pinocytosis can be experimentally induced with various solutes, and as much as one half the plasmalemma may be taken in during one cycle of pinocytosis (Chapman-Andresen, 1962). Ultrastructural aspects of pinocytosis were studied in detail by Chapman-Andresen and Nilsson (1967), who noted the exocytic vacuole membrane fusing with the plasmalemma to be devoid of glycocalyx at the time of fusion. It remains unclear as to how and when such denuded plasma membrane acquires its glycocalyx.

Various other aspects of pinocytosis have been extensively studied by W. Stockem and co-workers and by R. D. Prusch, and readers are referred to their chapters in this volume for details.

V. MEMBRANES OF SPECIALIZED VACUOLES

Apart from membranes of the usual cell organelles, amoebae may have specialized membranes or vacuoles, some of which are transient, but others persist for many cell generations. Amoebae may be induced to form new membranes, for example, by microsurgically caused injury. Amoebae may also contain specialized vesicles or vacuoles such as newly forming vacuole membranes to surround foreign objects introduced into the cytoplasm by direct injection and vesicles containing endosymbionts (symbiosomes).

A. Newly Formed Membranes

Wohlfarth-Bottermann (1964) reviewed earlier evidence for the membrane formation de novo under various conditions. According to Szubinska (1971,1978), injured amoebae form "new membranes" to heal the wounds, and the new membrane is initially thicker than the existing plasma membrane. Amoebae also appear to form new membranes when returned to atmospheric pressure following exposure to high pressures (Landau, 1965). Nachmias (1966) observed a rapid formation of large amounts of new plasmalemma as prechilled amoebae were returned to room temperature. In this case, the new plasmalemma appeared to be derived by interpolation of ground cytoplasm and not by fusion of preexisting vacuoles. Flickinger (1974) found rough endoplasmic reticulum membranes to participate in the repair of mechanically damaged nuclear envelopes.

When foreign objects such as bacteria or *Chlorella* are injected into the amoeba cytoplasm, initially they lie free in the amoeba's cytoplasm without any barrier sequestering them (Fig. 5) (Jeon and Lorch, 1982). However, within 2–6 hr after injection, such objects become surrounded by discontinuous, nontrilamellar membrane-like structures that later change into continuous vacuole membranes. These structures are not associated with any preexisting membranes, and the vacuole membrane apparently arises de novo by a yet unknown mechanism, suggesting that amoeba's cytoplasm has a pool of membrane precursors in a form not detectable as typical membranes.

B. Symbiosome Membranes

The symbiosome is one of several membranous organelles that do not fuse with amoeba lysosomes (Jeon and Jeon, 1976b). Established endosymbionts, including parasites, avoid destruction by their host cells by being resistant to digestive enzymes (Han and Jeon, 1980) and/or by preventing the fusion between symbiosomes and lysosomes (Steinhoff et al., 1989; Goren and Mor, 1990; Ishibashi and Arai, 1990; Joiner et al., 1990; Buchmeier and Heffron, 1991; Jeon, 1992).

The avoidance of lysosomal fusion suggests that the membranes of symbiosomes which initially arise from the plasmalemma during infection are different from those of the plasmalemma or phagolysosomes. In the case of the xD strain of *Amoeba proteus* harboring X-bacteria (Jeon, 1992), the symbiosome membrane has been found to contain a unique 96 kDa protein (Ahn et al., 1990) and lipopolysaccharides (Choi and Jeon, 1992b). While the origin of the 96 kDa protein is not known, lipopolysaccharides are produced by the bacterial endosymbionts. The symbiosome-specific membrane components are suspected to play a role in

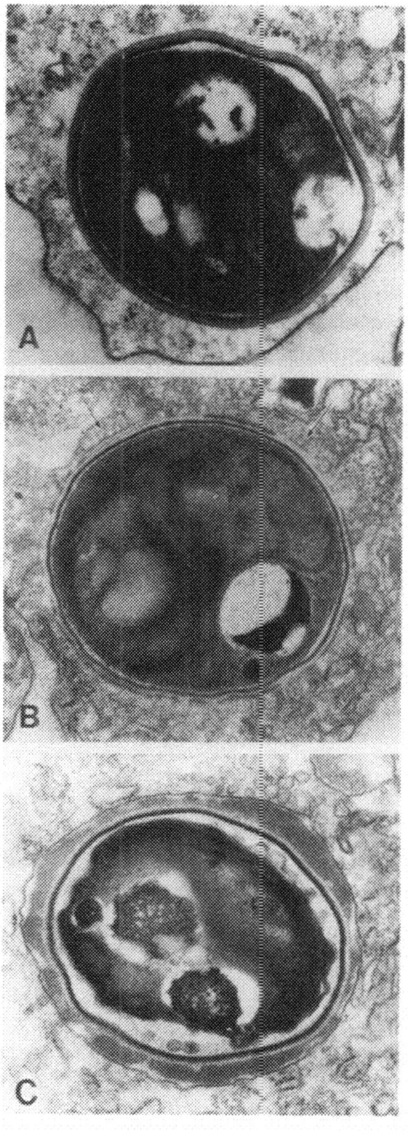

Figure 5. Electron micrographs to show an apparent de nove membrane formation around injected algae. **A**, injected alga lying free in the amoeba cytoplasm; **B**, alga fixed 6 hr after injection, partially surrounded by pieces of membrane-like structures (arrows); **C**, alga fixed 8 hr after injection, completely surrounded by a vacuole membrane. × 14,000. After Jeon and Lorch (1982).

the prevention of lysosomal fusion as a "fusion-blocking" agent(s) (Choi and Jeon, 1992b; Jeon, 1992).

As symbionts multiply and the number of symbiosomes increases, keeping up with the host cell growth, more symbiosome membrane will have to be produced. However, since components of symbiosome membranes are unique and not found on any other cellular membranes, it is not clear if such membranes are turned over or recycled.

VI. CONCLUDING REMARKS

In this chapter, we have reviewed what is known about the membrane traffic in large, free-living amoebae. While there are obvious membrane activities associated with normal locomotion and movement-related phenomena, very little is known about membrane turnover and recycling. This has been due largely to the technical difficulties of studying membrane traffic in amoebae arising from the presence of a thick glycocalyx and the lack of suitable markers for the plasma membrane components. Some markers for the carbohydrate portion of membrane glycoproteins or lipids and specific antibodies are available, but their use has been limited. It is hoped that more useful markers will be found and that many advantages amoebae offer for the study of membrane trafficking will be explored in the near future.

ACKNOWLEDGMENTS

I thank Dr. M. A. Handel for reading the manuscript. The work in our laboratory has been supported by grants from the U.S. National Science Foundation.

REFERENCES

Ahn, G. S., Choi, E. Y., and Jeon, K. W. (1990). A symbiosome-membrane-specific protein in symbiont-bearing *Amoeba proteus* as studied with a monoclonal antibody. *Endocyt. Cell Res.* 7: 45–50.

Ahn, T. I. and Jeon, K. W. (1982). Structural and biochemical characteristics of the plasmalemma and vacuole membranes in amoebae. *Exp. Cell Res.* 137: 253–268.

Ahn, T. I. and Jeon, K. W. (1985). Recycling of membrane proteins during endo- and exocytosis in amoebae. *Exp. Cell Res.* 160: 54–62.

Allen, H. J., Ault, C., Winzler, R. J., and Danielli, J. F. (1974). Chemical characterization of the isolated cell surface of *Amoeba. J. Cell Biol.* 60: 26–38.

Allen, H. J. and Winzler, R. J. 1973). The chemistry of amoeba surface. In: *The Biology of Amoeba* (Jeon, K. W. ed.), pp. 451–466, Academic Press, New York.

Bell, L. G. E. (1961). Surface extension as the mechanism of cellular movement and cell division. *J. Theoret. Biol.* 1: 104–106.

Bell, L. G. E. and Jeon, K. W. (1963). Locomotion of *Amoeba proteus. Nature* 198: 675–676.

Brown, M. S., Anderson, R. W., and Goldstein, J. L. (1983). Recycling receptors: the round-trip itinerary of migrant membrane proteins. *Cell* 32: 663.

Buchmeier, N. A. and Heffron, F. (1991). Inhibition of macrophage phagosome–lysosome fusion by *Salmonella typhimurium. Infect. Immun.* 59: 2232–2238.

Chapman-Andresen, C. (1962). Studies on pinocytosis in amoebae. *C. R. Trav. Lab. Carlsberg* 33: 73–264.

Chapman-Andresen, C. (1972). Membrane activity in freshwater amebae. *J. Protozool.* 19: 225–231.

Chapman-Andresen, C. (1973). Endocytotic processes. In: *The Biology of Amoeba* (Jeon, K. W., ed.), pp. 319–348, Academic Press, New York.

Chapman-Andresen, C. and Holter, H. (1960). The uptake of fluorescent albumin by pinocytosis in *Amoeba proteus. J. Biophys. Biochem. Cytol.* 3: 288–291.

Chapman-Andresen, C. and Nilsson, J. R. (1967). Studies on endocytosis in amoebae: the distribution of pinocytically ingested dyes in relation to food vacuoles in *Chaos chaos*. II. Electron microscopic observations using Alcian blue. *C. R. Trav. Lab. Carlsberg* 36: 189–207.

Choi, E. Y. and Jeon, K. W. (1989). The presence of a spectrin-like protein on symbiosome membrane of symbiont-bearing *Amoeba proteus* as studied with monoclonal antibodies. *Endocyt. Cell Res.* 6: 99–108.

Choi, E. Y. and Jeon, K. W. (1992a). Role of spectrin in *Amoeba proteus*, as studied by microinjection of anti-spectrin monoclonal antibodies. *Exp. Cell Res.* 199: 174–178.

Choi, E. Y. and Jeon, K. W. (1992b). Bacterial-endosymbiont-derived lipopolysaccharides on amoeba symbiosome membranes. *J. Protozool.* 39: 205–210.

Choi, E. Y., Kim, K. J., and Jeon, K. W. (1991). Lysosome–phagosome fusion in amoebae, as studied using lysosome-membrane-specific monoclonal antibodies. *J. Cell Biol.* 115: 229a (abstract).

Choi, E. Y., Kim, K. J., and Jeon, K. W. (1992). Lysosomal membrane proteins of *Amoeba proteus*, as studied with monoclonal antibodies. *J. Protozool.* 39: 671–677.

Christiansen, R. G. and Marshall, J. M. (1965). A study of phagocytosis in the ameba *Chaos chaos. J. Cell Biol.* 25: 443–457.

Couillard, P. and Bourdear, M. (1983). The contractile vacuole of *Amoeba proteus*: some of the consequences of the microcoalescence hypothesis. *Can. J. Zool.* 61: 2669–2673.

Daniels, E. W. (1973). Ultrastructure. In: *The Biology of Amoeba* (Jeon, K. W., ed.), pp. 125–169, Academic Press, New York.

Daniels, E. W. and Breyer, E. P. (1967). Ultrastructure of the giant amoeba *Pelomyxa palustris. J. Protozool.* 14: 167–179.

De Bruyn, P. P. H. (1947). Theories of amoeboid movement. *Quart. Rev. Biol.* 22: 1–24.

Flickinger, C. J. (1973). Cellular membranes of amoebae. In: *The Biology of Amoeba* (Jeon, K. W., ed.), pp. 171–199, Academic Press, New York.

Flickinger, C. J. (1974). The role of endoplasmic reticulum in the repair of amoeba nuclear envelopes damaged microsurgically. *J. Cell Sci.* 14: 421–437.

Flickinger, C. J. (1975). The relation between the Golgi apparatus, cell surface and cytoplasmic vesicles in amoebae studied by electron microscope radioautography. *Exp. Cell Res.* 96: 189–201.

Flickinger, C. J. (1984). Altered labelling of the cell surface and intracellular organelles with $^{[3]}$H-mannose in enucleated amoebae. *J. Cell Sci.* 68: 83–94.

Flickinger, C. J. and Read, G. A. (1982). Distribution of phospholipids labeled with $^{[3]}$H-choline and relationship between membranous organelles in amoebae, as studied by electron-microscopic radioautography. *Cell Tissue Res.* 222: 523–530.

Goldacre, R. J. (1961). The role of the cell membrane in the locomotion of amoebae, and the source of the motive force and its control by feedback. *Exp. Cell Res. Suppl.* 8: 1–16.

Goldacre, R. J. and Lorch, I. J. (1950). Folding and unfolding of protein molecules in relation to cytoplasmic streaming, amoeboid movement and osmotic work. *Nature* 166: 497.

Goren, M. B. and Mor, N. (1990). Influence of phagosomal contents on the apparent inhibition of phagosome–lysosome fusion mediated by polyanionic substances in mouse peritoneal macrophages. *Biochem. Cell Biol.* 68: 24–32.

Grębecki, A. (1987). Velocity distribution of the anterograde and retrograde transport of extracellular particles by *Amoeba proteus. Protoplasma* 141: 126–134.

Grębecki, A. (1990). Dynamics of the contractile system in the pseudopodial tips of normally locomoting amoebae, demonstrated in vivo by video-enhancement. *Protoplasma* 154: 98–111.
Griffin, J. L. and Allen, R. D. (1960). The movement of particles to the surface of amoeba in relation to current theories of amoeboid movement. *Exp. Cell Res.* 20: 619–622.
Han, J. H. and Jeon, K. W. (1980). Isolation and partial characterization of two plasmid DNAs from endosymbiotic bacteria in *Amoeba proteus*. *J. Bacteriol.* 141: 1466–1469.
Hare, J. F. (1990). Mechanism of membrane protein turnover. *Biochim. Biophys. Acta* 1031: 71–90.
Ishibashi, Y. and Arai, T. (1990). Specific inhibition of phagosome–lysosome fusion in murine macrophages mediated by *Salmonella typhimurium* infection. *FEMS Microbiol. Immunol.* 2: 35–43.
Jeon, K. W. (1992). Macromolecules involved in the amoeba–bacteria symbiosis. *J. Protozool.* 39: 199–204.
Jeon, K. W. and Bell, L. G. E. (1964). Behavior of cell membrane in relation to locomotion in *Amoeba proteus*. *Exp. Cell Res.* 33: 531–539.
Jeon, K. W. and Jeon, M. S. (1976a). Scanning electron microscope observations of *Amoeba proteus* during phagocytosis. *J. Protozool.* 23: 83–86.
Jeon, K. W. and Jeon, M. S. (1976b). Endosymbiosis in amoebae: recently established endosymbionts have become required cytoplasmic components. *J. Cell. Physiol.* 89: 337–344.
Jeon, K. W. and Lorch, I. J. (1982). The formation of vacuole membranes in the presence and absence of cell nucleus in amoebae. *Exp. Cell Res.* 141: 351–356.
Joiner, K. A., Fuhrman, S. A., Miettinen, H. M., Kasper, L. H., and Mellman, I. (990). *Toxoplasma gondii*: fusion competence of parasitophorous vacuoles in Fc receptor-transfected fibroblasts. *Science* 249: 641–646.
Komnick, H., Stockem, W., and Wohlfarth-Bottermann, K. E. (1973). Cell motility: mechanism in protoplasmic streaming and ameboid movement. *Int. Rev. Cytol.* 34: 169–249.
Landau, J. V. (1965). High hydrostatic pressure effects on *Amoeba proteus*. *J. Cell Biol.* 24: 332–336.
McKanna, J. A. (1973). Membrane recycling: vesiculation of the amoeba contractile vacuole at systole. *Science* 179: 88–90.
Mast, S. O. (1926). Structure, movement, locomotion, and stimulation in amoeba. *J. Morphol. Physiol.* 41: 347–424.
Muller, W. A., Steinman, R. M., and Cohn, Z. A. (1980). The membrane proteins of the vacuolar system. I. Analysis by a novel method of intralysosomal iodination. *J. Cell Biol.* 86: 292–303.
Nachmias, V. T. (1966). A study by electron microscopy of the formation of new surface by *Chaos chaos*. *Exp. Cell Res.* 43: 583–601.
O'Neill, C. H. (1964). Isolation and properties of the cell surface membrane of *Amoeba proteus*. *Exp. Cell Res.* 35: 477–496.
O'Neill, C. H. and Wolpert, L. (1961). Isolation of the cell membrane of *Amoeba proteus*. *Exp. Cell Res.* 24: 592–594.
Opas, M. (1981). Effects of induction of endocytosis on adhesiveness of *Amoeba proteus*. *Protoplasma* 107: 161–169.
Pappas, G. D. (1959). Electron microscope studies on amoebae. *Ann. N.Y. Acad. Sci.* 78: 448–473.
Preston, T. M. and King, C. A. (1978). An experimental study of the interaction between the soil amoeba *Naegleria gruberi* and a glass substrate during amoeboid locomotion. *J. Cell Sci.* 34: 145–158.
Rösel von Rosenhof, A. J. (1755). Der kleine Proteus. Der Monat.-herausgeg. *Insect.-Belust.* 3: 622.
Roth, L. E., Jeon, K., and Stacey, G. (1988). Homology in endosymbiotic systems: the term "Symbiosome." In: *Molecular Genetics of Plant–Microbe Interactions* (Palacios, R. and Verma, D. P. S., eds.), pp. 220–225, APS Press, St. Paul, MN.
Sanders, E. J. and Bell, L. G. E. (1970). Some physiological and morphological effects of polylysine on *Amoeba proteus*. *J. Cell Sci.* 7: 739–753.
Steinhoff, U., Golecki, J. R., Kazda, J., and Kaufmann, S. H. (1989). Evidence for phagosome–lysosome fusion in *Mycobacterium leprae*-infected murine Schwann cells. *Infect. Immun.* 57: 1008–1010.

Stockem, W. (1972). Membrane-turnover during locomotion of *Amoeba proteus. Acta Protozool.* 11: 83–101.

Szubinska, B. (1971). "New membrane" formation in *Amoeba proteus* upon injury of individual cells: electron microscope observations. *J. Cell Biol.* 49: 747–772.

Szubinska, B. (1978). Closure of the plasma membrane around microneedle in *Amoeba proteus. Exp. Cell Res.* 111: 105–115.

Wise, G. E. (1972). Connections between cisternae of the Golgi apparatus and the granular endoplasmic reticulum in *Amoeba proteus. Z. Zellforsch. mikrosk. Anat.* 126: 431–436.

Wise, G. E. and Flickinger, C. J. (1970). Relation of the Golgi apparatus to the cell coat in amoebae. *Exp. Cell Res.* 61: 13–23.

Wohlfarth-Bottermann, K. E. (1964). Cell structure and their significance for ameboid movement. *Int. Rev. Cytol.* 16: 61–131.

Wohlfarth-Bottermann, K. E. and Stockem, W. (1966). Pinocytose und Bewegung von Amoeben. II. Permanente und induzierte Pinocytose by *Amoeba proteus. Z. Zellforsch. mikrosk. Anat.* 73: 444–474.

Wollman, S. H. (1989). Turnover of plasma membrane in thyroid epithelium and review of evidence for the role of micropinocytosis. *Eur. J. Cell Biol.* 50: 247–256.

Wolpert, L. and O'Neill, C. H. (1962). Dynamics of the membrane of *Amoeba proteus* studied with labelled specific antibody. *Nature* 196: 1261–1266.

FOOD UPTAKE AND DIGESTION IN AMOEBAE

Wilhelm Stockem and Melpo Christofidou-Solomidou

Advances in Cell and Molecular Biology of Membranes, Volume 2B
Membrane Traffic in Protozoa, pages 371–407

ISBN: 1-55938-628-2

I. INTRODUCTION

Considerable progress in the past decade has extensively enhanced our view of the mechanisms related to the incorporation and intracellular processing of endocytotic contents. A series of recent reviews summarized and described the present knowledge on this topic in a large variety of cells (Rodman et al., 1990; Griffiths and Gruenberg, 1991; Smythe and Warren, 1991). Accordingly, the digestive system is composed of a heterogeneous population of endosomal and lysosomal compartments containing extracellular or intracellular material, a mixture of up to 40 acid hydrolases, and some highly specific membrane proteins. Three different cytotic processes can be distinguished with respect to the origin of the lysosomal contents, namely, receptor-mediated endocytosis (Goldstein et al., 1985), phagocytosis (Wright and Silverstein, 1986), and autophagy (Glauman et al., 1981). Receptor-mediated endocytosis is the most intensely studied mechanism whereby a highly specific uptake of extracellular material such as low-density lipoproteins, transferrin, immunoglobulins, peptide hormones, and others is accomplished. Under normal conditions, the ligand–receptor complex is enclosed first in coated vesicles and then targeted via an uncoated stage to early (peripheral) endosomes (Geuze et al., 1983). Early endosomes have a characteristic tubulovesicular morphology as well as a slightly acidic pH and serve as the compartment for uncoupling of receptor and ligand (CURL). While the receptor often recycles to the plasma membrane, the ligand is carried to late (perinuclear) endosomes or prelysosomes (endolysosomes) that are multivesicular or multilamellar in appearance (Gruenberg and Howell, 1989). Prelysosomes contain low levels of digestive enzymes and represent the penultimate stage of the degradative pathway followed by the occurrence of typical (secondary) lysosomes, which are the strongly acidic compartment of final digestion (Kornfeld and Mellman, 1989). So far, two fundamental models have been proposed to describe the intracellular fate of vacuoles resulting from receptor-mediated endocytosis. The first model is that early and late endosomes represent preexisting and independent compartments which communicate by recycling transport vesicles or tubules (Griffiths and Gruenberg, 1991). The second model states that late endosomes derive from early endosomes by successive maturation (Stoorvogel et al., 1991). Although specialized cells, such as MDCK cells or hepatoma cells, may have developed two different pathways, it remains to be determined whether one or both of them are required for endosomal processing in all higher multicellular organisms.

In addition to the specific endocytotic incorporation of proteins, the engulfment of particulate matter by phagocytosis represents another important receptor-mediated process (Wright and Silverstein, 1986). Examples demonstrating the significance of phagocytosis at least for mammalian cells are the uptake of cell debris, various microorganisms, and aged red blood cells by monocytes and macrophages as well as neutrophils and eosinophils. All these phagocytotic cells express a set of particular plasma membrane receptors for immunoglobulins, complement factors, mannose- and galactose-terminated cell surface oligosaccharides, and other glycosylation end products in the mucous layer (Silverstein et al., 1989). The intracellular fate of phagosomes differs from the typical endosomal pathway in that they obviously receive their digestive enzymes by direct fusion with either late endosomes or preexisting secondary lysosomes (Cohn and Wiener, 1963). On the other hand, formation of phagolysosomes can be inhibited by ingested pathogenic microorganisms such as *Mycobacterium*, *Rickettsia*, *Toxoplasma*, *Trypanosoma*, *Leishmania*, *Legionella*, *Chlamydia*, or *Histoplasma*, thus resulting in intracellular parasitism (Horwitz, 1988). Meanwhile, it is also established that the mode of phagosome formation shows some fundamental variations as compared to receptor-mediated endocytosis because it reflects a localized response between the ligand of the target particle and the receptor of the phagocyte, which proceeds according to a zipper mechanism (Griffin et al., 1975).

A third source for the origin of lysosomal contents results from the process of autophagy (Dunn, 1990a). Eukaryotic cells continuously degrade cytoplasmic molecules and cell organelles such as proteins and mitochondria by an ER-mediated sequestration into autophagosomes. According to recent studies, autophagic vacuoles mature into digestive stages in a similar way to that of phagosomes by fusing with either late endosomes or preexisting secondary lysosomes (Dunn, 1990b). The rate of autophagosome formation can be stimulated by starvation (Nilsson, 1984; Mortimor et al., 1988) or by microinjection of exogenous material (Dice et al., 1987) but is distinctly inhibited during experimentally suppressed cell growth (Papadopoulos and Pfeifer, 1987).

In addition to the described possibilities of intralysosomal digestion, there exists also a way of nonlysosomal degradation of proteins in the cytoplasmic matrix or within cellular organelles such as the nuclear envelope, ER, nucleoplasm, and mitochondria (Hruban et al., 1991). Short-lived or abnormal proteins are degraded within the cytosol either by Ca^{2+}-activated proteases, the calpains, or by ubiquitin-dependent proteolysis; on the other hand, most long-lived proteins are sequestered first by autophagy and then digested within lysosomes.

In opposition to the extensive literature regarding cytosis and the endosomal pathway in metazoa and especially in mammalian cells, only a few laboratories have addressed these topics in protozoa by applying modern techniques of cell biology. For instance, rather sporadic information exists about the mode of food uptake and intracellular digestion in sporozoa. Some coccids like *Plasmodium* or *Eimeria* are known to develop during their erythrocytic stages a cytostome-like membrane invagination (micropore) which ingests the cytoplasm of red cells and forms food vacuoles (Aikawa et al., 1966). However, there are differences of opinion as to whether this mechanism should be classified as pinocytosis or phagocytosis.

More detailed information has been gathered about the feeding process in flagellates. Whereas typical pinocytotic activity was observed only in a few species such as *Opalina* (Münch, 1970), *Giardia* (Bockman and Winborn, 1968), *Crithidia* (Brooker, 1971), and *Trypanosoma* (Langreth and Balber, 1975) upon the experimental addition of thorotrast or ferritin, most autotrophic and heterotrophic flagellates investigated under natural conditions were shown to ingest food particles and organisms by phagocytosis (Aaronson, 1973; Eeckhout, 1973). In some cases, these cells possess special differentiations like the projecting collar of choanoflagellates (Laval, 1971) or the food-catching filopodia (peduncles) of dinoflagellates, to which nutritive particles adsorb before they are phagocytozed (Schnepf and Elbrächter, 1992).

Comparable results were obtained with ciliates in that pinocytosis has been demonstrated only for some suctorians like *Tokophrya* (Rudzinska, 1980), and, more recently, for *Paramecium* (Allen et al., 1992). The majority of ciliated protozoa, however, are again highly specialized phagocytotic cells that possess a particular organelle (cytostom) for the uptake of food (Hausmann, 1980). A rather extensive literature describes the endocytotic pathway and digestive system in *Paramecium* (Fok and Allen, 1988; see chapters by these authors in this volume), where about twenty different hydrolytic enzymes have been identified in four stages (I–IV) of lysosomal vacuoles (Fok and Paeste, 1982).

A vast diversity of information exists also for rhizopods. Besides in true amoebae, which are the subject of this review and will be discussed more thoroughly in the following sections, cytotic processes as well as the intracellular fate of endosomes and phagosomes were also profoundly studied in other rhizopods, such as foraminifera (Lee and Anderson, 1991), radiolaria (Anderson, 1983), and heliozoa (Hausmann and Patterson, 1982; Linnenbach et al., 1983). Moreover, a detailed description of the nutritive physiology of these organisms and of other protozoa is given in numerous chapters of this book.

II. AMOEBOID MOVEMENT AND FOOD UPTAKE

True amoebae express different morphodynamic forms during active locomotion and food uptake. Giant or freshwater amoebae such as *Chaos chaos, Pelomyxa palustris*, *Amoeba proteus*, or *Amoeba dubia* belong to the "proteus" group (Andresen, 1973), that is, they are mostly polytactic and move by formation and retraction of large pseudopodia (Fig. 1a). In conjunction with normal movement, the proteus-amoebae exhibit a continuous but low turnover of membrane resulting from weak, permanent pinocytosis at the uroid and simultaneous exocytosis at the front (Stockem and Klopocka, 1988). However, during intense pinocytosis or phagocytosis and the initial stages of digestion, giant amoebae stop to locomote and become spheric (Fig. 1b).

Small lobose amoebae such as *Acanthamoeba castellanii*, *Entamoeba histolytica*, *Hartmaniella rhysodes*, *Hyalodiscus rubicundus*, *Naegleria gruberi*, or *Vanella (Hyalodiscus) simplex* are assigned to the "limax" group (Vickerman, 1962), that is, they are predominantly orthotactic by the expression of a broad or fan-like lobopodium and distinct separation of hyaline ectoplasm from granular endoplasm (Fig. 1c). In most limax-amoebae, active locomotion is also combined with the uptake of external fluid and particulate matter; however, phagocytotic

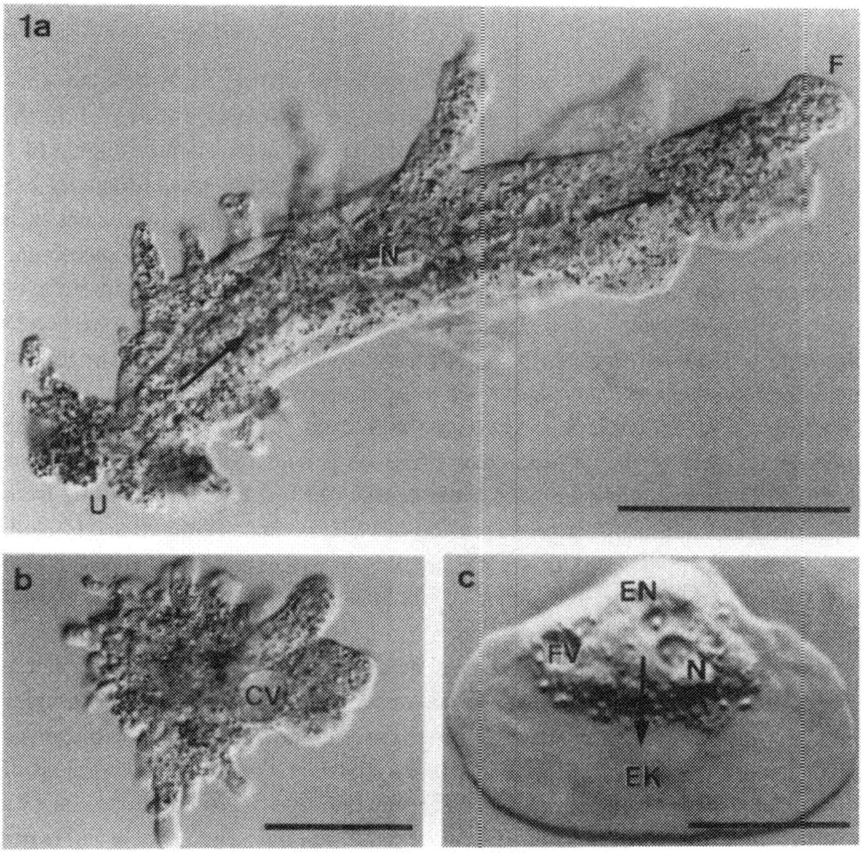

Figure 1. DIC micrographs of a polypodial *Amoeba proteus* (**a**) and a lobopodial *Vanella simplex* (**c**) during normal locomotion (arrows indicate moving direction), and (**b**) a resting *A. proteus* during food uptake. EK = ectoplasm, EN = endoplasm, F = front, FV = food vacuole, N = nucleus, CV = contractile vacuole, U = uroid. (a, b) are unpublished, (c) is reproduced from Hausmann et al. (1972) by permission of Springer-Verlag (Heidelberg). Bars: (a, b) 100 μm, (c) 20 μm.

activities with typical changes in external morphology and cessation of movement as observed in giant amoebae normally represent the exceptional case.

Besides proteus- and limax-amoebae there exists a third group of rhizopods exhibiting radiosa or reticulopodial cell forms. Many of these amoebae such as *Netzelia tuberculata* (Anderson, 1989), *Vampyrella lateritia* (Hülsmann, 1983), *Thecamoeba spaeronucleolus* (Haberey, 1973b), *Nuclearia delicatula* (Cann, 1986), or *Difflugiella* (Griffin, 1972) have developed very special strategies of food ingestion and, hence, will not be dealt with here.

A. Fluid-Phase Endocytosis

The permanent pinocytotic activity exhibited by locomoting freshwater amoebae at the uroid (Wolpert and O'Neil, 1962; Bruce and Marshall, 1965; Wohlfarth-Bottermann and Stockem, 1966) is characterized by the constant formation of 1 to 3 short membrane invaginations and the generation of a few small (diameter 2–5 µm) endosomes (Fig. 2a). Studies with inert, nonadsorptive marker substances such as Aerosil (amorphous silicon dioxide) have described this kind of membrane incorporation as typical fluid-phase endocytosis, that is, the concentration of marker within the endosomes (Fig. 2b) corresponds to the marker concentration in the culture medium (Stockem, 1966). Reliable conclusions regarding the importance of permanent pinocytosis for the normal nutritive metabolism in proteus-amoebae are so far difficult to draw, although fluid-phase endosomes enter the digestive system by fusion with preexisting lysosomes (Stockem, 1977). Most probably the measured rate of membrane uptake (0.15–0.2%/min) resulting from this pinocytotic activity in *Amoeba proteus* and *Chaos chaos* (Wolpert and O'Neil, 1962; Stockem, 1973) is much too low not only to cover the need of food but also to play any role in the generation of motive force for amoeboid movement (Stockem and Klopocka, 1988).

Distinctly higher rates of continuous membrane incorporation during active locomotion have been measured in limax-amoebae. Accordingly, *Vanella simplex* incorporates 0.2%/min or more of the plasma membrane by permanent pinocytosis (Hausmann et al., 1972), and *Acanthamoeba castellanii* can attain even values of 3–15%/min (Bowers and Olzewski, 1983). Nevertheless, it remains to be clarified whether this cytotic activity represents normal fluid-phase endocytosis and is not aimed at a rather specific uptake of adsorbed external matter.

Figure 2. DIC (**a**, **c**, **e**) and electron micrographs (**b**, **d**, **f**) to demonstrate different forms of pinocytosis and phagocytosis as well as different endosomes and phagosomes in *Amoeba proteus*. (**a**) Permanent (fluid phase) pinocytosis (arrowheads point to pinocytotic channels, arrows to single endosomes). FV= food vacuole. (**b**) Fluid-phase endosome containing randomly distributed Aerosil. (**c**) Induced (adsorptive) pinocytosis (arrows point to large tubular endosomes, arrowhead to a pinocytotic channel). Notice the high contrast of the plasma membrane in (c) due to the adsorption of egg white as compared with (a). (**d**) Adsorptive endosome containing egg white concentrated on the mucous layer (arrowheads). (**e**) Phagocytosis of a *Tetrahymena* (arrowhead; arrows indicate elongation of two pseudopodia during food-cup formation). (**f**) Phagosome containing an undigested *Tetrahymena* (T). (b, d, e, f) are unpublished, (a, c) are reproduced from Stockem (1970) by permission of Verlag Georg Fromme and Co (Wien). Bars: (a, also valid for c) 20 µm, (b, also valid for d) 1 µm, (e) 50 µm, (f) 5 µm.

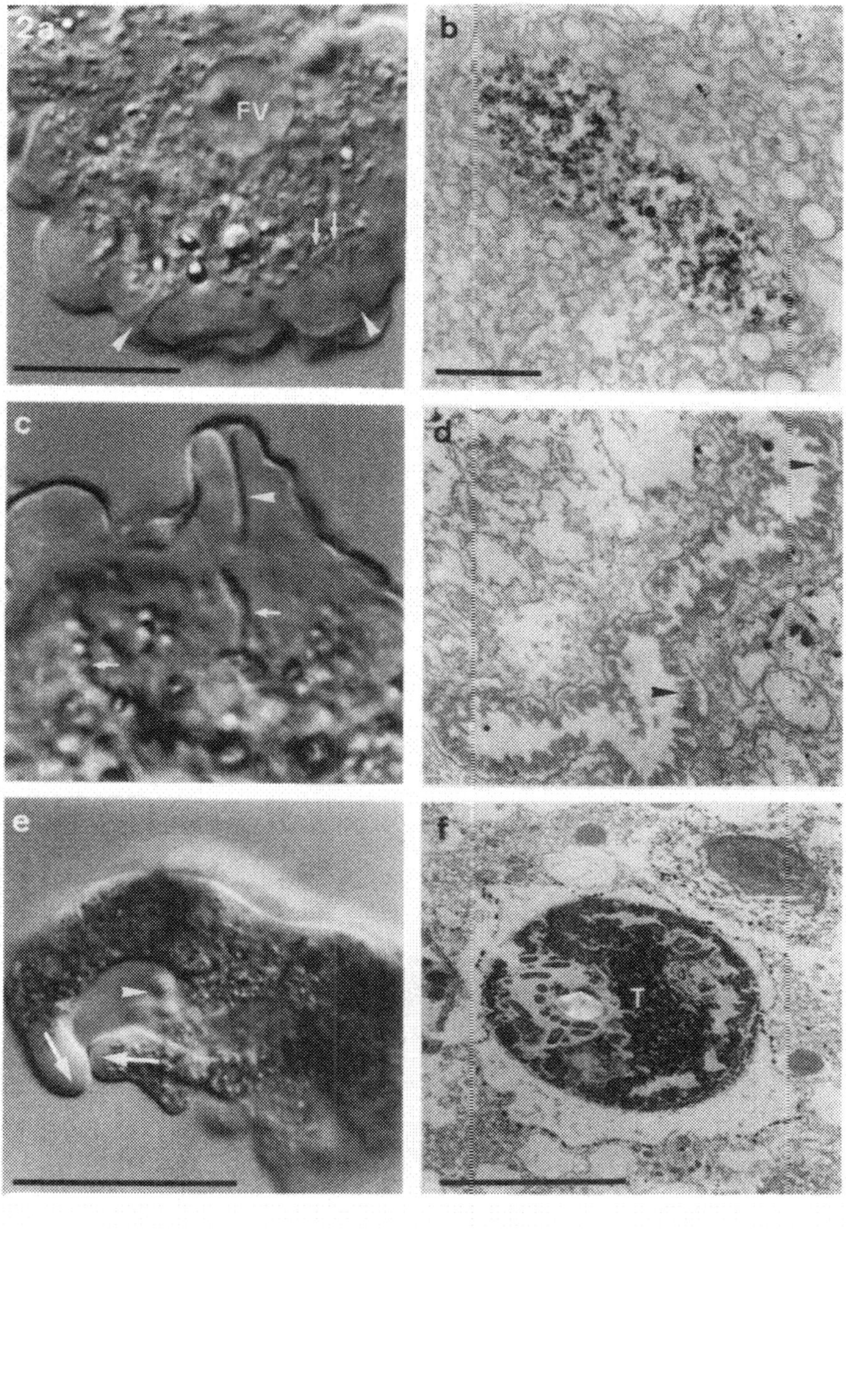
2a
FV
b
c
d
e
f
T

B. Adsorptive Endocytosis

Cationic substances such as inorganic salts, amino acids, proteins, dyes, or others are adsorbed first to the plasma membrane and then induce intense pinocytotic activity at the entire cell surface of giant amoebae (Chapman-Andresen, 1962; Josefsson, 1975; Stockem, 1977). Due to the chemical nature of the applied substances as many as 100 pinocytotic channels ranging from 10 to 60 μm in length and 0.5 to 5 μm in diameter are formed within 15–30 min (Klein et al., 1988; Stockem and Klein, 1988). The elongated endosomes constricted from these invaginations can measure up to 50 μm (Fig. 2c) and contain the cationic material distinctly concentrated at the mucous layer of the plasma membrane (Fig. 2d). Biochemical studies have shown that acid mucopolysaccharides with inositol-bound phosphate polymers are essential components of the negatively charged cell surface (Allen et al. 1974, 1976). By means of these binding sites external substances such as ferritin (Braatz-Schade and Stockem, 1973) or monovalent

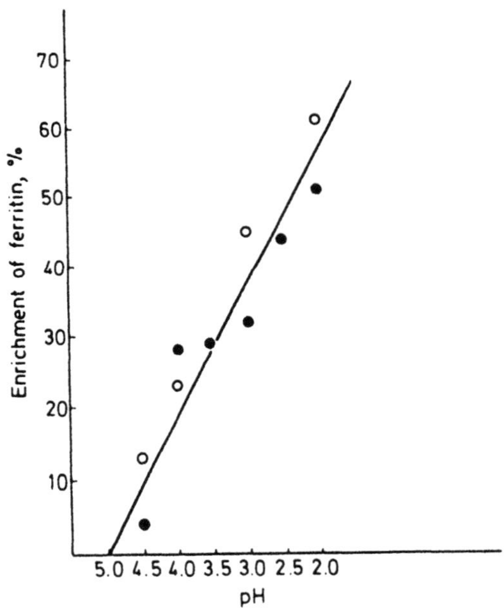

Figure 3. Diagram demonstrating the pH-dependent adsorption of ferritin on the mucous layer of *Amoeba proteus* by measuring the density (o) and counting the number (•) of ferritin particles. Reproduced from Braatz-Schade and Stockem (1973) by permission of Springer-Verlag (Heidelberg).

cations (Hendil, 1971) can be enriched 17 to 21-fold over the original concentration in the culture medium (Fig. 3). Hence, induced pinocytotic activity of giant amoebae shares distinct similarities with the so-called adsorptive endocytosis of metazoan cells, however without attaining the same degree of specificity as receptor-mediated endocytosis. Although the intensity of ingestion during induced pinocytosis can be increased up to a membrane incorporation rate of 2.25%/min (Stockem, 1973), only a smaller part of the formed endosomes is targeted to the lysosomal system (Christofidou-Solomidou et al., 1989; Christofidou-Solomidou and Stockem, 1992), that is, this kind of endocytosis seems not to be important for nutrition. Moreover, the majority of endosomes is rapidly released from the cell (Kukulies et al., 1986) and thus escapes intracellular digestion. Accordingly, induced pinocytosis rather represents a survival mechanism by which giant amoebae can overcome extracellular stress situations (Christofidou-Solomidou et al., 1989) and, thereby, exhibits certain similarities with the process of surface cleaning as displayed by macrophages or lymphocytes (De Petris, 1977). In this connection, it remains unexplained why typical adsorptive pinocytosis upon the application of cations has not been observed in limax-amoebae.

C. Receptor-Mediated Phagocytosis

The most common mechanism of food uptake in carnivorous or herbivorous amoebae of the proteus- and limax-group is to capture motile prokaryotic and eukaryotic animal cells or to ingest nonmotile plant material by phagocytotic activity (Christiansen and Marshall, 1965; Haberey, 1973a; Prusch and Minck, 1985). Amoeboid phagocytosis always represents the localized and specific response to a distinct chemotactic or mechanotactic stimulus (Seravin, 1986) and starts with the oriented formation of pseudopodia that engulf food material by flowing around it (Fig. 2e). Many amoebae show a clear preference for definite prey organisms, for example, *Amoeba proteus* and *Chaos chaos* for *Colpidium, Tetrahymena*, *Paramecium*, or *Chilomonas, Vanella simplex* and *Dictyostelium discoideum* for different bacteria, *Entamoeba histolytica* for *Crithidia* or cell debris, *Acanthamoeba castellanii* for erythrocytes, and other specialists such as *Thecamoeba* and *Vampyrella* for algae (Kitching, 1956; Rabinovitch and de Stefano, 1971a,b; Braun et al., 1972; Westphal et al., 1972; Haberey, 1973b; Hülsman, 1983). There is evidence that the division rate of amoebae is distinctly influenced by the nature of ingested food, as shown for *Amoeba proteus* with a highest value of 13.5 hr when fed on *Colpidium* and a lowest one of 130 hr when fed on *Chilomonas* (Andresen, 1956; Salt, 1968). The number and size of food vacuoles formed by giant amoebae within a few hours also depend on the nature of the prey cells and can range from 70 specimens of *Tetrahymena* to 100 of *Chilomonas* in phagosomes with diameters between 10 and 40 μm (Fig. 2f). This intense food capture corresponds to a maximum rate of plasma membrane turnover of an average of 1%/min and emphasizes the outstanding significance of phagocytosis for the nutrition of large amoebae.

Small amoebae of the limax group can internalize plasma membrane at comparably high rates by phagocytosis. For instance, Ryter and Chastellier (1977) have shown that *Dictyostelium* amoebae are stimulated by yeast to ingest cell surface portions equal to 40% of the total area. A similar phagocytotic affinity to yeast cells, modified erythrocytes, and *Crithidia* cells with ingestion rates of more than 50 particles/hr was also described for *Acanthamoeba* (Rabinovitch and de Stefano, 1971a, b; Bowers and Olzewski, 1983) and for *Entamoeba histolytica* (Westphal et al., 1972).

Although molecular biological studies designed to isolate and characterize specific phagocytotic binding sites on the plasma membrane of amoebae are so far lacking, more recent investigations on *Dictyostelium* and *Amoeba proteus* clearly show phagocytosis to be a receptor-mediated process (Vogel et al., 1980; Prusch and Wells, 1990). One of the phagocytotic receptors proposed for *Amoeba proteus* specifically binds a signal tripeptide (NFMLP) and uses extracellular Ca^{2+} to control the level of membrane uptake (Prusch and Vedovatti, 1989). In addition, there are indications that prostaglandins obviously play a signal-coupling role during this type of receptor-mediated phagocytosis (Prusch et al., 1989) and that the receptors have similar binding kinetics and are recycled in the same way as described for metazoan cells (Prusch and Wells, 1990; see also chapters by Prusch and Jeon in this volume).

III. GENERAL MORPHOLOGY OF THE VACUOLAR APPARATUS

The morphological appearance of the vacuolar apparatus in giant amoebae is mainly characterized by the presence of numerous phagosomes and endosomes (Fig. 4a). Whereas phagosomes are always easy to identify as large prey-containing vacuoles, the smaller endosomes have a more inconspicuous content, unless marker substances have been applied. Early and late endosomes can be distinguished because of the more regular shape, fuzzy content, and electron-dense granules in the periphery of the late endosomal stage (Fig. 4b). The histochemical demonstration of acid phosphatase activity (Fig. 5a, b) indicates that the majority of endosomal and phagosomal vacuoles in both proteus-(Chapman-Andresen and

---

Figure 4. Semithin section (**a**) of *Amoeba proteus* (phase contrast) showing the cytoplasmic distribution of endosomes (arrows) and phagosomes (T) containing freshly fed *Tetrahymena* (arrowheads point to mitochondria). (**b**) Electron micrograph of *A. proteus* showing the general morphology of the vacuolar apparatus. 1 = phagosome, 2 = early endosomes, 3 = late endosomes, 4 = endosymbiontic vacuole, 5 = crystal vacuoles. Arrows point to exocytotic vacuoles. M = mitochondria. (a) is unpublished, (b) is reproduced from Stockem (1969) by permission of Springer-Verlag (Heidelberg). Bars: (a) 10 μm, (b) 5 μm.

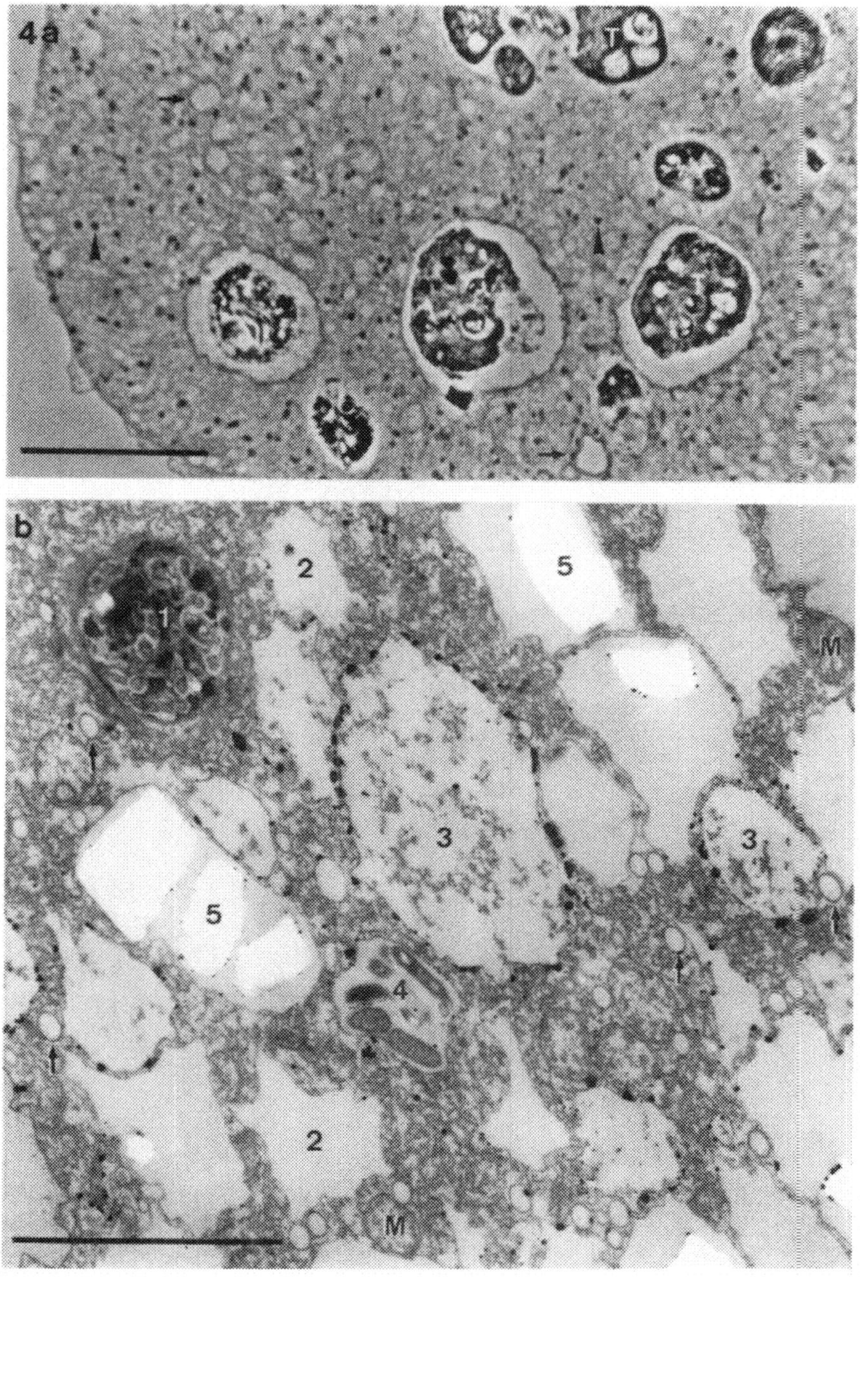
4a
T
b
1
2
5
M
3
3
5
4
2
M

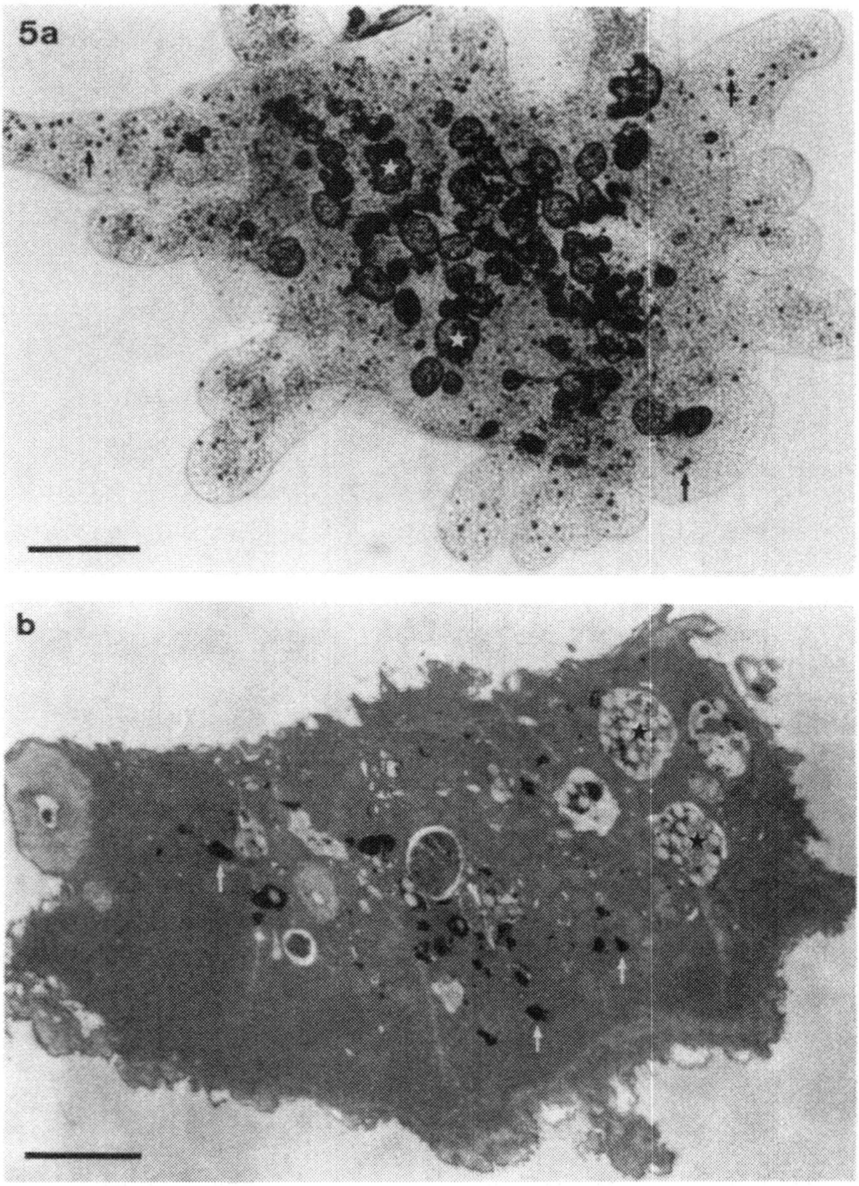

Figure 5. Bright field (**a**) and electron micrograph (**b**) to demonstrate the presence of endolysosomes (arrows) and phagolysosomes (stars) in *Amoeba proteus* (a) and *Vanella simplex* (b) by the Gomori method for acid phosphatase. *A. proteus* was fed with *Tetrahymena* and *V. simplex* with latex beads. (a) is reproduced from Stockem and Mitarbeiter (1977) by permission of Franckh'sche Verlagshandlung (Stuttgart), (b) is from Hausmann and Stockem (1973) by permission of Wissenschaftliche Verlagsgesellschaft (Stuttgart). Bars: (a) 20 μm, (b) 3 μm.

Lagunoff, 1966; Wise and Flickinger, 1970; Hausmann and Stockem, 1973) and limax-amoebae (Hausmann and Stockem, 1972) are lysosomes in different stages of digestion (see Section VI). Another population of minor vesicles such as exocytotic vacuoles, primary lysosomes, and acidosomes which partly belong to the digestive system, are more difficult to detect in light and electron micrographs and will be dealt with later in the review (see Sections VI.A, B; IX).

Besides the different lysosomal compartments, giant amoebae exhibit two further classes of vacuoles that are not implicitly related to the digestive system. The first class includes the group of endosymbiontic vacuoles (Fig. 4b). These vacuoles contain endosymbiontic bacteria and are segregated from phagosomes in which ingested microbes have survived the initial stages of digestion (Jeon, 1983). In the course of a series of complex interactions between the symbionts and the host, the endosymbiontic vacuoles lose their ability to fuse with lysosomes so that an irreversible endosymbiosis is established with various effects on amoeboid morphology and physiology. The second class comprises the group of crystal vacuoles (Fig. 4b) and is usually found in all giant amoebae (Bovee and Jahn, 1973); they contain a stored nitrogenous waste product (triuret) originating from the nucleic acid metabolism (Grunbaum et al., 1959). Although it was recently observed that crystal vacuoles occasionally fuse with endosomes (M. Christofidou-Solomidou, unpublished observations), their normal mode of generation as well as their role in the digestive process remains to be elucidated.

IV. ENDOSOME FORMATION AND MICROFILAMENTS

Staining of giant amoebae with fluorescent phalloidin or with anti-actin antibodies reveals the existence of a continuous system of microfilaments beneath the plasma membrane (Stockem et al. 1983b, 1984). Accordingly, endosome formation proceeds stepwise by the repeated detachment and regeneration of the submembraneous actin layer parallel to the elongation of small hyaline pseudopodia and channel-like invaginations (Klein and Stockem, 1979; Klein et al., 1988; Christofidou-Solomidou et al., 1989). As a result, several stratified actin layers form that are clearly visible in both living (Fig. 6a,b) and fixed amoebae (Fig. 6c, d). Endosome constriction always starts at the base of channels and occurs at places where the stratified layers come into contact with the membrane. The area of contact between the lower actin layers and the cytoplasmic face of the basal channel region (Fig. 6c) determines the endosomal size, in that extended contact areas result in large endosomes and vice versa (Klein et al., 1988). The ultrastructural organization of the actin layer along the channels (Fig. 6f) and around the endosomes (Fig. 6g) clearly differs. Microfilaments forming the layer of the channels exhibit a strictly parallel orientation, thus pointing to stress forces between the base and the tip of the channels. In contrast, microfilaments in the layer surrounding endosomes are disorganized, a characteristic feature of isotonic conditions. Various

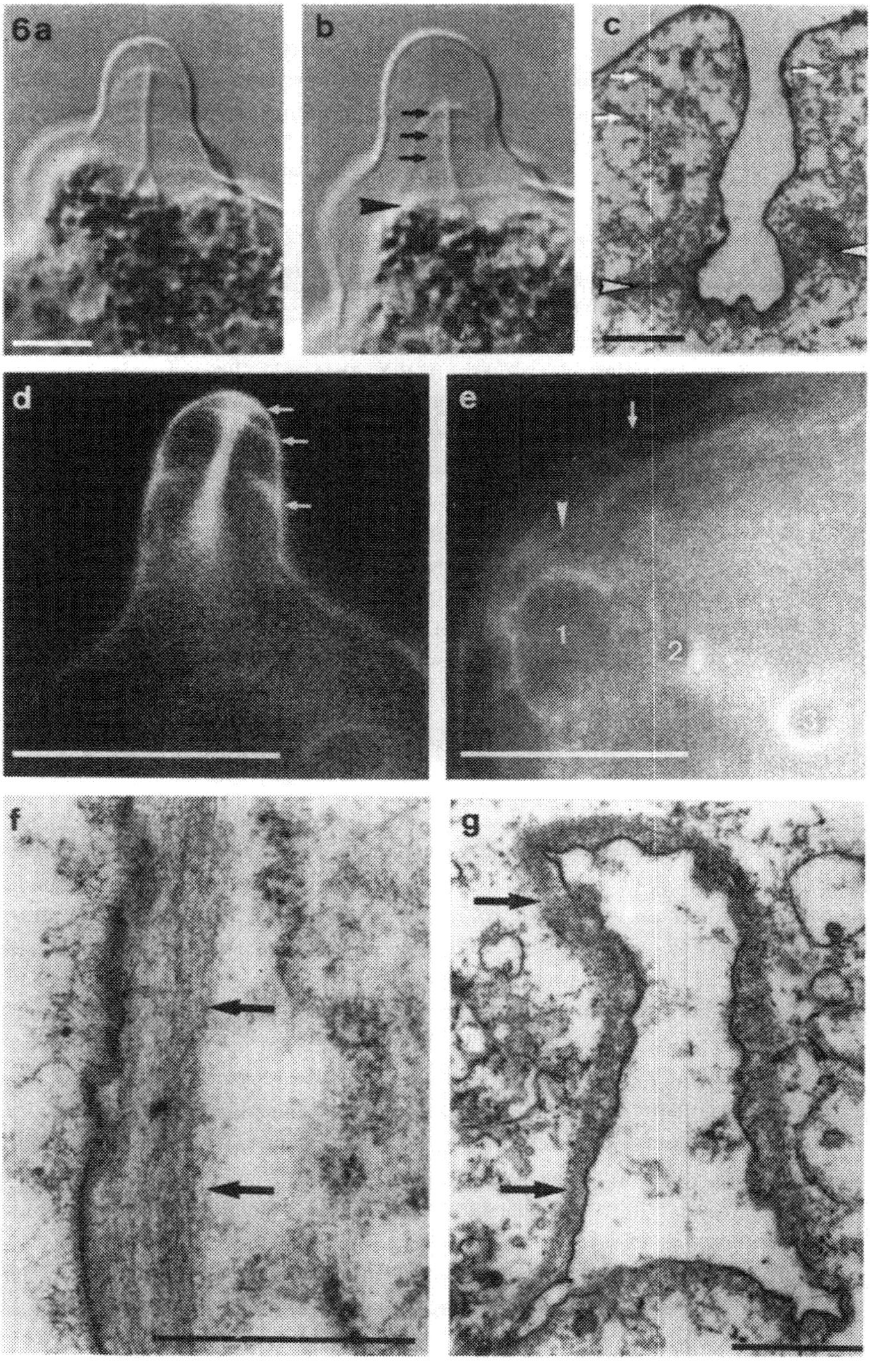
6a
b
c
d
e
1
2
f
g

Figure 6. DIC (**a**, **b**), immunofluorescence (anti-actin antibodies) (**d**, **e**), and electron micrographs (**c**, **f**, **g**) to demonstrate the organization of the microfilament system during induced pinocytosis in *Amoeba proteus*. The arrows in (b–d) point to the stratified actin layers after successive detachment from the plasma membrane. The arrowheads in (b, c, e) label the lowest actin layer determining the size of endosomes. Endocytotic channels (f, arrows) and endosomes (e, 1–3; g, arrows) are surrounded by actin layers differing in microfilament orientation and density. (a–c, f, g) are reproduced from Klein et al. (1988) by permission of Springer-Verlag (Vienna), (d) is from Stockem et al. (1983b) by permission of Wissenschaftliche Verlagsgesellschaft (Stuttgart), (e) is from Christofidou-Solomidou et al. (1989) by permission of Gustav Fischer Verlag (Stuttgart). Bars: (a, also valid for b) 25 μm, (c) 5 μm, (d) 50 μm, (e) 25 μm, (f) 0.5 μm, (g) 2 μm.

---

degrees in density of the microfilament network correlate with changes in endosomal diameter and hence point to different stages of endosomal contraction.

External Ca^{2+} plays a significant role in the process of endocytotic membrane incorporation in that cationic substances displace surface-bound calcium (Josefsson, 1975; Prusch and Hannafin, 1979; Kukulies et al., 1986). By this means, subsequent changes in permeability of the plasma membrane and induced transients in intracellular Ca^{2+} activate the actin layer to generate motive force for the initial stages of endocytosis (Gollnick et al., 1991). In a recent study on *Amoeba proteus*, Prusch and Minck (1985) have shown that similar Ca^{2+} transients are essential for receptor-mediated phagocytosis by again stimulating the actin system to produce the dynamic forces necessary for food-cup formation (Jeon and Jeon, 1983; Stockem et al., 1983a).

In contrast to giant amoebae, only a few data exist concerning the association of microfilaments with phagosomal or endosomal processes in small amoebae. The involvement of microfilaments in phagosome generation has been demonstrated only for *Dictyostelium* (Fukui and Imamoto, 1985). Altogether, it seems evident that microfilaments are more important for the adsorptive pinocytosis in large amoebae than for receptor-mediated endocytosis in mammalian cells.

V. ENDOSOMAL PATHWAYS AND PROCESSING

Recent results have shown that the mode of membrane internalization in giant amoebae obviously influences the intracellular pathway and processing of endosomal vacuoles (Christofidou-Solomidou et al., 1989). Accordingly, the fate of endosomes formed by induced pinocytosis distinctly differs from the endosomal pathway following permanent pinocytosis or phagocytosis. Endosomes resulting from weak induced pinocytotic activity (Fig. 7, Ia and Ib) accumulate within 30 min of active locomotion near the former uroid and fuse there to one large endosomal compartment. Afterward, the heteroendosome either releases its condensed content into the external environment by normal exocytosis (pathway Ia)

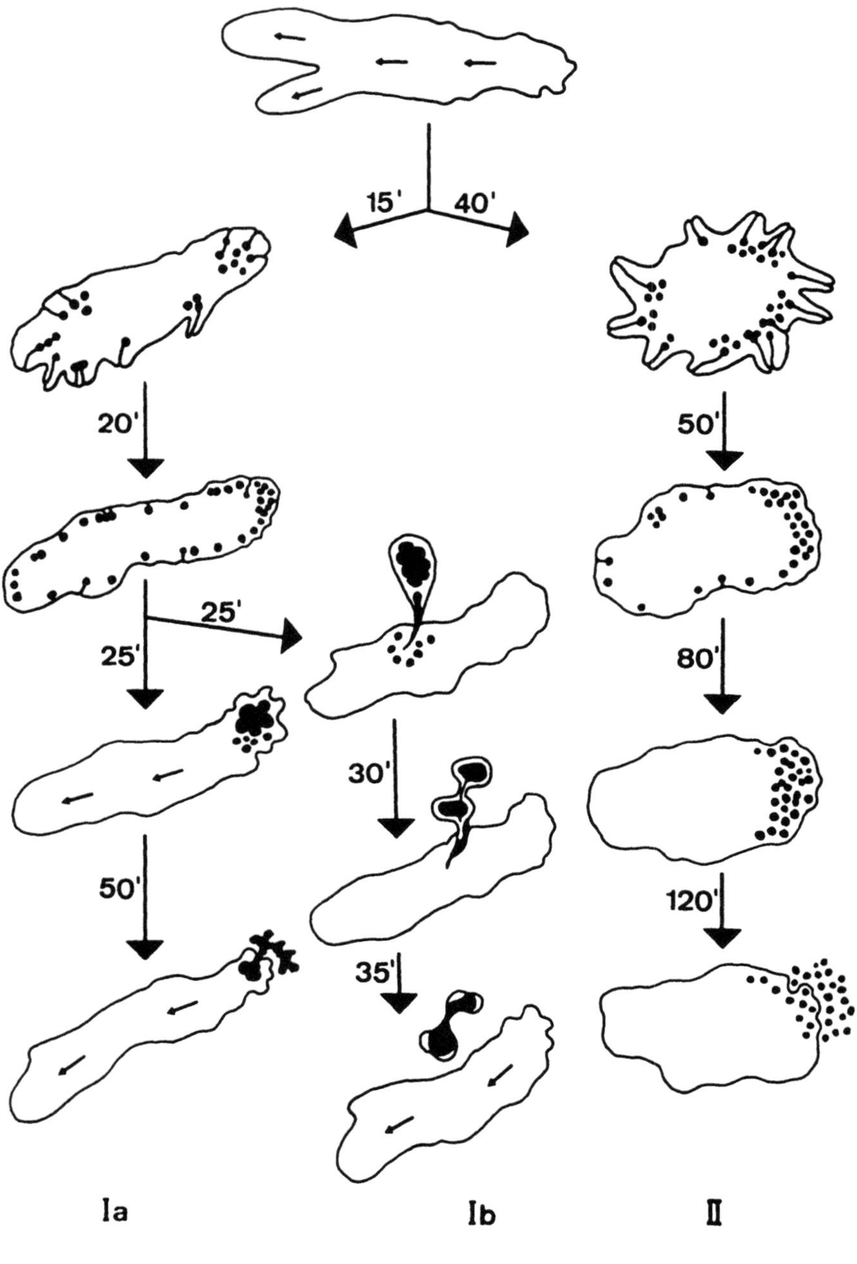

Figure 7. Schematic summary of the stages of induced (adsorptive) pinocytosis, endosome gathering and fusion, exocytosis and cell constriction in *Amoeba proteus*. Ia and Ib = weak pinocytosis, II = strong pinocytosis. Time after incubation with the marker is given in minutes. For further explanations see Section V. From Christofidou-Solomidou et al. (1989) by permission of Gustav Fisher Verlag (Stuttgart).

or it is pinched off in bulk together with a small portion of cytoplasm (pathway Ib). Endosomes resulting from intense induced pinocytotic activity also gather in a discrete cell area but do not undergo mutual fusion (Fig. 7, II). In this case, the exocytotic removal of the endosomal contents occurs in a progressive fashion and needs much more time as compared to weak pinocytosis. The different endosomal pathways in amoebae resemble similar findings on the intracellular fate of sequentially formed endosomes in chinese hamster ovary cells (Salzman and Maxfield, 1988). Correspondingly, in contrast to late endosomes only early endosomes can fuse with each other. This would explain that endosomes formed by amoebae during weak and short pinocytosis can coalesce to large compartments, whereas others formed during intense pinocytosis over a much longer period of time represent a heterogeneous population of early and late endosomes which are impeded to fuse. Results obtained with cell-free assays also confirm this suggestion (Gruenberg and Howell, 1989).

VI. INTRACELLULAR DIGESTION AND SORTING

The intracellular fate of food vacuoles has been studied in a large variety of different protozoan species (Kitching, 1956). Based on the comprehensive studies on *Paramecium* (see chapters by Allen and Fok as well as by Fok and Allen in this volume), a general concept for the endosomal pathway in phagocytotically active protists has been developed, according to which the maturation or processing period between phagosome formation and release is divided into four different stages (Allen and Staehelin, 1981; Fok and Paeste, 1982). Several characteristic parameters such as vacuolar age, size, membrane morphology, pH, and AcPase content allow us to distinguish the stages DV-I (0–6 min, pH 6.0–6.5, no AcPase), DV-II (4–10 min, pH 3, no AcPase), DV-III (8–20 min, pH 7, with AcPase), and DV-IV (20 min and older, pH 7, no AcPase). The following description of intracellular digestion and sorting in amoebae refers to this concept, although some representatives of the proteus group show peculiarities in which they differ from *Paramecium* and other protozoa.

A. Acidification of Food Vacuoles

The majority of endosomes, phagosomes, and lysosomes in amoebae prove to have an acidic content when stained with acridine orange as a suitable indicator of pH (Fig. 8a). In a fluorometrical investigation with FITC-RNAase, McNeil et al. (1983) demonstrated that surface binding of this marker is associated with a slight external decrease of pH from 7.0 to 6.1. Within the first 10 min, early FITC-RNAase-containing endosomes show a pH of 5.4, and after 30–40 min late endosomes and lysosomes have attained a pH of 5.0, that is, they are less acidic than *Paramecium* phagosomes with a pH of 3.0 (Fok and Allen, 1988). Since the pH of the surrounding cytoplasm was measured to be 6.75 (Heiple and Taylor,

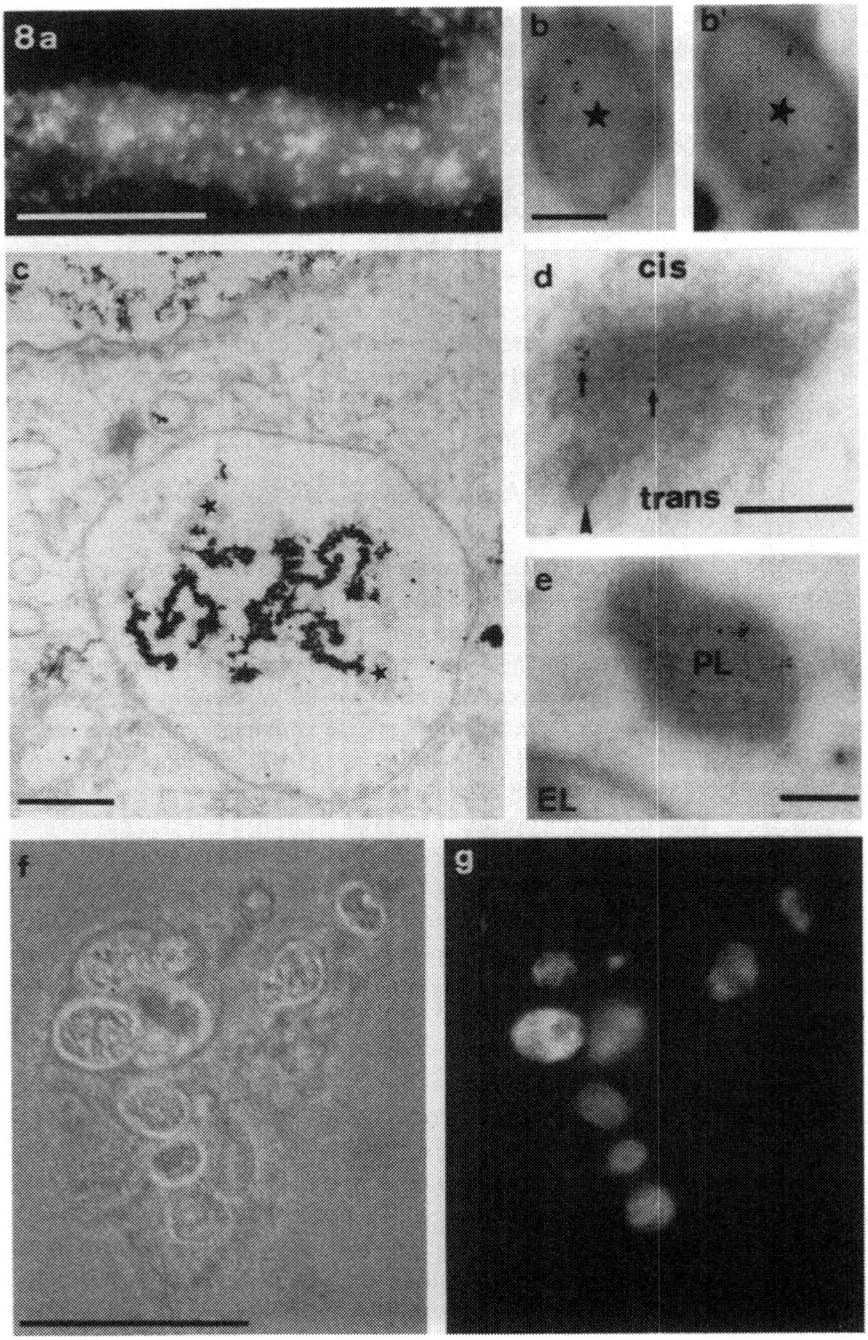
8a
b
b'
c
d
cis
trans
e
PL
EL
f
g

Figure 8. Fluorescence (**a**) and electron micrographs (**b**, **b**,') of *Amoeba proteus* treated with acridine orange (**a**) and with the DAMP-method (**b**, **b'**) to demonstrate acidic compartments such as lysosomes and acidosomes (stars). (**c**) Aggregated marker complex (Au_{12}-BSA-TRITC) in the center of an endosome in *A. proteus* and beginning separation of Au_{12} and BSA-TRITC (stars). (**d**, **e**) *A. proteus* treated with anti-cathepsin D/GAR-Au_5 to demonstrate localization of this enzyme in dictyosomes (d, arrows) and primary lysosomes (PL). The arrowhead in (d) points to Golgi vesicles. EL = endolysosome. (**f**) Light and (**g**) fluorescence micrograph of freshly fed *A. proteus* containing many phagosomes with *Tetrahymena* (f), which stain distinctly positive for cathepsin D (g). (a) is unpublished, (b, b′, d–g) are reproduced from Christofidou-Solomidou and Stockem (1992) by permission of Gustav Fisher Verlag (Stuttgart), (c) is from Christofidou-Solomidou et al. (1989) by permission of Gustav Fischer Verlag (Stuttgart). Bars: (a) 100 μm, (b, b′, d, e,) 0.2 μm, (c) 1 μm, (f, also valid for g) 100 μm.

1980), acidification of endosomes has to be an active process. More recently, Christofidou-Solomidou and Stockem (1992) detected a population of small vesicles (200–500 nm) randomly distributed in the cellular matrix or gathered in the periphery of digestive vacuoles, which obviously contain a strong acidic milieu as shown by the DAMP-method (Fig. 8b, b′). DAMP is a weak base which diffuses freely through biomembranes in the unprotonated state. In cell organelles with acidic pH, it becomes positively charged and trapped so that it can be detected by antibodies. DAMP-positive acidic vesicles of amoebae have a strong similarity to the so-called acidosomes of *Paramecium* (Allen and Fok, 1983) and are obviously important for acidification, since they fuse with food vacuoles. Corresponding prelysosomal acidic vesicles were also described in amoebae of the acellular slime mold *Dictyostelium* (Padh et al., 1989). On the other hand, early endosomes and phagosomes can coalesce with preexisting lysosomes to form an acidic compartment. The presence of an ATP-dependent proton pump or of some other cation exchange mechanism that could mediate acidification of vacuoles should also be explored. Nevertheless, the results on amoebae are in good agreement with observations on other protozoan (Kitching, 1956) or metazoan cells (Storrie et al., 1984). This supports the hypothesis that acidification precedes provision of food vacuoles with digestive enzymes, thus playing a significant role in the endosomal pathway. The physiological significance of acidification is diverse, for example, killing of prey organisms and a first hydrolysis of food material in phagosomes or uncoupling of adsorbed ligands from membrane binding sites in endosomes.

B. Provision of Food Vacuoles with Lysosomal Enzymes

About 20 different lysosomal enzymes have been detected in different species of giant and small amoebae (Table 1). Studies using the Gomori-method (Fig. 9a, b) and more recent results obtained with immunological techniques (Fig. 8d) point to the importance of the Golgi apparatus in providing *Chaos chaos*, *Amoeba*

Table 1. Lysosomal Enzymes in Amoebae as Summarized and Described by Müller et al. (1962), Kudo (1966), Chapman-Andresen (1973, 1976), Hausmann and Stockem (1973), Nisbet (1984) and McLaughlin and Aley (1985).

Enzyme	*Species*	*pH*
Acid proteases (EC 3.4)	*Amoeba proteus*	3.8–4.0
	Chaos carolinensis	3.5–4.0
	Pelomyxa palustris	4.0
	Acanthamoeba	3.0
	Entamoeba	—
	Entamoeba histolytica	—
Thiamine pyrophosphatase (EC 2.7.6.2)	*Amoeba proteus*	7.2
Acid phosphatase (EC 3.1.3.2)	*Amoeba proteus*	5.0–6.5
	Amoeba discoides	—
	Chaos chaos	5.5
	Hyalodiscus simplex	—
	Acanthamoeba	5.0
	Entamoeba	—
Acid phosphomonoesterase (EC 3.1.4.1)	*Chaos chaos*	4.5
Arylsulfatase (EC 3.1.6.1)	*Chaos carolinensis*	5.4
	Chaos chaos	5.5
	Hyalodiscus simplex	.5
	Amoeba proteus	5.5
Proteinase (EC 3.4)	*Entamoeba histolytica*	—
Cathepsin B (EC 3.4.22.1)	*Entamoeba histolytica*	—
Cathepsin D (EC 3.4.23.5)	*Amoeba proteus*	—
α-Glucosidase (EC 3.2.1.20)	*Chaos carolinensis*	3.8
	Acanthamoeba palestinensis	4.0
	Acanthamoeba	4.0
β-Glucosidase (EC 3.2.1.21)	*Acanthamoeba palestinensis*	3.6
N-acetylglucosaminidase (EC 3.2.2)	*Entamoeba*	—
Amylase (EC 3.2.1)	*Amoeba proteus*	6.0
Peptidase (EC 3.4.4)	*Chaos chaos*	7.4
	Pelomyxa palustris	7.4
Dipeptidase (EC 3.4.3)	*Amoeba proteus*	7.4
Esterase (EC 3.1)	*Entamoeba*	—
Carboxylic esterhydrolase (EC 3.1.1.1)	*Amoeba proteus*	7.0–8.7
	Amoeba dubia	7.0–8.7
	Amoeba discoides	—
	Chaos chaos	7.0–8.1
Lipase (EC 3.1.1.3)	*Chaos chaos*	7.0
	Amoeba proteus	7.0–8.7
	Amoeba discoides	7.0–8.7
Lipolytic substance	*Amoeba dubia*	—
Diastatic enzyme	*Pelomyxa palustris*	—
Pepsin-like enzyme (EC 3.4.4.1)	*Pelomyxa palustris*	—
Succinate dehydrogenase (EC 1.3.99.2)	*Pelomyxa palustris*	—
RNAase (EC 3.1.4.22)	*Entamoeba*	—

proteus, *Vanella simplex*, *Dictyostelium discoideum*, or *Acanthamoeba castellanii* with hydrolytic enzymes for intracellular digestion (Wise and Flickinger, 1970; Ashworth, 1973; Hausmann and Stockem, 1973; Ryter and Bowers, 1976). Within 10–20 min, primary lysosomes with a diameter of 50–500 nm (Chapman-Andresen and Lagunoff, 1966) and a rather high concentration of acid hydrolases (Fig. 8e) gather around endosomes and phagosomes (Fig. 9c, e) to deliver their lytic content into these vacuoles by fusion (Fig. 9d, f). This could be demonstrated by investigating amoebae either treated with TRITC-BSA for pinocytosis (Fig. 10b, b′) or fed with *Tetrahymena* for phagocytosis (Fig. 8f). Cathepsin D as the major acid proteinase of most animal cells with a specificity resembling that of pepsin (Barret, 1969) was detected for the first time in all phagosomes of *Amoeba proteus* at 15–30 min after their formation (Fig. 8g). On the contrary, amoebae treated with TRITC-BSA exhibit a different staining pattern in that many endosomes are positive for both cathepsin D and BSA (Fig. 10b, b′), whereas others either contain BSA or cathepsin D. This favors the idea that, after induced pinocytosis, only some endosomes fuse with primary or secondary lysosomes, and that others escape provision with acid hydrolases to avoid the digestive pathway (see also Section V). Vacuoles containing exclusively lysosomal enzymes but not ingested substrate may represent preexisting lysosomes as an alternative compartment to primary lysosomes for the origin of acid hydrolases; this was shown not only for amoebae (Hausmann and Stockem, 1973) but also for different metazoan cells (Straus, 1963).

C. Segregation of Lysosomal Contents

Some protozoa have developed the ability to distinguish between different engulfed substances by intracellular segregation, thus possessing a mechanism which resembles the specific sorting of endosomal contents in metazoan cells (Goldstein et al., 1985). After feeding mixtures of proteins (e.g., egg white) and inert materials (e.g., Aerosil), the amoebae ingest both substances and transport them together in the same endosome (Stockem, 1966). However, 30 min later smaller vacuoles are found which contain either egg white or Aerosil, but not both. The same phenomenon was observed in acellular slime molds (Stockem and

---

Figure 9. Electron micrographs to demonstrate the histochemical localization of acid phosphatase activity by the Gomori method in *Amoeba proteus* (**a**, **e**, **f**) and *Vanella simplex* (**b–d**). Acid phosphatase activity is present in dictyosomes (**a**, **b**), in primary lysosomes surrounding phagosomes (**c**, arrows) and endosomes (e, arrows) as well as in lysosomes containing latex beads (**d**) and a partly digested *Tetrahymena* (**f**). B = bacteria. (c) is unpublished, (a, b, d–f) are reproduced from Hausmann and Stockem (1972, 1973) by permission of Wissenschaftliche Verlagsgesellschaft (Stuttgart). Bars: (a, b) 0.1 µm, (c, d) 0.5 µm, (e, f) 2 µm.

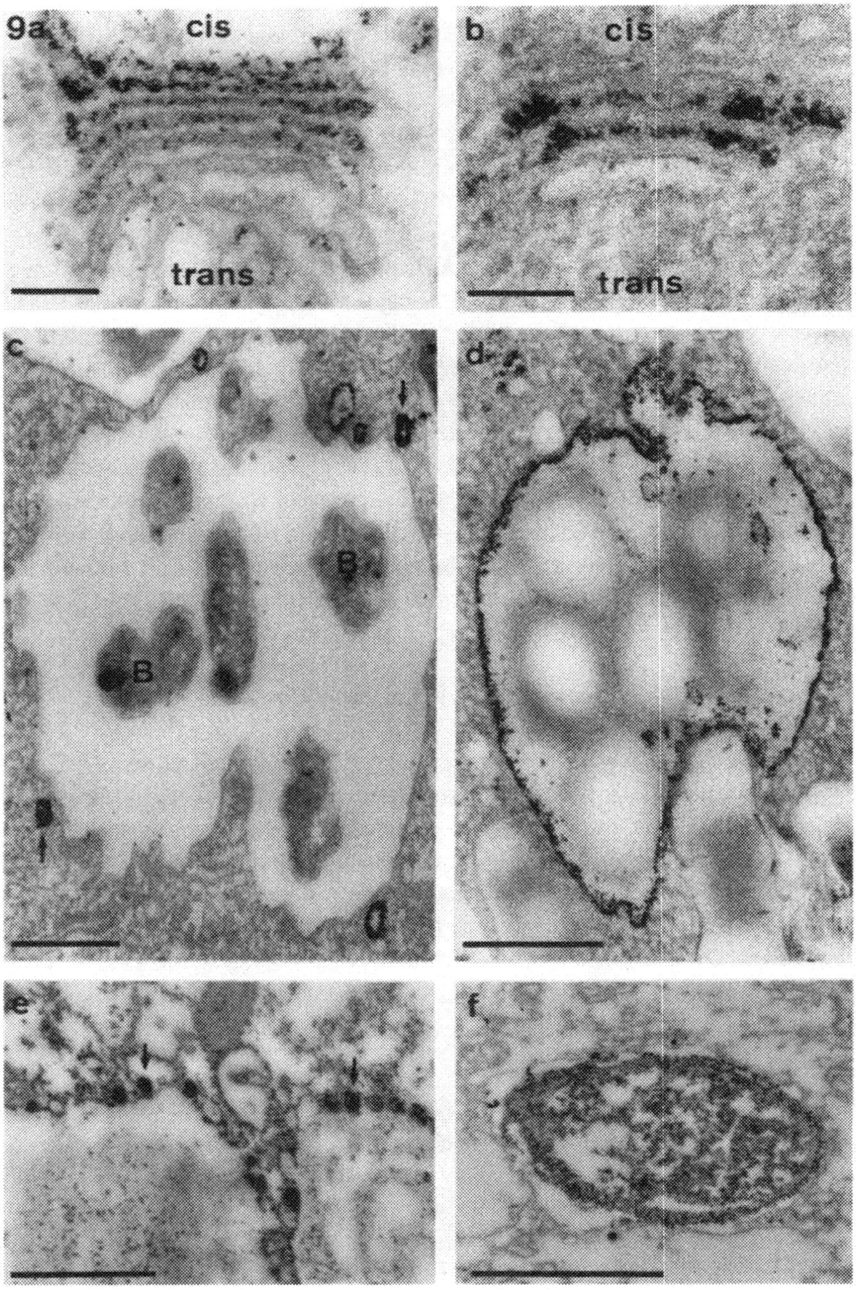
9a
cis
trans
b
cis
trans
c
B
B
d
e
f

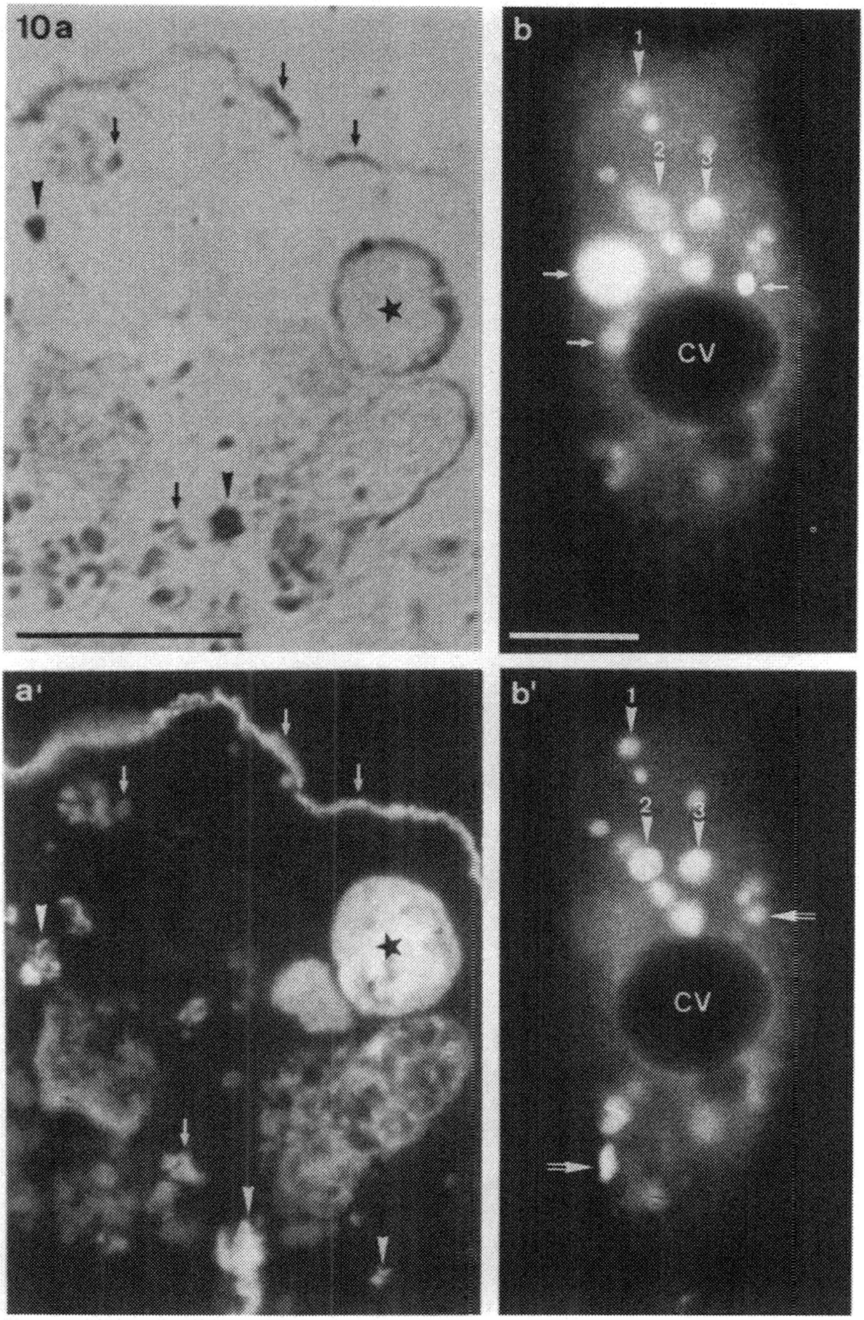
10a
b
1
2
3
CV
a'
b'
1
2
3
CV

Figure 10. Bright field (**a**) and fluorescence micrograph (**a′**) of the same semithin section of *Amoeba proteus* treated with Au_{12}-BSA for 50 min. Au_{12}-particles can be localized by their red color (intense contrast in a) and BSA by fluorescence after staining with anti-BSA antibodies (a′). Arrows = cell surface areas and endosomes showing colocalization of Au_{12} and BSA. Arrowheads (black in a) = endosomes containing only Au_{12}, arrowheads (white in a′) = endosomes containing only BSA. Stars (in a and a′) = identical endosome with diverse internal distribution of Au_{12} and BSA. (**b**, **b′**) *A. proteus* double-stained with anti-cathepsin D and GAR-FITC and photographed at FITC excitation (b) as well as with anti-BSA-TRITC and photographed at TRITC excitation (b′). Some vacuoles are positive for both, cathepsin D and BSA (numbered arrowheads), whereas others show presence of either cathepsin D (arrows in b) or BSA (double arrows in b′). CV = contractile vacuole. (a, a′) are reproduced from Christofidou-Solomidou et al. (1989) by permission of Gustav Fisher Verlag (Stuttgart), (b, b′) from Christofidou-Solomidou and Stockem (1992) by permission of Gustav Fischer Verlag (Stuttgart). Bars: (a, also valid for a′) 25 μm, (b, also valid for b′) 50 μm.

11

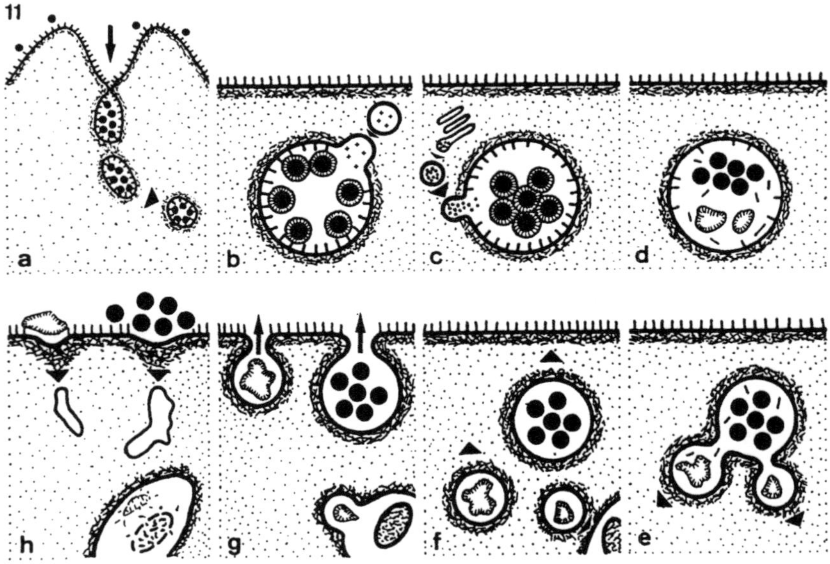

Figure 11. Schematic summary of the endosomal pathway in *Amoeba proteus*. The arrows and arrowheads indicate the direction of bulk transport and membrane flow, respectively. For detailed description see Section VI. Reproduced from Christofidou-Solomidou and Stockem (1992) by permission of Gustav Fisher Verlag (Stuttgart).

Stiemerling, 1976) and has been further studied in *Amoeba proteus* by using a complex protein-gold-marker (Christofidou-Solomidou et al., 1989). The question whether this complex persists or disintegrates into Au and TRITC-BSA was analyzed by combined light (Fig. 10a) and fluorescence microscopy (Fig.10a′). Accordingly, at the cell surface and in some endosomal vacuoles the Au and BSA are colocalized, whereas two further vacuolar populations contain either BSA but no Au or Au but no BSA. Other endosomes seem to be in an intermediate state of segregation because both markers show a diverse distribution pattern within the same vacuole, that is, they are already separated but not segregated to different compartments. Although molecular events underlying such processes in protozoa are still unknown, endosomal sorting could be significant for the fact that many amoebae (Bowers and Olszewski, 1983) and ciliates (Kitching, 1956; Fok and Allen, 1988) discriminate internally between digestible and indigestible material; this has far-reaching consequences for cell metabolism.

D. Chronology of the Digestive Cycle

The temporal sequence of events characterizing the endosomal pathway in amoebae is summarized in Fig. 11 (see also Christofidou-Solomidou and Stockem, 1992). This chronology shows many similarities to the digestive cycle of other protozoa, especially to that of *Paramecium* (Allen and Staehelin, 1981). Within the first 5 min after their constriction from the plasma membrane (a) endosomes and phagosomes fuse with so-called acidosomes, thus attaining a pH between 6.0 and 5.5 (b). This stage is comparable to DV-I and II in *Paramecium* and causes mortification of phagocytozed prey organisms or release of endocytozed ligands from their membrane-binding sites. After 10–20 min, primary lysosomes derived from the Golgi apparatus and containing high concentrations of digestive enzymes coalesce with acidic food vacuoles (c) to initiate dissociation of different ligand components (d) or to start hydrolysis of prey material. The resulting vacuoles correspond to DV-III in *Paramecium* and undergo further processing (e) insofar as digestible (BSA) and indigestible contents (Au) are targeted to different lysosomal compartments (f). For the most part, indigestible contents are released by exocytosis within 30–60 min, whereas vacuoles with digestible material either fuse with preexisting lysosomes or also discharge their contents by exocytosis (g). Similar to the DV-IV in *Paramecium*, there are strong indications that the membrane of defecation vacuoles is not permanently incorporated into the cell surface (h) but recycled to the cytoplasmic membrane pool (see Section VIII). This is also in agreement with biochemical studies on induced phagocytosis in amoebae (Ahn and Jeon 1982, 1985) indicating that single membrane proteins are continuously altered and recycled during the endosomal pathway (see chapter by Jeon in this volume). Finally, it should be mentioned that food vacuoles resulting from permanent pinocytosis or phagocytosis can also take a shorter pathway by directly fusing with preexisting lysosomes (Fig. 11a, b, f–h).

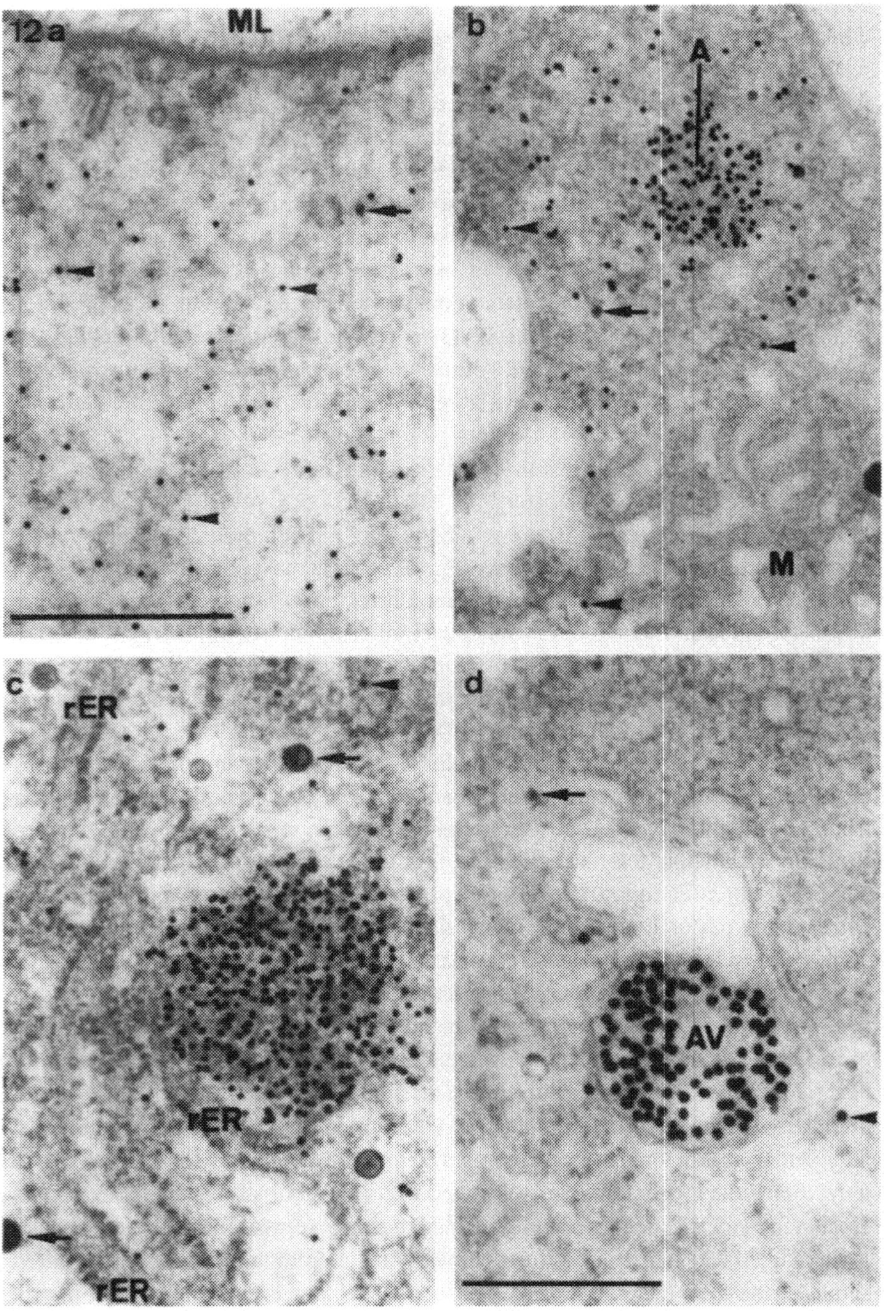
12 a
ML
b
A
M
c
rER
rER
rER
d
AV

Figure 12. Electron micrographs to demonstrate the autophagosomal sequestration of microinjected TRITC-BSA-Au_{16} by *Amoeba proteus.* (**a**) Uniform distribution of the marker (arrowheads) in the cytosol, (**b**) condensation of the marker within small aggregates (A), (**c**) forming of a secluded autophagosomal compartment by rough ER-cisternae around aggregates, (**d**) autophagosome (AV) containing a marker aggregate. Arrows = electron-dense granules of unknown nature often present in *A. proteus.* M = mitochondrium. (a–d) are reproduced from Sodeik et al. (1989) by permission of Gustav Fischer Verlag (Stuttgart). Bars: (a, also valid for b, c) 0.5 μm, (d) 0.25 μm.

VII. AUTOPHAGOCYTOSIS

Unlike heterophagocytosis (uptake of exogenous material), the process of autophagocytosis is aimed at the sequestration and lysosomal digestion of endogenous cell components (e.g., mitochondria) or foreign substances experimentally transferred into living cells. Previous studies have already shown that bacteria and algae microinjected into giant amoebae are sequestered from the cytosol by a membrane suggested to be formed de novo (Jeon and Lorch, 1982). In a more recent study the mechanism and mode of autophagosome formation was analyzed in *Amoeba proteus* with the aid of a microinjected Au–BSA complex (Sodeik et al., 1989). In accordance with data obtained on metazoan cells (see Section I), the results of this investigation show that the autophagic process of amoebae can be divided into four stages: cytoplasmic aggregation, autophagosome formation, lysosomal digestion, and exocytotic release. Immediately after microinjection the Au–BSA particles distribute randomly in the cytoplasmic matrix (Fig. 12a). However, within 45 min discrete aggregates are formed consisting of numerous Au–BSA particles embedded in a homogeneous material (Fig. 12b). Later on, the aggregates are partly enclosed by single cisternae of the rough ER (Fig. 12c), and phagosomes form by mutual fusion of the cisternal elements (Fig. 12d). The autophagosomes are then further processed as described for typical heterophagosomes, that is, they fuse with preexisting lysosomes and release their digested content by defecation (see Section VIII).

VIII. DEFECATION OF FOOD VACUOLES

Defecation of food vacuoles was extensively studied in *Paramecium* (Allen and Wolf, 1974) and also documented for other protozoa (Kitching, 1956; Hausmann, 1980) including giant amoebae (Andresen, 1956; Stockem, 1970). Accordingly, the defecation vacuole first approaches the internal cell surface (Fig. 13a) and then coalesces with the plasma membrane (b) to release its content into the external environment (c). All these stages have been proved for *Amoeba proteus* (Andresen and Holter, 1945; Stockem, 1977) and *Vanella simplex* (Hausmann and Stockem,

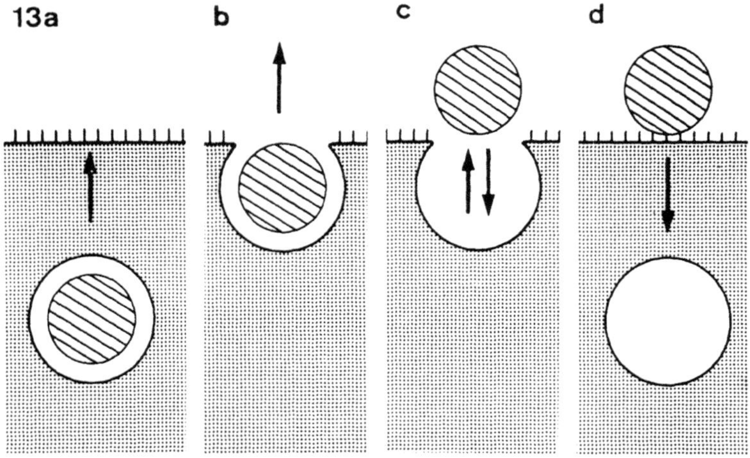

Figure 13. Schematic summary of the suggested course of defecation in amoebae. For detailed description of stages (a–d) see Section VIII. Reproduced from Hausmann and Stockem (1972) with permission of Wissenschaftliche Verlagsgesellschaft (Stuttgart).

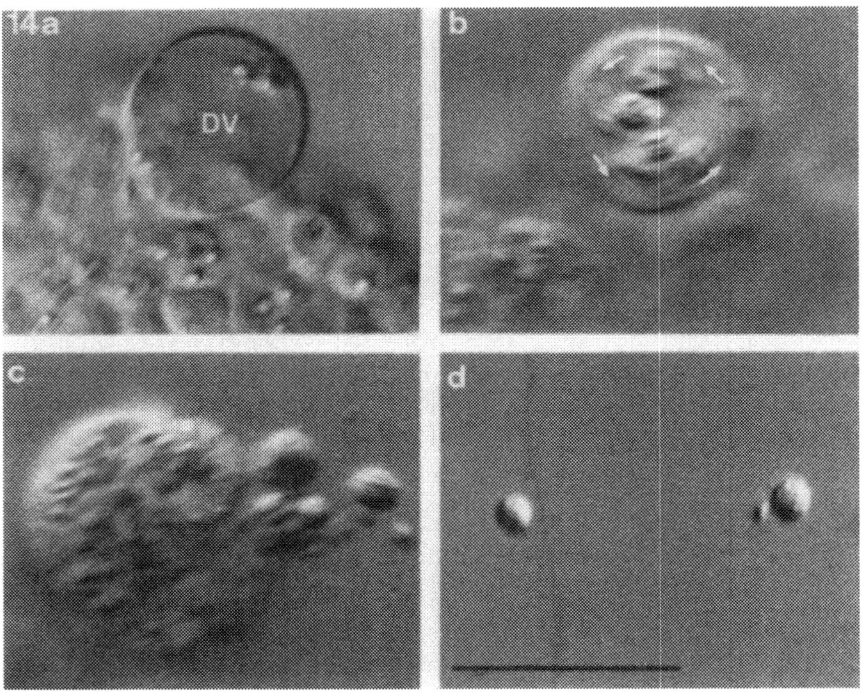

Figure 14. DIC micrographs to demonstrate the release of a defecation vacuole (**a**, DV) by *Amoeba proteus*. Time distance between single pictures = 2 sec. Arrows (in **b**) indicate intense rotation of food residues during fusion of the vacuole with the plasma membrane. Reproduced from Stockem (1970) by permission of Verlag Georg Fromme and Co (Viennna). Bar: (**d**, also valid for **a–c**) 40 μm.

1972) by *in vivo* documentation (Fig. 14) and by electron microscopy. However, the final step of defecation, the transport of the vacuolar membrane back into the cytoplasm, is still hypothetical and analyzed in detail only for *Paramecium*, where the membrane is recycled from the cytoproct to the cytostome via small, discoidal vesicles (Allen and Fok, 1980).

IX. EXOCYTOSIS AND CELL SURFACE REGENERATION

Most cells showing intense incorporation of plasma membrane during the endocytotic or phagocytotic uptake of substances have developed a mechanism to retrieve the lost cell surface areas by exocytosis. In contrast to the above mentioned direct recycling of lysosomal membrane portions via discoidal vesicles in *Paramecium*, several investigations have demonstrated that small and giant amoebae synthesize new membrane, together with mucous material, by the classical secretory pathway (Stockem, 1969; Flickinger 1975, 1982; Hausmann, 1975). Acid mucopolysaccharides, a characteristic component of the mucous layer, can be traced back to distal cisternae of dictyosomes, from where small vesicles already exhibiting the characteristic morphological appearance of the plasma membrane are pinched off. These vesicles, referred to as exocytotic vacuoles (compare also Fig. 4), are transported to the cell surface and fuse with the plasma membrane (Fig. 15). Nachmias (1966) and Hausmann (1975) succeeded in the direct demonstration of this exocytotic process by irreversibly labeling the old mucous layer at the cell surface of *Amoeba proteus* and *Vanella simplex* so that new membrane material became identifiable because of different morphology and contrast (Fig. 16). The intensity of the exocytotic process is in a range corresponding to that of endocytosis (Nachmias, 1966) and thus enables the amoebae to continuously keep their cell surface area at a constant level (Stockem, 1977; Stockem and Klopocka, 1988).

15

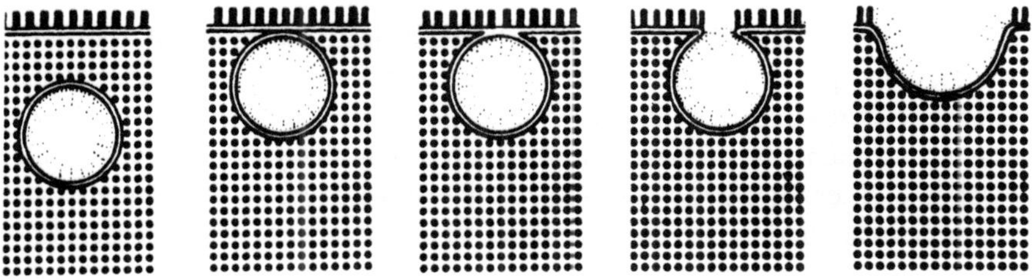

Figure 15. Schematic summary of the course of plasma membrane regeneration by exocytosis in amoebae. For detailed description see Section IX. Reproduced from Hausmann (1975) by permission of Editions du Centre National de la Recherche Scientifique (Gap).

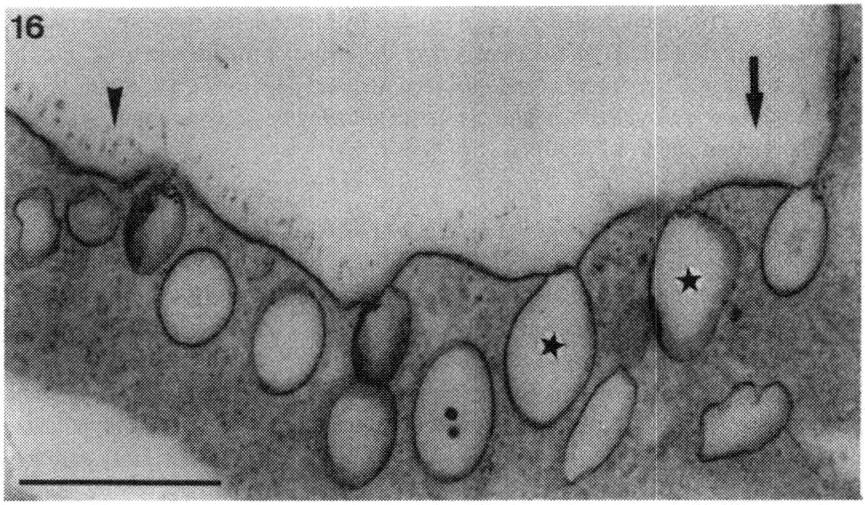

Figure 16. Electron micrograph demonstrating the presence of exocytotic vesicles in *Vanella simplex.* Vesicles have been fixed before or during exocytosis (stars) and contain unlabeled mucous layer. The mucous layer of the plasma membrane is partly more labeled (arrowhead) and partly less labeled (arrow). Reproduced from Hausmann (1975) by permission of Editions du Centre National de la Recherche Scientifique (Gap). Bar: 0.5 μm.

X. CONCLUSIONS AND OUTLOOK

Our present knowledge about the mode of food uptake and course of intracellular digestion in amoebae has remarkably increased during the last decade. Especially, with regard to some recent publications, it seems to be established that there are not only distinct conformities but also several characteristic dissimilarities between the lysosomal concept of amoebae and other protozoan or metazoan cells.

Amoebae have developed several types of membrane flow mechanisms comparable to the fluid-phase endocytosis, adsorptive endocytosis, and receptor-mediated phagocytosis of mammalian cells. Nevertheless, the mode of endosome formation and processing shows distinct differences due to the regulation of membrane–cytoskeleton contacts and functional interactions with the lysosomal system. A certain part of the endosomal population always eludes intracellular digestion and segregates its contents into the external medium by prompt exocytosis or cellular budding. On the other hand, a large number of endosomes, mainly derived from fluid-phase endocytosis or phagocytosis and containing hydrolyzable ligands, escape rapid exocytotic release; they are acidified and provided with digestive enzymes, thus contributing to the nutrient pool of the cell. Altogether, the endosomal pathway, at least of giant amoebae, largely corresponds to the model of Stoorvogel et al. (1991), which states that the single vacuolar stages of digestion derive from each other by successive maturation. Preexisting lysosomes play a significant role in providing food vacuoles with acid hydrolases. However, there is no indication that early and late endosomes also represent preexisting and independent compartments in amoebae or that these compartments would communicate by a membrane shuttle via transport vesicles, as postulated by Griffiths and Gruenberg (1991) for cultured cells. A recycling of obviously existing endocytotic or phagocytotic membrane receptors between certain endosomal stages and the cell surface is likewise questionable for amoebae, because early endosomes fail to show the typical polymorphic appearance as a necessary precondition for such a recycling process. On this account, the digestive system in amoebae seems to be rather primordial when compared with the endosomal pathway of metazoan cells. However, to answer this question conclusively much more experimental work must be done.

ABBREVIATIONS

AcPase:	acid phosphatase
Au_5, Au_{12}, Au_{16}:	coloidal gold with particle diameters of 5, 12, and 15 nm
BSA:	bovine serum albumin
CURL:	compartment of uncoupling receptor and ligand
DAMP:	3-(2,4-dinitroanilino)-3-amino-*N*-methyldipropylamine
DIC:	differential interference contrast
DV:	digestive vacuole
ER:	endoplasmic reticulum
FITC:	fluoresceine isothiocyanate
GAR:	goat anti-rabbit antibodies
MDCK:	Madin Darby Canine Kidney
TRITC:	tetramethyl rhodamine isothiocyanate

ACKNOWLEDGMENT

The authors wish to thank Dr. Fred Minnear (Albany Medical College, New York, USA) for reading the manuscript and helpful suggestions as well as Mrs. Antje Schuster and Mrs. Elisabeth Kraemer for excellent technical assistance.

REFERENCES

Aaronson, S. (1973). Digestion in phytoflagellates. In: *Lysosomes in Biology and Pathology* (Dingle, J.T., ed.), pp. 18–37, North-Holland, Amsterdam.

Ahn, T. I. and Jeon, K. W. (1982). Structural and biochemical characteristics of the plasmalemma and vacuole membranes in amoebae. *Exp. Cell Res.* 137: 253–268.

Ahn, T. I. and Jeon, K. W. (1985). Recycling of membrane proteins during endo- and exocytosis in amoebae. *Exp. Cell Res.* 160: 54–62.

Aikawa, M., Hepler, P. K., Huff, G. G., and Sprinz, H. (1966). The feeding mechanism of avian malarial parasites. *J. Cell Biol.* 28: 355–373.

Allen, H. J., Winzler, R. J., and Danielli, J. F. (1974). Chemical characterization of the isolated cell surface of *Amoeba. J. Cell Biol.* 60: 26–38.

Allen, H. J., Winzler, R. J., and Danielli, J. F. (1976). Stimulation of ^{3}H glucosamin incorporation into cell surface of *Amoeba* by pinocytosis. *Exp. Cell Res.* 100: 408–411.

Allen, R. D. and Fok, A. K. (1980). Membrane recycling and endocytosis in *Paramecium* confirmed by horseradish peroxidase pulse-chase studies. *J. Cell Sci.* 45: 131–145.

Allen, R. D. and Fok, A. K. (1983). Nonlysosomal vesicles (acidosomes) are involved in phagosome acidification in *Paramecium. J. Cell Biol.* 97: 566–570.

Allen, R. D. and Staehelin, L. A. (1981). Digestive system membranes: freeze–fracture evidence for differentiation and flow in *Paramecium. J. Cell Biol.* 89: 9–20.

Allen, R. D., Schroeder, C. C., and Fok, A. K. (1992). Endosomal system of *Paramecium*: coated pits to early endosomes. *J. Cell Sci.* 101: 449–461.

Allen, R. D. and Wolf, R. W. (1974). The cytoproct of *Paramecium caudatum*: structure and function, microtubles, and fate of food vacuole membranes. *J. Cell Sci.* 14: 611–631.

Anderson, O. R. (1983). Trophic activity and nutrition. In: *Radiolaria* (Anderson, O. R., ed.), pp. 175–188, Springer-Verlag, New York.

Anderson, O. R. (1989). Some observations of feeding behavior, growth, and test particle morphology of a silica-secreting testate amoeba *Netzelia tuberculata* (Wallich) (Rhizopoda, Testacea) grown in laboratory culture. *Arch. Protistenk.* 137: 211–221.

Andresen, N. (1956). Cytological investigations on the giant amoeba *Chaos chaos* L. *C. R. Trav. Lab. Carlsberg* 29: 435–555.

Andresen, N. (1973). General morphology. In: *The Biology of Amoeba* (Jeon, K. W., ed.), pp. 99–120, Academic Press, New York.

Andresen, N. and Holter, H. (1945). Cytoplasmic changes during starvation of the amoeba *Chaos chaos. C. R. Trav. Lab. Carlsberg* 25: 107–146.

Ashworth, J. M. (1973). The lysosomes of the cellular slime mould *Dictyostelium discoideum.* In: *Lysosomes in Biology and Patholgy* (Dingle, J. T., ed.), pp. 38–48, North-Holland, Amsterdam.

Barret, A. J. (1969). Properties of lysosomal enzymes. In: *Lysosomes in Biology and Pathology* (Dingle, J. T. and Fell, H. B., eds.), pp. 245–312, North-Holland, Amsterdam.

Bockman, D. E. and Winborn, W. B. (1968). Electron microscopic localization of exogenous ferritin within vacuoles of *Giardia muris. J. Protozool.* 15: 26–30.

Bovee, E. C. and Jahn, T. L. (1973). Taxonomy and phylogeny. In: *The Biology of Amoeba* (Jeon, K. W., ed.), pp. 37–82, Academic Press, New York.

Bowers, B. and Olzewski, T. E. (1983). *Acanthamoeba* discriminates internally between digestible and indigestible particles. *J. Cell Biol.* 97: 317–322.

Braatz-Schade, K. and Stockem, W. (1973). Pinocytosis and locomotion of amoebae. X; The significance of the mucous layer for the initial phase of the induced endocytosis in *Amoeba proteus. Z. Zellforsch. mikrosk. Anat.* 137: 97–109.

Braun, V., Hantke, K., and Wolff, H. (1972). Degradation of the murein-lipoprotein complex of *Escherichia coli* cell walls by *Dictyostelium* amoebae. *Eur. J. Biochem.* 27: 116–125.

Brooker, B. E. (1971). The fine structure of *Crithidia fasciculata* with special reference to the organelles involved in the ingestion and digestion of protein. *Z. Zellforsch. mikrosk. Anat.* 116: 532–563.

Bruce, D. L. and Marshall, J. M. (1965). Some ionic and bioelectric properties of the amoeba *Chaos chaos. J. Gen. Physiol.* 49: 151–178.

Cann, J. P. (1986). The feeding behavior and structure of *Nuclearia delicatula* (Filosea: Aconchulinida). *J. Protozool.* 33: 392–396.

Chapman-Andresen, C. (1962). Studies on pinocytosis in amoebae. *C. R. Trav. Lab. Carlsberg* 33: 73–264.

Chapman-Andresen, C. (1973). Endocytotic processes. In: *The Biology of Amoeba* (Jeon, K. W., ed.), pp. 319–345, Academic Press, New York.

Chapman-Andresen, C. (1976). Lysosomes in the large freshwater Amoebae. In: *Proceedings of the Ninth Congress of the Nordic Society on Cell Biol.* (Bierring, F., ed.), pp. 113–119, Odense University Press.

Chapman-Andresen, C. and Lagunoff, D. (1966). The distribution of acid phosphatase in the amoeba *Chaos chaos L. C. R. Trav. Lab. Carlsberg* 35: 419–436.

Christiansen, R. G. and Marshall (1965). A study of phagocytosis in the amoeba *Chaos chaos. J. Cell Biol.* 254: 443–457.

Christofidou-Solomidou, M., Brix, K., and Stockem, W. (1989). Induced pinocytosis and cytoskeletal organization in *Amoeba proteus*—a combined fluorescence and electron microscopic study. *Eur. J. Protistol.* 24: 336–345.

Christofidou-Solomidou, M. and Stockem, W. (1992). Induced pinocytosis and endosomal pathways in *Amoeba proteus. Eur. J. Protistol.* 28: 56–70.

Cohn, Z. A. and Wiener, E. (1963). The particulate hydrolases of macrophages. II. Biochemical and morphological responses to particle ingestion. *J. Exp. Med.* 118: 1009–1020.

De Petris, S. (1977). Distribution and mobility of plasma membrane components on lymphocytes. In: *Dynamic Aspects of Cell Surface Reviews* (Poste, G. L. and Nicolson, G. L., eds.), pp. 643–728, North-Holland, Amsterdam.

Dice, J. F., Backer, J. M., Chiang, H. L., and Goff, S. (1987). Lysosomal degradation of microinjected proteins. *Biochem. Soc. Trans.* 15: 824–826.

Dunn, W. A. (1990a). Studies on the mechanisms of autophagy: formation of the autophagic vacuole. *J. Cell Biol.* 110: 1923–1933.

Dunn, W. A. (1990b). Studies on the mechanisms of autophagy: maturation of the autophagic vacuole. *J. Cell Biol.* 110: 1935–1945.

Eeckhout, Y. (1973). Digestion and lysosomes in zooflagellates. In: *Lysosomes in Biology and Pathology* (Dingle, J. T., ed.), pp. 3–17, North-Holland, Amsterdam.

Flickinger, C. J. (1975). The relation between the Golgi apparatus, cell surface and cytoplasmic vesicles in amoebae studied by electron microscope radioautography. *Exp. Cell Res.* 96: 189–201.

Flickinger, C. J. (1982). Redistribution of material labelled with [^{3}H]mannose in amoebae induced to undergo pinocytosis. *J. Cell Sci.* 58: 79–93.

Fok, A. K. and Allen, R. D. (1988). The lysosome system. In: *Paramecium* (Görtz, H. D., ed.), pp. 301–322, Springer-Verlag, Berlin.

Fok, A. K. and Paeste, R. M. (1982). Lysosomal enzymes of *Paramecium caudatum* and *Paramecium tetraurelia. Exp. Cell Res.* 139: 159–169.

Fukui, Y. and Imamoto, N. (1985). A cortical cytoskeleton associated with phagosomes disclosed by a monoclonal antibody against capped membrane of *Dictyostelium. Protoplasma* 128: 173–183.

Geuze, H. J., Slot, J. W., Strous, G. J., Lodish, H. F., and Schwartz, A. L. (1983). Intracellular site of asialoglycoprotein receptor–ligand uncoupling: double label immunoelectron microscopy during receptor-mediated endocytosis. *Cell* 32: 277–287.

Goldstein, J. L., Brown, M. S., Anderson, R. G. W., Russel, D. W., and Schneider, W. J. (1985). Receptor-mediated endocytosis: concepts emerging from LDL receptor system. *Annu. Rev. Cell Biol.* 1: 1–39.

Gollnick, F., Meyer, R., and Stockem, W. (1991). Visualization and measurement of calcium transients in *Amoeba proteus* by fura-2 fluorescence. *Eur. J. Cell Biol.* 55: 262–271.

Griffin, F. M., Griffin, J. A., Leider, J. E., and Silverstein, S. C. (1975). Studies on mechanism of phagocytosis. I. Requirements for circumferential attachment of particle-bound ligands to specific receptors on the macrophage plasma membrane. *J. Exp. Med.* 142: 1263–1282.

Griffin, J. L. (1972). Movement, fine structure, and fusion of pseudopods of an enclosed amoeba, *Difflugiella sp. J. Cell Sci.* 10: 563–583.

Griffiths, G. and Gruenberg, J. (1991). The arguments for pre-existing early and late endosomes. *Trends Cell Biol.* 1: 5–9.

Gruenberg, J. E. and Howell, K. E. (1989). Membrane traffic in endocytosis: insights from cell-free assays. *Annu. Rev. Cell Biol.* 5: 453–481.

Grunbaum, B. W., Möller, K. M., and Thomas, R. S. (1959). Cytoplasmic crystals of the amoebae: *Amoeba proteus* and *Chaos chaos*. *Exp. Cell Res.* 18: 385–389.

Haberey, M. (1973a). Food cup formation of *Amoeba proteus*. *Research Film* 8: 9–20.

Haberey, M. (1973b). Phagocytosis of *Oscillatoria* by *Thecamoeba sphaeronucleolus*. II. Electronmicroscopical observations. *Arch. Protistenk.* 115: 111–124.

Hausmann, E. (1975). Die Regeneration der Zelloberfläche bei *Hyalodiscus simplex*. *Protistologica* 11: 135–147.

Hausmann, E. and Stockem, W. (1972). Pinocytosis and locomotion of amoebae. VIII. Endocytosis and intracellular digestion in *Hyalodiscus simplex*. *Cytobiologie* 5: 281–300.

Hausmann, E. and Stockem, W. (1973). Pinocytosis and locomotion of amoebae. XI. The intracellular digestive system of different kinds of amoebae. *Cytobiologie* 7: 55–75.

Hausmann, E., Stockem, W., and Wohlfarth-Bottermann, K. E. (1972). Pinocytosis and locomotion of amoebae. VII. Quantitative studies on membrane turnover in *Hyalodiscus simplex*. *Z. Zellforsch. mikrosk. Anat.* 127: 270–286.

Hausmann, K. (1980). Cytosen bei Ciliaten. *Biol. i. u. Zeit* 10: 137–147.

Hausmann, K. and Patterson, D. J. (1982). Pseudopod formation and membrane production during prey capture by a heliozoon (feeding by *Actinophrys*, II). *Cell Motil. Cytoskel.* 2: 9–24.

Heiple, J. M. and Taylor, D. L. (1980). Intracellular pH in single motile cells. *J. Cell Biol.* 86: 885–890.

Hendil, K. B. (1971). Ion exchange properties of glycocalyx of amoeba *Chaos chaos* and its relation to pinocytosis. *C. R. Trav. Lab. Carlsberg* 38: 187–211.

Horwitz, M. A. (1988). Intracellular parasitism. *Curr. Op. Immunol.* 1: 41–46.

Hruban, Z., Dubien, L. H., Glagov, S., and Getz, G.S. (1991). Nonlysosomal cytoplasmic degradation. *J. Submicros. Cytol. Pathol.* 23: 491–500.

Hülsmann, N. (1983). Development and mode of nutrition in *Vampyrella lateritia* (Rhizopoda). In: *Verzeichnis der wissenschaftlichen Filme. Biologie* (IWF, ed.), p. 6, Institut wiss. Film, Göttingen.

Jeon, K. W. (1983). Integration of bacterial endosymbionts in amoebae. *Int. Rev. Cytol.* 14: 29–47.

Jeon, K. W. and Jeon, M. S. (1983). Generation of mechanical forces in phagocytosis amoebae: light and electron microscopic study. *J. Protozool* 30: 536–538.

Jeon, K. W. and Lorch, I. J. (1982). The formation of vacuole membranes in the presence and absence of cell nucleus in amoebae. *Exp. Cell Res.* 141: 351–356.

Josefsson, J. O. (1975). Studies on the mechanism of induction of pinocytosis in *Amoeba proteus*. *Acta Physiol. Scand.* 432: 1–65.

Kitching, J. A. (1956). Food vacoules. *Protoplasmatologia* 3: 1–54.

Klein, H. P., Köster, B., and Stockem, W. (1988). Pinocytosis and locomotion of amoebae. XVIII. Different morphodynamic forms of endocytosis and microfilament organization in *Amoeba proteus*. *Protoplasma* 2: 76–87.

Klein, H. P. and Stockem, W. (1979). Pinocytosis and locomotion of amoebae. XII. Dynamics and motive force generation during induced pinocytosis in *Amoeba proteus*. *Cell Tiss. Res.* 197: 263–279.

Kornfeld, S. and Mellman, I. (1989). The biogenesis of lysosomes. *Annu. Rev. Cell Biol.* 5: 483–525.

Kudo, R. R. (1966). *Protozoology*. (Thomas, C. C., ed.), pp. 117–177, Thomas Books, Springfield.

Kukulies, J., Ackermann, G., and Stockem, W. (1986). Pinocytosis and locomotion of amoebae. XIV. Demonstration of two different receptor sites on the cell surface of *Amoeba proteus*. *Protoplasma* 131: 233–243.

Langreth, S. G. and Balber, A. E. (1975). Protein uptake and digestion in bloodstream and culture forms of *Trypanosoma brucei*. *J. Protozool.* 22: 40–53.

Laval, M. (1971). Ultrastructure et mode de nutrition du choanoflagellé *Salpingoeca pelagica*, sp. nov., comparison avec les choanocytes des spongiaires. *Protistologica* 7: 325–336.

Lee, J. J. and Anderson, O. R. (1991). Cytology and fine structure. In: *Biology of Foraminifera* (Lee, J. J. and Anderson, O. R., eds.), pp. 7–40, Academic Press, San Diego.

Linnenbach, M., Hausmann, K., and Patterson, D. J. (1983). Ultrastructural studies on the food vacuole cycle of a Heliozoon (feeding by *Actinophrys*, III). *Protoplasma* 115: 43–51.

McLaughlin, J. and Aley, S. (1985). The biochemistry and functional morphology of the *Entamoeba*. *J. Protozool.* 32: 221–240.

McNeil, P. L., Tanasugran, L., Meigs, J. B., and Taylor, D. L. (1983). Acidification of phagosomes is initiated before lysosomal enzyme activity is detected. *J. Cell Biol.* 97: 692–702.

Mortimore, G. E., Lardeux, B. R., and Adams, C. E. (1988). Regulation of microautophagy and basal protein turnover in rat liver: effect of short-term starvation. *J. Biol. Chem.* 263: 2506–2512.

Müller, M., Tóth, J., and Törö, I. (1962). Studies on feeding and digestion in Protozoa. IV. Acid phosphatase and nonspecific esterase activity of food vacuoles in *Amoeba proteus*. *Acta Biol. Hung.* 13: 105–116.

Münch, R. (1970). Food uptake by endocytosis in *Opalina ranarum*. *Cytobiologie* 2: 108–122.

Nachmias, V. (1966). A study by electron microscopy of the formation of new surface by *Chaos chaos*. *Exp. Cell Res.* 43: 583–601.

Nilson, J. R. (1984). On starvation-induced autophagy in *Tetrahymena*. *Carlsberg Res. Commun.* 49: 323–340.

Nisbet, B. (1984). *Nutrition and Feeding Strategies in Protozoa*, pp. 185–201, Croom Helm, London.

Padh, H., Lavasa, M., and Steck, T. L. (1989). Prelysosomal acidic vacuoles in *Dictyostelium discoideum*. *J. Cell Biol.* 108: 865–874.

Papadopoulus, T. and Pfeifer, U. (1987). Protein turnover and cellular autophagy in growing and growth-inhibited 3T3 cells. *Exp. Cell Res.* 171: 110–121.

Prusch, R. D., Goette, S. M., and Haberman, P. (1989). Prostglandins may play a signal-coupling role during phagocytosis in *Amoeba proteus*. *Cell Tiss. Res.* 255: 553–557.

Prusch, R. D. and Hannafin, J. (1979). Sucrose uptake by pinocytosis in *Amoeba proteus* and the influence of external calcium. *J. Gen. Physiol.* 74: 523–535.

Prusch, R. D. and Minck, D. R. (1985). Chemical stimulation of phagocytosis in *Amoeba proteus* and the influence of external calcium. *Cell Tiss. Res.* 242: 557–564.

Prusch, R. D. and Vedovatti, P. A. (1989). Calcium movements associated with peptide induced phagocytosis in *Amoeba proteus*. *J. Comp. Physiol.* 159: 287–292.

Prusch, R. D. and Wells, S. M. (1990). Binding characteristics of receptors for phagocytosis on the surface of *Amoeba proteus*. *Cell Tiss. Res.* 261: 445–450.

Rabinovitch, M. and de Stefano, M. J. (1971a). Phagocytosis of erythrocytes by *Acanthamoeba* sp. *Cell Tiss. Res.* 64: 275–284.

Rabinovitch, M. and de Stefano, M. J. (1971b). Antibody and plant agglutinins stimulate phagocytosis of erythrocytes by *Acanthamoeba*. *Nature* 234: 414–415.

Rodman, J. S., Mercer, R. W., and Stahl, P. D. (1990). Endocytosis and transcytosis. *Curr. Opin. Cell Biol.* 2: 664–672.

Rudzinska, M. A. (1980). Internalization of macromolecules from the medium in Suctoria. *J. Cell. Biol.* 84: 172–183.

Ryter, A. and Bowers, B. (1976). Localization of acid phosphatase in *Acanthamoeba castellanii* with light and electron microscopy during growth and after phagocytosis. *J. Ultrastruct. Res.* 57: 309–321.

Ryter, A. and de Chastellier, C. (1977). Morphometric and cytochemical studies of *Dictyostelium discoideum* in vegetative phase: digestive system and membrane turnover. *J. Cell Biol.* 75: 200–217.

Salt, G. W. (1968). The feeding of *Amoeba proteus* on *Paramecium aurelia*. *J. Protozool.* 15: 275–280.

Salzman, N. H. and Maxfield, F. R. (1988). Intracellular fusion of sequentially formed endocytotic compartments. *J. Cell Biol.* 106: 1083–1091.

Schnepf, E. and Elbrächter, M. (1992). Nutritional strategies in dinoflagellates: a review with emphasis on cell biological aspects. *Eur. J. Protistol.* 28: 3–24.

Seravin, L. M. (1986). The role of mechanical and chemical stimulation on the induction of phagocytotic reactions in *Amoeba proteus* and *Amoeba dubia. Acta Protozool.* 6: 97–107.

Silverstein, S. C., Greenberg, S., di Virgilio, F., and Steinberg, T. H. (1989). Phagocytosis. In: *Fundamental Immunology* (Paul, W. E., ed.), pp. 1–18, Raven Press, New York.

Smythe, E. and Warren, G. (1991). The mechanism of receptor-mediated endocytosis. *Eur. J. Biochem.* 202: 689–699.

Sodeik, B., Brix, K., and Stockem, W. (1989). Sequestration of microinjected molecular probes from the cytoplasm of *Amoeba proteus. Eur. J. Protistol.* 25: 75–84.

Stockem, W. (1966). Pinocytose und Bewegung von Amöben. I. Mitteilung. Die Reaktion von *Amoeba proteus* auf verschiedene Markierungssubstanzen. *Z. Zellforsch. mikrosk. Anat.* 74: 372–400.

Stockem, W. (1969). Pinocytosis and locomotion of amoebae. III. The function of the Golgi apparatus of *Amoeba proteus* and *Chaos chaos. Histochemistry* 18: 217–240.

Stockem, W. (1970). Untersuchungen mit dem Differential-Interferenz-Kontrast über Morphologie und Cytosen von *Amoeba proteus. Mikroskopie* 24: 332–344.

Stockem, W. (1973). Pinocytosis and locomotion of amoebae. IV. Quantitative investigations on permanent and induced endocytosis of *Amoeba proteus. Z. Zellforsch. mikrosk. Anat.* 136: 433–446.

Stockem, W. (1977). Endocytosis. In: *Mammalian Cell Membranes* (Jamieson, G. A. and Robinson, D. M., eds.), vol. 5, pp. 151–195, Butterworths, London.

Stockem, W., Hoffmann, H. U., and Gruber, B. (1983a). Dynamics of the cytoskeleton in *Amoeba proteus*. I. Redistribution of microinjected fluorescein-labeled actin during locomotion, immobilization and phagocytosis. *Cell Tiss. Res.* 232: 79–96.

Stockem, W. and Klein, H. P. (1988). Pinocytosis and locomotion of amoebae. XVII. Influence of different cations on induced pinocytosis in *Amoeba proteus. Eur. J. Protistol.* 23: 317–326.

Stockem, W. and Klopocka, W. (1988). Ameboid movement and related phenomena. *Int. Rev. Cytol.* 112: 137–183.

Stockem, W. and Mitarbeiter (1977). Nahrungsaufnahme beim Pantoffeltier *Paramecium caudatum. Mikrokosmos* 7: 202–206.

Stockem, W., Naib-Majani, W., and Wohlfarth-Bottermann, K. E. (1984). Preservation and phallotoxin-staining of the microfilament system in *Amoeba proteus. Cell Biol. Int. Rep.* 8: 207–213.

Stockem, W., Naib-Majani, W., Wohlfarth-Bottermann, K. E., Osborn, M., and Weber, K. (1983b). Pinocytosis and locomotion of amoebae. XIX. Immunocytochemical demonstration of actin and myosin in *Amoeba proteus. Eur. J. Cell Biol.* 29: 171–178.

Stockem, W. and Stiemerling, R. (1976). Intracellular segregation of endocytotically ingested substances. *Cytobiologie* 13: 158–162.

Stoorvogel, W., Strous, G. J., Geuze, H. J., Oorschot, V., and Schwartz, A. L. (1991). Late endosomes derive from early endosomes by maturation. *Cell* 65: 417–427.

Storrie, B., Pool, R., Sachdeva, M., Maurey, K. M., and Oliver, C. (1984). Evidence for both prelysosomal and lysosomal intermediates in endocytotic pathways. *J. Cell Biol.* 98: 108–115.

Straus, W. (1963). Comparative observations on lysosomes and phagosomes in kidney and liver of rats after administration of horseradish peroxidase. In: *Lysosomes* (de Reuck, A. V. S. and Cameron, M. P., eds.), pp. 151–175, Ciba Foundation Symposium, Cameron.

Vickerman, K. (1962). Patterns of cellular organisation in limax amoebae: an electron microscope study. *Exp. Cell Res.* 26: 497–519.

Vogel, G., Thilo, L., Schwarz, H., and Steinhart, R. (1980). Mechanism of phagocytosis in *Dictyostelium discoideum*: phagocytosis is mediated by different recognition sites as disclosed by mutants with altered phagocytotic properties. *J. Cell Biol.* 86: 456–465.

Westphal, A., Michel, R., and Mühlpfordt, H. (1972). Elektronenmikroskopische Befunde bei der Phagozytose von *Crithidia sp.* durch *Entamoeba histolytica. Tropenmed. Parasitol.* 23: 117–127.

Wise, G. E. and Flickinger, C. J. (1970). Cytochemical staining of the Golgi apparatus in *Amoeba proteus. J. Cell Biol.* 46: 620–626.

Wohlfarth-Bottermann, K. E. and Stockem, W. (1966). Pinocytose und Bewegung von Amöben. II. Mitteilung. Permanente und induzierte Pinocytose bei *Amoeba proteus*. *Z. Zellforsch. mikrosk. Anat.* 73: 444–474.

Wolpert, L. and O'Neil, C. H. (1962). Dynamics of the membrane of *Amoeba proteus* studied with labelled specific antibody. *Nature* 196: 1261–1265.

Wright, S. D. and Silverstein, S. C. (1986). Overview: the function of receptors in phagocytosis. In: *Handbook of Experimental Immunology* (Weirs, D. M., ed.), pp. 1–14, Blackwell Scientific, Oxford.

THE LYSOSOMAL SYSTEM IN MALARIA PARASITES

Christian Slomianny

I. INTRODUCTION

The malaria parasite, *Plasmodium*, has a complex life cycle shared between the vector Anopheline and the mammalian host. During this cycle, it presents different growing and dividing phases. The most important one is the intraerythrocytic phase, which produces the well-known clinical signs of rhythmic fever. At that level, the parasite matures within a cell in which hemoglobin is the major cytosolic protein. This stage, called trophozoite, is characterized by an intense and complex endocytosis of the erythrocytic stroma (Slomianny et al., 1985). The endocytic vesicles and vacuoles are pinched off from the parasitic plasma membrane and their

Advances in Cell and Molecular Biology of Membranes, Volume 2B
Membrane Traffic in Protozoa, pages 409–418

ISBN: 1-55938-628-2

content is progressively degraded, leading to the formation of black-pigmented crystals which accumulate within residual vacuoles. These vacuoles are released into the bloodstream after parasite division (schizogony) and host cell lysis.

Since the parasite has limited capacity to synthetize amino acids or to take them up directly from the plasma (Sherman and Tanigoshi, 1970)—although recent work tends to show the existence of a duct between the parasite and the plasma (Sherman and Zidovetski, 1992)—hemoglobin is thought to be its main source of nutrients, as shown by the incorporation of labeled hemoglobin (Theakston et al., 1970). Moreover, the parasite produces amino acids, the composition of which resembles that of hemoglobin (Zarchin et al., 1986). This production increases with parasite age. It has also been estimated that more than 25% of the hemoglobin content of the host cell could be degraded by the trophozoite during its growing phase (Roth et al., 1986). Nevertheless, this process could be not only for nutritional purposes but also in order to produce sufficient space, since the parasite releases large amounts of amino acids into the bloodstream (Charet et al., 1983).

These results demonstrate that the malaria parasite necessarily possesses a proteolytic system which is able to break down hemoglobin into amino acids. This means also that there occur active proteases in the parasite, the origin of which could be double: endogenous, that is, synthetized de novo, or exogenous, since the host cell still contains this kind of enzyme.

In this chapter I shall try to show the different stages of endocytosis, then of degradation of the host stroma, leading to the formation of residual pigment. I will focus mainly on the intraerythrocytic phase. Little information is available on the digestive process of the other growing and feeding phases of the parasite (hepatic phase in the mammalian host, then sporogenesis in the Anopheline), these stages being more difficult to study. I will also discuss the origin of proteolytic enzymes and the existence of a true lysosomal system in the malaria parasite. See also chapter by Duszenko and Seyfang on trypanosomes in this volume.

II. THE DIGESTIVE SYSTEM

Endocytosis occurs by two routes: peripheral pinocytosis, especially in young stages, and a cytostomal system which is more or less important depending on the species (Slomianny, 1990). Micropinocytosis (Fig. 1a) is represented by the budding of small, smooth-surfaced vesicles (60–100 nm) from the plasma membrane. Young stages of malaria parasites show a large number of micropinocytic vesicles. These stages are very similar in all the species studied. However, these vesicles are less numerous in *P. berghei* RC than in other species (*P. berghei* N, *P. chabaudi, P. yoelii nigeriensis,* and *P. falciparum*), possibly because this species is very resistant to chloroquine or perhaps due to a very particular type of exocytosis of pigment vesicles into the cytoplasm of the host reticulocyte (the other species invade preferentially mature erythrocytes).

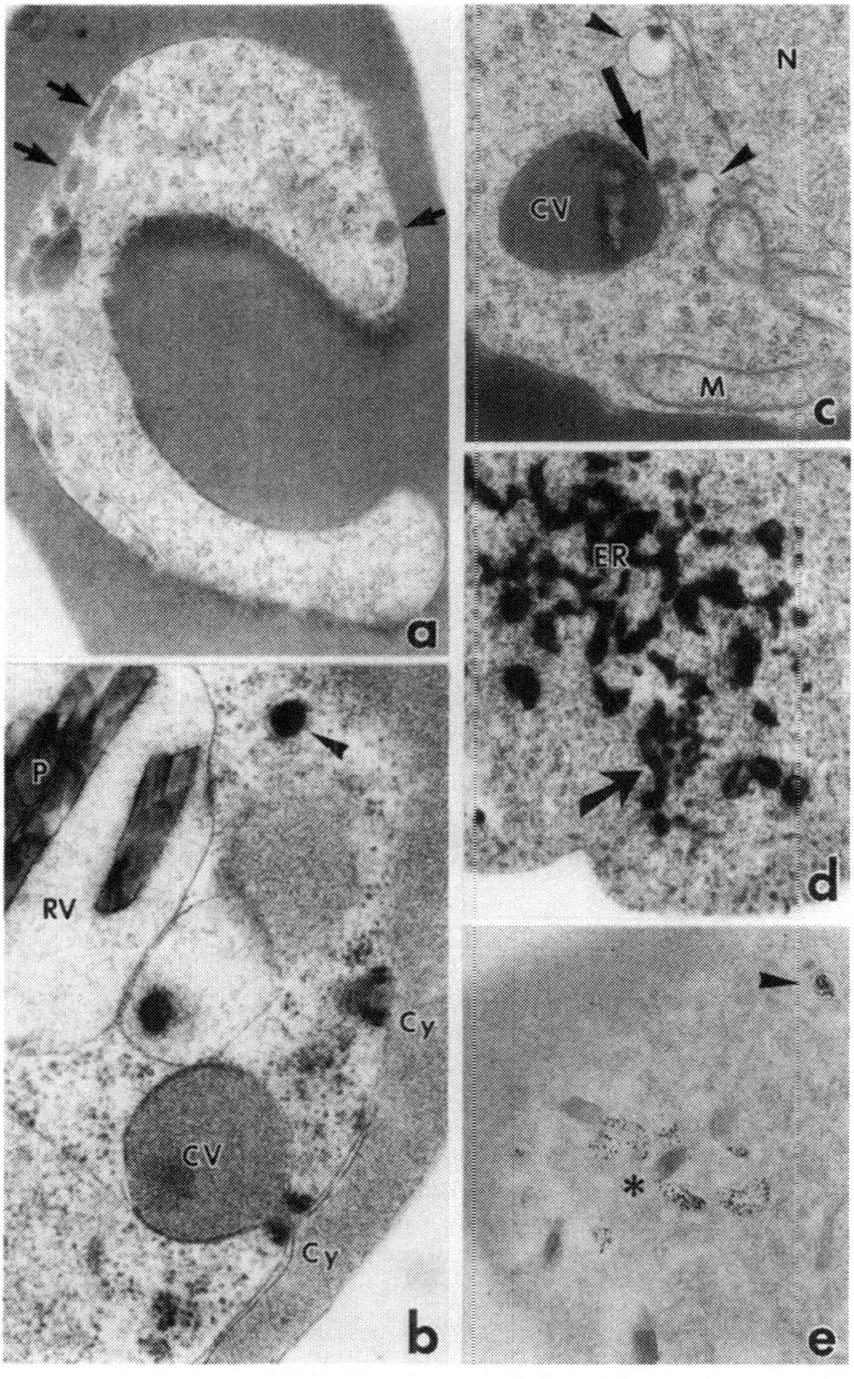
a
b
c
d
e
P
RV
Cy
CV
N
M
ER
*

Figure 1. (**a**) Young cup-shaped trophozoite of *P. falciparum* showing intensive micropinocytosis (arrows). × 27,000. (**b**) Middle trophozoite of *P. falciparum* showing cytostomal feeding. The two limiting membranes are invaginated through cytostomes (Cy) and form cytostomal vacuoles (CV). After the erythrocyte stroma is degraded, the digestive vacuoles are gathered into the residual vacuole (RV). The undegraded moiety of the hemoglobin is stocked as pigment crystals (P). A lysosome-like vesicle (arrowhead) is observed near the cytostomal vacuoles. × 51,000. (**c**) Trophozoite of *P. falciparum*. Several lysosomal vesicles (arrowheads) are shown in the vicinity of a cytostomal vacuole in the course of digestion. One of them is just touching the cytostomal vacuole (arrow). M = section through the unique mitochodrion, N = nucleus. × 27,000. (**d**) The zinc-iodide-osmium technique heavily contrasts some intracellular compartments of *P. berghei* trophozoite, in particular the endoplasmic reticulum (ER). The arrow points to a zone of intense budding, corresponding to an atypical Golgi apparatus. × 47,000. (**e**) Trophozoite of *P. yoelii nigeriensis*. Acid phosphatase is shown in vesicles just endocytosed by micropinocytosis (asterisk) and in lysosomal vesicles (arrowhead). Unpublished micrographs. × 51,000.

In older stages, a cytostomal system is formed (Fig. 1b), whereas micropinocytosis tends to vanish. The oldest stages display only a cytostomal system. This mechanism of endocytosis was first described as a ring structure (Rudzinska and Trager, 1959; Aikawa et al., 1966) resulting from plasmalemma differentiation. The erythrocyte stroma is engulfed through this structure, leading to the formation of either an elongated vacuole or a rounded vacuole whose diameter is about 300–500 nm. Other vacuoles can then bud off from the elongated structure, leading to a very complicated system which takes up a great deal of room in the parasitic cytoplasm.

Nevertheless, these two endocytic processes always lead to the formation of vesicles or vacuoles that are surrounded by two membranes. The outer membrane is derived from the plasma membrane of the parasite, whereas the inner one comes from the parasitophorous vacuole which always surrounds the parasite.

Ultrastructural evidence for hemoglobin degradation includes the disappearance of the inner membrane, the progressive decrease of electron density of the vesicular or vacuolar content, and the gradual appearance of malarial pigment called hemozoin. In *P. falciparum* (FCR_3 strain), the pigment-containing vacuoles gather into one or two large residual vacuoles early in the life cycle, whereas in murine species this phenomenon takes place just before schizogony. However, in both cases, the large residual pigment vacuoles, whenever formed, are always a dumping place where vesicles or vacuoles can fuse after hemoglobin digestion has been completed. At the end of the intraerythrocytic phase, all the pigment produced by the parasite is deposited into a big residual vacuole. This is ejected into the blood plasma after host cell lysis and merozoite release.

Even though the endocytic mechanism is well established, the hemoglobin degradation process itself remains rather obscure. Also the formation of the crystalline particles is poorly understood. A likely explanation is that proteases degrade the protein moiety of hemoglobin, releasing the heme residue.

III. LYSOSOMAL SYSTEM

The digestion of hemoglobin takes place in vacuoles and vesicles which share many characteristics with lysosomes of mammalian cells (Choi and Mego, 1987). As in lysosomes of normal cells (de Duve and Wattiaux, 1966), the inner pH of such vacuoles is acidic, between pH 5.0 and pH 5.4 (Homewood et al., 1982; Yayon et al., 1984; Krogstadt et al., 1985). This acidic pH within the digestive system is most likely maintained by an ATP-energized proton pump similar to those characterized in mammalian lysosomes, as shown on the vacuolar membrane of *P. falciparum* (Choi and Mego, 1988).

One of the hydrolytic enzymes known to be usually associated with lysosomes is acid phosphatase (AP). Work on this subject is scarce and has yielded contradictory results. Scorza et al. (1972) investigated three acid hydrolases in the intraerythrocytic stages of three malarial species. Of these three hydrolases, only arylsulfatase, at low pH, could be demonstrated. These results contradict those of other authors (Sen Gupta et al., 1955; Aikawa and Thompson, 1971; Slomianny and Prensier, 1990), who found AP activity in constant association with malarial particles in the food vacuole of *P. berghei* and *P. gallinaceum*. The close relationship between food vacuoles and endoplasmic reticulum led these authors to suggest the transfer of AP from the reticulum to the food vacuole. Acid phosphatase was not detected in the intraerythrocytic stages but was demonstrated in ookinetes (Freyvogel and Hunter, 1969). This enzyme is not strictly relevant to hemoglobin catabolism. Since it is localized mainly in early stages of endocytic vesicles (Fig. 1e), it could play a role in the degradation of the inner membrane of these vesicles and indicate an acidification of vesicle contents, a prerequisite for protease activity. It is interesting to note that some authors did not find any nonproteolytic lysosomal activity in the pigment vacuoles, in contrast to some others. An explanation could be that the different authors did not work on the same stages, as this activity is present only in freshly endocytosed vesicles where the inner membrane is being degraded.

The presence of acid proteases able to degrade hemoglobin has been demonstrated in various malaria species, such as *P. berghei* (Levy and Chou, 1973; Sato et al., 1987), *P. knowlesi* (Cook et al., 1961; Hempelmann and Wison, 1980), *P. lophurae* (Sherman and Tanigoshi, 1983), *P. yoelii nigeriensis* (Aissi and Charet, 1983), and *P. falciparum* (Gyang et al., 1982; Vander Jagt et al., 1986; Rosenthal et al., 1987, 1988; Bailly et al., 1991). A complex proteolytic system has been proposed involving cathepsin-D, an endoarylamidase and finally an aminopeptidase (Aissi et al., 1983). Aminopeptidase has been found in the malaria parasite (Charet et al., 1983; Vander Jagt et al., 1984), as revealed also by electron microscopy using enzyme-cytochemical techniques in endocytic vesicles (Slomianny et al., 1983). More recent work (Goldberg et al., 1990) also suggests that there are multiple and diverse proteases at work in the vacuoles. The initial event could be the generation of large fragments by aspartic protease activity, inhibited by pepstatin (Goldberg et al., 1991). A cystein protease could also act in the vacuoles, as inhibitors block hemozoin accumulation and arrest cell growth (Rosenthal et al., 1988).

The characterization and the purification of these enzymes have been partially done. But a major problem in the studies is to know whether the parasite uses the erythrocyte enzyme (Levy et al., 1974), as for cathepsin-D, or whether it synthetizes these enzymes itself. If it uses exogenous enzymes, another question arises: How can it activate them? A response to this question may be obtained by studying the complete lysosomal system. Indeed, the presence of a Golgi apparatus with vesicles budding off as well as the observation of the fusion of lysosomes with other vacuoles, could be a good argument in favor of a complete lysosomal system in *Plasmodium.*

Very few authors actually observed a lysosomal system or a Golgi apparatus in the parasite, and the results were rather contradictory. Few of the morphologists who have studied the fine structure of the asexual stages of the various species of *Plasmodium* have mentioned or even described lysosome-like structures (Garnham et al., 1962; Warhurst and Hockley, 1967; Homewood et al., 1972). Furthermore, organelles with the typical structure of the Golgi complex have not been reported for malarial parasites, except by Rudzinska and Trager (1959), who interpreted certain vesicles in the vicinity of the nucleus as a Golgi apparatus. These findings have not been confirmed by other investigators in the same or in related species of the parasite.

Nevertheless, more recently a very atypical Golgi-like apparatus has been shown (Slomianny and Prensier, 1990) by the use of techniques contrasting distinct subcellular compartments (Fig. 1c). The Golgi apparatus is active during the whole phase of growth and not only during the phase of synthesis of the apical complex of the merozoite as stated before (Rudzinska and Trager, 1959). The budding of vesicles from the endoplasmic reticulum has also been seen in exoerythrocytic forms of *P. berghei* (Meis et al., 1981, 1983, 1986). This apparatus is atypical, because it does not consist of flattened cisternae but of vesicles that show a structure very similar to those of the coated vesicles of a true Golgi complex. To our knowledge, no fusion with other vacuoles was shown up to now. Only close vicinity (Fig. 1c) was observed, but not real fusion (Slomianny and Prensier, 1990). This is perhaps a very rapid phenomenon not observable with the techniques employed.

The hypothesis on the endogenous origin of the proteolytic enzymes is supported by some other facts. For example, AP and aminopeptidase activity has never been detected in the stroma of the erythrocyte itself (Aikawa and Thompson, 1971; Slomianny, 1990). Some enzymes do not exist in the host (aminopeptidase, AP). For the other enzymes (cathepsin-D, for example) the possibility exists that the enzymatic activity seen in the parasite may be due to enzymes ingested by endocytosis of the erythrocytic stroma, a suggestion supported by the fact that the labeling is generally found near malarial pigment. This kind of enzyme might be inactive in the erythrocytic stroma because it is in a neutral or slightly basic environment, and it could be then activated in the acidic digestive compartment of the parasite. The response could be given by the discovery of genes coding for some enzymes, for example, for cystein protease, which shares many properties with cathepsine-L (Rosenthal and Nelson, 1992).

IV. CONCLUSIONS AND OUTLOOK

Plasmodium has developed a very unusual digestive system due to its growth in a parasitophorous vacuole. By micropinocytosis and cytostomal feeding, hemoglobin is engulfed along with part of the parasitophorous membrane and with some other host stroma components (Fig. 2). Two kinds of enzymes are then active: hydrolases (which are not directly related to host stroma degradation but necessary to degrade the membrane surrounding the hemoglobin bulk) and proteases. These enzymes are successively active, first the acid phosphatase, then the different proteases. Amino acids are generated from the globin moiety. The heme residue then accumulates as pigment. The amino acids can either be used for new protein synthesis or released from the parasite (Fig. 3).

The endogenous origin of some proteases can be derived from their absence from the host cells, as well as from their detection in lysosomal vesicles whose origin is a very atypical Golgi apparatus. Unfortunately, to our knowledge, fusion of endocytic vacuoles or vesicles with these lysosomes has not yet been observed. This process must be very rapid and perhaps cryofixation and cryosubstitution might solve this problem. The endogenous origin of some enzymes involved in the degradation process will also have to be analyzed by molecular genetics.

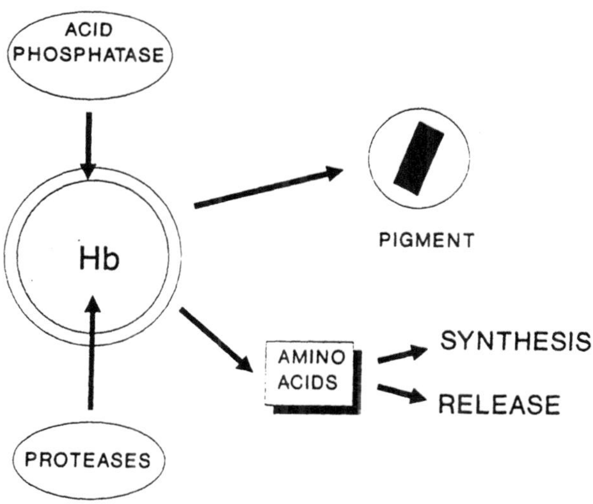

Figure 2. Endocytosis and degradation of host stroma by *Plasmodium.* Hb = hemoglobin. For more details, see text.

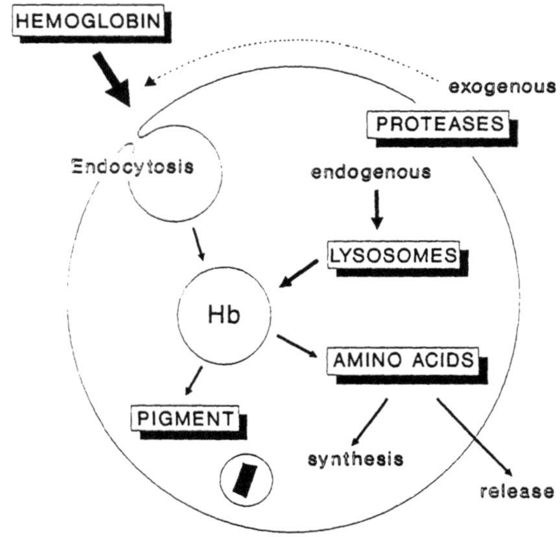

Figure 3. The lysosomal system of *Plasmodium*. Hb = hemoglobin. For more details, see text.

REFERENCES

Aikawa, M., Hepler, P. K., Huff, C. G., and Sprinz, H (1966). The feeding mechanism of avian malarial parasite. *J. Cell Biol.* 28: 355–373.

Aikawa, M. and Thompson, F. E. (1971). Localization of acid phosphatase activity in *Plasmodium berghei* and *Plasmodium gallinaceum*: an electron microscopic observation. *J. Parasitol.* 53: 603–610.

Aissi, E., Charet, P., Bouquelet, S., and Biguet, J. (1983). Endoprotease in *Plasmodium yoelii nigeriensis*. *J. Comp. Biochem. Physiol.* 74B: 559–566.

Bailly, E., Savel, J., Mahouy, G., and Jaureguiberry, G. (1991). *Plasmodium falciparum*: isolation and characterization of a 55-kDa protease with a cathepsine D-like activity from *P. falciparum*. *Exp. Parasitol.* 72: 278–284.

Charet, P., Aissi. E., Bouquelet, S., and Biguet, J. (1983). Aminopeptidase in rodent *Plasmodium*. *Comp. Biochem. Physiol.* 65B: 515–524.

Charet, P., Prensier, G., and Slomianny C. (1983). Comparative study of amino acid production in erythrocytes parasitized in *Plasmodium sp. and Babesia hylomysci. Comp. Biochem. Physiol.* 75B: 347–353.

Choi, I. and Mego, J. L. (1987). Intravacuolar proteolysis in *Plasmodium falciparum* digestive vacuoles is similar to intralysosomal proteolysis in mammalian cells. *Biochim. Biophys. Acta* 926: 170–176.

Choi, I. and Mego, J. L. (1988). Purification of *Plasmodium falciparum* digestive vacuoles and partial characterization of the vacuolar membrane ATPase. *Mol. Biochem. Parasitol.* 31: 71–78.

Cook, L., Grant, P. T., and Kermack, W. O. (1961). Proteolytic enzymes of the erythrocytic forms of rodent and simian species malaria plasmodia. *Exp. Parasitol.* 11: 372–379.

de Duve, C. and Wattiaux, R. (1966). Function of lysosomes. *Annu. Rev. Physiol.* 28: 435–492.

Freyvogel, J. A. and Hunter, R. L. (1969). Esterases and acid phosphatases in sporogonic stages of *Plasmodium. Acta Trop.* 26: 15–28.

Garnham, P. C. C., Bird, R. G., and Baker, J. R. (1962). Electron microscope study of motile stages of malaria parasites. III. The ookinete of *Haemamoeba* and *Plasmodium*. *Trans. R. Soc. Trop. Med. Hyg.* 56: 116–120.

Goldberg, D. E., Slater, A. F. G., Beavis, R., Chait, B., Cerami, A., and Henderson, G. B. (1991). Hemoglobin degradation in the human malaria pathogen *Plasmodium falciparum*: a catabolic pathway initiated by a specific aspartic protease. *J. Exp. Med.* 73: 961–969.

Goldberg, D. E., Slater, A. F. G., Cerami, A., and Henderson, G. B. (1990). Hemoglobin degradation in the malaria parasite *Plasmodium falciparum*: an ordered process in a unique organelle. *Proc. Natl. Acad. Sci. U.S.A.* 87: 2931–2935.

Gyang, F. N., Poole, B., and Trager, W. (1982). Peptidase from *Plasmodium falciparum* cultured *in vitro*. *Molec. Biochem. Parasitol.* 5: 263–273.

Hempelmann, E. and Wison, R. J. M. (1980). Endopeptidase from *Plasmodium knowlesi*. *Parasitology* 80: 323–330.

Homewood, C. A., Warhurst, D. C., Peters, W., and Baggaley, V. C. (1972). Lysosomes, pH and antimalarial action of chloroquine. *Nature* 235: 50–52.

Krogstadt, D. J., Schlesinger, P. H., and Gluzman, I. Y. (1985). Antimalarials increase vesicle pH in *Plasmodium falciparum*. *J. Cell Biol.* 101: 2302–2309.

Levy, M. R. and Chou, S. C. (1973). Activity and some properties of an acid proteinase from normal and *Plasmodium berghei*-infected red cells. *Exp. Parasitol.* 59: 1064–1070.

Levy, M. R., Siddiqui, W. A., and Chou, S. C. (1974). Acid protease activity in *Plasmodium falciparum* and *Plasmodium knowlesi* and ghosts of their respective host red cell. *Nature* 247: 546–549.

Meis, J. F. G. M., Verhaeve, J. P., Jap, P. H. K., Hess, F., and Meuwissen, J. H. E. T. (1981). An ultrastructural study of developing stages of exoerythrocytic *Plasmodium berghei* in rat hepatocytes. *Parasitology* 82: 195–204.

Meis, J. F. G. M., Verhaeve, J. P., Jap, P. H. K., and Meuwissen, J. H. E. T. (1983). Ultrastructural studies of a vesicle system associated with endoplasmic reticulum in exoerythrocytic forms of *Plasmodium berghei*. *J. Protozool.* 30: 111–114.

Meis, J. F. G. M., Verhaeve, J. P., Jap, P. H. K., Hollingdale, M. R., and Meuwissen, J. H. E. T. (1986). Organization of the exoerythrocytic stage of the rodent malaria parasite *Plasmodium berghei*: a cytochemical study. *J. Submicrosc. Cytol.* 18: 755–760.

Rosenthal, P. J., Kim, K., McKerrow, J. H., and Leech, J. H. (1987). Identification of three stage-specific proteinases of *Plasmodium falciparum*. *J. Exp. Med.* 166: 816–821.

Rosenthal, P. J., McKerrow, J. H., Aikawa, M., Nagasawa, H., and Leech, J. H. (1988). A malarial cystein proteinase is necessary for hemoglobin degradation by *Plasmodium falciparum*. *J. Clin. Invest.* 82: 1560–1566.

Rosenthal, P. J. and Nelson, R. G. (1992). Isolation and characterization of a cystein proteinase gene of *Plasmodium falciparum*. *Mol. Biochem. Parasitol.* 51: 143–152.

Roth, E. F., Brotman, D. S., Vanderberg, J. P., and Schulman, S. (1986). Malarial pigment-dependent error in the estimation of hemoglobin content in *Plasmodium falciparum*-infected red cells: implications for metabolic and biochemical studies of the erythrocytic phases of malaria. *Am. J. Trop. Med. Hyg.* 35: 906–911.

Rudzinska, M. A. and Trager, W. (1959). Phagotrophy and two new structures in the malaria parasite *Plasmodium berghei*. *J. Biophys. Biochem. Cytol.* 6: 103–122.

Sato, K., Fukabori, Y., and Suzuki, M. (1987). *Plasmodium berghei*: a study of globinolytic enzyme in erythrocyte parasite. *Zbl. Bakt. Hyg.* A264: 487–495.

Scorza, J. V., De Souza, C., and Monteiro, M. G. C. (1972). Cytochemical observation of three acid hydrolases in blood stages of malaria parasites. *Ann. Trop. Med. Parasitol.* 66: 167–172.

Sen Gupta, P. C., Ray, H. N., and Chaudhury, R. N. (1955). A cytochemical study of *Plasmodium berghei* (Vincke and Lips, 1948). *Ann. Trop. Med. Parasitol.* 49: 273–277.

Sherman, I. W. and Tanighoshi, L. (1970). Incorporation of ^{14}C-amino acids by malaria (*Plasmodium lophurae*). IV. In vivo utilization of host cell haemoglobin. *Int. J. Biochem.* 1: 635–637

Sherman, I. W. and Tanighoshi, L. (1983). Purification of *Plasmodium lophurae* cathepsin D and its effects on erythrocyte membrane proteins. *Mol. Biochem. Parasitol.* 8: 207–226.
Sherman, I. W. and Zidovetzki, R. (1992). A parasitophorous duct in *Plasmodium*-infected red blood cells. *Parasitol. Today* 8: 2–3.
Slomianny, C. (1990). Three-dimensional reconstruction of the feeding process of the malaria parasite. *Blood Cells* 16: 369–378.
Slomianny, C., Charet, P., and Prensier, G. (1983). Ultrastructural localization of enzymes involved in the feeding process in *Plasmodium chabaudi* and *Babesia hylomysci*. *J. Protozool.* 30: 376–382.
Slomianny, C. and Prensier, G. (1990). A cytochemical study of the lysosomal system of different species of malaria parasites. *J. Protozool.* 37: 465–470.
Slomianny, C., Prensier, G., and Charet, P. (1985). Ingestion of erythrocytic stroma by *Plasmodium chabaudi* trophozoites: ultrastructural study by serial sectioning and 3-dimensional reconstruction. *Parasitology* 90: 579–588.
Theakston, R. D. G., Fletcher, K. A., and Maegraith, B. G. (1970). The use of electron microscope autoradiography for examining the uptake and degradation of haemoglobin by *Plasmodium falciparum*. *Ann. Trop. Med. Parasitol.* 64: 63–71.
Vander Jagt, D. L., Baack, B. R., and Hunsaker, L. A. (1984). Purification of an aminopeptidase from *Plasmodium falciparum*. *Mol. Biochem. Parasitol.* 10: 45–54.
Vander Jagt, D. L., Hunsaker, L. A., and Campos, N. M. (1986). Characterization of a hemoglobin-degrading, low molecular weight protease from *Plasmodium falciparum*. *Mol. Biochem. Parasitol.* 18: 389–400.
Warhurst, D. C. and Hockley, D. J. (1967). Mode of action of chloroquine on *Plasmodium berghei* and *Plasmodium cynomolgi*. *Nature* 214: 935–936.
Yayon, A., Cabantchik, Z. I., and Ginsburg, H. (1984). Identification of the acidic compartment of *Plasmodium falciparum*-infected human erythrocytes as the target of the antimalarial drug chloroquine. *Eur. Mol. Biol. Organ.* J. 3: 2695–2700.
Zarchin, S., Krugliak, M., and Ginsburg, H. (1986). Digestion of the host erythrocyte by malaria parasites is the primary target for quinoline-containing antimalarials. *Biochem. Pharmacol.* 35: 2435–2442.

MEMBRANE AND MICROTUBULE DYNAMICS IN HELIOZOA

Toshinobu Suzaki and Yoshinobu Shigenaka

I. INTRODUCTION

The actinophryid heliozoan *Echinosphaerium* and *Actinophrys* extend a number of needle-like axopodia, each of which contains several hundred microtubules as the cytoskeletal component (Roth et al., 1970; Shigenaka et al., 1971; Sugiyama et al., 1992). The axonemal microtubules have an intense birefringence that is clearly visualized when the organism is observed under a polarizing microscope. The proximal parts of the axonemes penetrate into the endoplasm, where they terminate either freely or in close association with the outer membrane of the nuclear

Advances in Cell and Molecular Biology of Membranes, Volume 2B
Membrane Traffic in Protozoa, pages 419–432

ISBN: 1-55938-628-2

envelope. The axonemal microtubules have an intrinsic physiological property that allows them to disassemble quite readily when the organism is treated with various physical or chemical factors such as Ca^{2+}, low temperature, and mitotic inhibitors. Thus, a number of investigations have been carried out using heliozoan axopodia as a model of dis- and reassembly of microtubules *in vivo* (Tilney, 1968; Shigenaka et al., 1971, 1974; Schliwa, 1976). Interest has also been focused on the morphology of the microtubular axoneme, especially on the cross-bridges which connect and uniformly position the microtubules to form an elaborate double-helical configuration (Roth et al., 1970). By repeating retraction, contraction, and reextension of the axopodia, heliozoans carry out various activities such as cytokinesis (Suzaki et al., 1978), cell fusion (Shigenaka and Kaneda, 1979), cell locomotion (Watters, 1968; Suzaki et al., 1978), cell-to-substrate interactions (Suzaki et al., 1978), and food capture and ingestion (Suzaki et al., 1980; Shigenaka et al., 1982). In addition to these, two other types of motilities have been observed in heliozoan axopodia, which do not involve any bulk transformation of the axopodia themselves. One of them is a unidirectional surface motility (Bloodgood, 1978), and the other is a bidirectional movement of the intra-axopodial particles (Fitzharris et al., 1972; Shigenaka et al., 1974; Edds, 1975a). The latter fulfills most of the criteria for saltatory movement proposed by Rebhun (1972). Although little is known about the mechanism of these motilities, the fact that both of them are associated with the axopodial surface membrane has raised the possibility that they are mediated by similar mechanisms (Bardele, 1976; Bloodgood, 1978). However, recent critical comparison of the characteristics of these two types of motilities showed that they are quite different from each other (Suzaki and Shigenaka, 1982).

In this chapter, we will present and discuss the various aspects of membrane-associated motilities found in heliozoan axopodia. This review will also be concerned with the mechanism of microtubule disorganization in the axopodia through which the heliozoans carry out a variety of motile behavior.

II. INTRA-AXOPODIAL PARTICLE MOVEMENT

Transport of intracellular organelles is a ubiquitous feature of eukaryotic cells (Rebhun, 1972; Hyams and Stebbing, 1979; Schliwa, 1984). In many instances, microtubules have been postulated as important elements along which bidirectional particle transport takes place (Hayden and Allen, 1985; Koonce and Schliwa, 1985). In addition to these microtubule-associated movements, there are some other examples in which organelles are transported even in the absence of microtubules (reviewed in Hyams and Stebbings, 1979), which include particle movement in association with heliozoan cell surfaces.

Echinosphaerium and *Actinophrys* cells bear a number of microtubule-containing axopodia, in which numerous particles move up and down along the axopodial axis in a saltatory manner. These particles, or extrusomes (Patterson, 1979), can be seen as refractive spheres (0.2–0.5 μm in diameter) under the light microscope. The highest velocity of the particle movement is about 1.5 μm/sec. Since these organelles are known to discharge their contents when the cells capture food organisms, they are implicated in the process of prey adhesion and membrane supply for making food vacuoles (Suzaki et al., 1980; Patterson and Hausmann, 1981; Hausmann and Patterson, 1982). Extrusomes are also observed in close association with the cell body surface, where saltatory particle movement can also be observed. The particles are absent in the cell body. In *Echinosphaerium akamae*, each axopodium contains 60 to 90 particles, and the motile activity of the extrusomes is highest at about the 40 μm level from the axopodial base (Suzaki and Shigenaka, 1982).

Inside the axopodia, there are two types of filamentous structures, microtubules and contractile filaments, as shown in Fig. 1. The microtubular axoneme is located at the center of the axopodium as the main cytoskeleton. The contractile filaments, which are considered to be indispensable for a rapid axopodial contraction, run parallel to the axoneme throughout the axopodial length. The extrusomes are always observed to be attached closely to the axopodial surface membrane with a distance of 5–10 nm. Between the plasma membrane and the limiting membrane of the extrusomes, bridges are sometimes observed that connect the apposing membranes (Suzaki and Shigenaka, 1982).

Edds (1975a) showed that particle movement of the heliozoan *Echinosphaerium* still occurred in an artificial axopodium, where a glass microneedle was substituted for the microtubular axoneme, and that colchicine did not inhibit the motion of particles in either normal or artificial axopodia (Tilney, 1968; Edds, 1975a). Moreover, particle movement is observed also in the cortex of the heliozoan cell body, where microtubules are not present (Fitzharris et al., 1972; Suzaki and Shigenaka, 1982). These findings indicate that the microtubular axoneme is not involved in this motility system. When compared with the cell body surface and other portions in the axopodia, the particle movement is most active in the proximal region of axopodia, where most extrusomes are in close contact with either the microtubules or the contractile filaments (Suzaki and Shigenaka, 1982). Although neither microtubules nor the contractile filaments can be regarded as essential for particle motility, such a close proximity may imply that these structures could be involved in acceleration of the particle movement.

Microfilaments have been observed in the cytoplasm of *Echinosphaerium* (Tilney et al., 1966; Edds, 1975b; Suzaki et al., 1980) and *Actinophrys* (Hausmann and Patterson, 1982). They are, however, located either deep in the cytoplasm (Tilney et al., 1966) or in the food vacuole-forming pseudopodia (Suzaki et al., 1980; Hausmann and Patterson, 1982), without any spatial interaction with the moving extrusomes. Cytochalasin B does not affect the particle movement (Suzaki and Shigenaka, 1982). Therefore, it seems unlikely that the movement of heliozoan extrusomes is generated through the actomyosin system.

The intra-axopodial particle movement is inhibited by treatment with energy inhibitors (NaN_3, dinitrophenol, and KCN) or chloralhydrate (Suzaki and Shigenaka, 1982). Antimicrotubule drugs (colchicine or vinblastine) do not inhibit

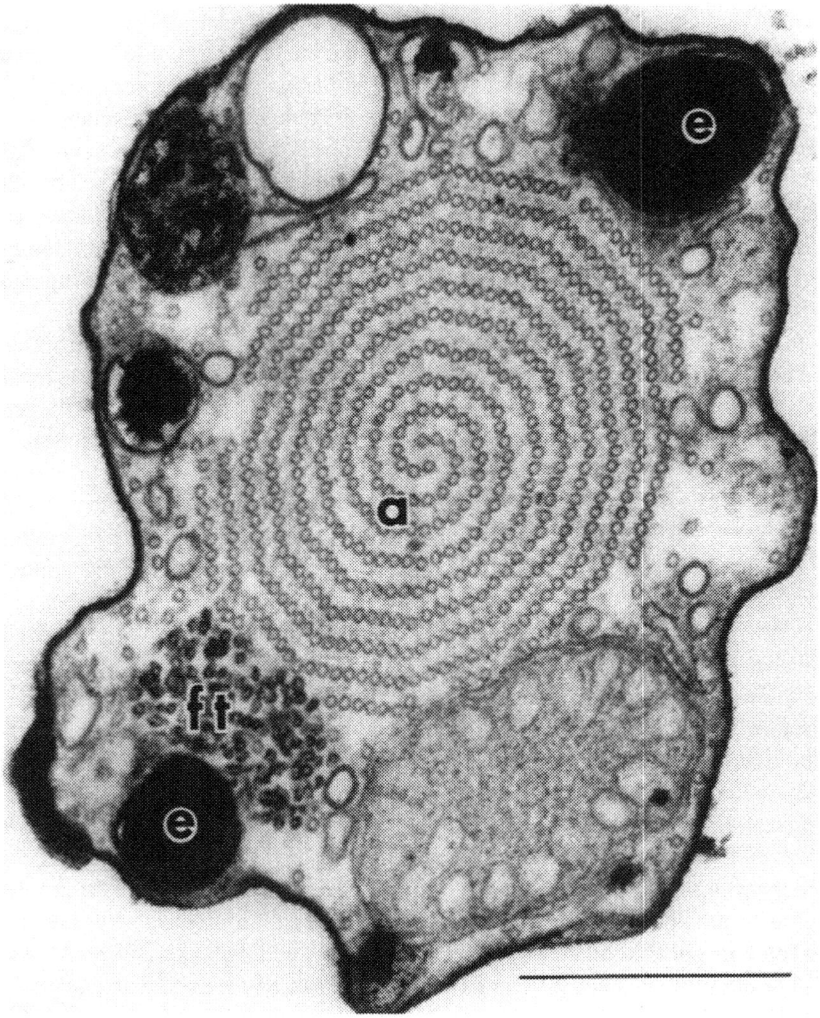

Figure 1. Electron micrograph of a cross section of an axopodium of *Echinosphaerium nucleofilum* showing double-helical arrangement of the axonemal microtubules (a), a bundle of the contractile filaments (ft), and extrusomes (e). Unpublished micrograph. Bar: 0.5 μm.

the motility at all, while axopodial retraction is induced by these treatments because of disassembly of the axonemal microtubules. The motility is temperature dependent and is inhibited after the cells are treated for 2 hr at 0°C. The extrusomes were found to be removed from contact with the plasma membrane when the movement was inhibited by treatment with 5 mM chloralhydrate for 10 min (Suzaki and Shigenaka, 1982). Other intra-axopodial structures including microtubules and contractile filaments were not affected by the treatment. This result indicates that an interaction between the extrusome membrane and that of the cell surface is indispensable for generating the movement of extrusomes.

Saltatory movement of axopodial particles has been reported also in *Heterophrys, Raphidiophrys,* and *Acanthocystis* (Bardele, 1976; Troyer, 1975). In *Heterophrys*, the movements of axopodial particles, called conicysts, are reported to depend upon a specialized fibrillar structure attached to the outer surface of the axopodial membrane (Davidson, 1976). In addition, extracellular markers such as adherent bacteria or gold particles also undergo saltation in a manner similar to the intracellular movement of the conicysts (Troyer, 1975). These observations suggest that an active interaction between extrusomes and the plasma membrane is the strongest candidate for the mechanism of intra-axopodial motility among different heliozoan species.

III. AXOPODIAL SURFACE MOTILITY

In addition to the motility of intra-axopodial particles, another type of motility, the axopodial surface motility, has been reported in *Echinosphaerium nucleofilum* (Bloodgood, 1978) and *E. akamae* (Suzaki and Shigenaka, 1982). Extracellular marker particles, such as polystyrene microspheres, attached to the axopodial surface exhibit unidirectional movement toward the axopodial tip. Carbon particles, carmine particles, squid ink, and some kind of bacteria have also been found to be transported along the axopodial surface. When a marker particle is attached to an axopodium, it begins to move centrifugally without any interruption until it reaches the most distal end of the axopodium. Velocity of the marker particle is quite constant over the axopodial length in the range of 1.6 to 1.9 μm/sec. In this respect, the axopodial surface motility is quite different from the intra-axopodial movement that is characterized by bidirectional saltation. The axopodial surface motility is energy independent and not affected by any of the inhibitors tested including those which inhibit intra-axopodial motility. Moreover, it was not even affected by low temperature (Suzaki and Shigenaka, 1982). These characteristics of the axopodial surface motility even give us an impression that the mechanism of this motility might be nonbiogenic in nature.

Similar cell surface motilities have been reported for *Chlamydomonas* flagella (Bloodgood, 1977; Bloodgood et al., 1979), cilia of sea urchin embryo (Bloodgood, 1980), and also for the centrohelidian heliozoan, *Heterophrys* (Troyer, 1975). In contrast to the surface motility of *Echinosphaerium*, all these surface motilities are not unidirectional, and they are characterized as saltatory. Effects of various types of environmental conditions have also been examined for *Chlamydomonas* flagellar surface motility (Bloodgood et al., 1979), which appeared to be quite different from those on the axopodial surface motility of *Echinosphaerium*. For *Chlamydomonas* flagella, surface motility is inhibited with several kinds of reagents, such as sodium citrate, cycloheximide, chloralhydrate, and also by cold treatment. Although the mechanism of axopodial surface motility remains to be explained, it seems evident that it is quite different from that of the surface motility of *Chlamydomonas* flagella.

Axopodial surface motility plays important roles in the bulk gliding displacement of the cell. When the cell is held between two glass plates, the cell begins to exhibit gliding cell movement (Watters, 1968). The velocity of the cell displacement is similar to that of marker particles along the axopodial surface, and both these motilities are ascribed to the relative sliding between the outer surface of axopodial membrane and substrates (Suzaki and Shigenaka, 1982). In addition to the gliding movement, there are two other fundamental cell functions associated with the axopodial surface motility. One is concerned with the conveyance of excretory materials which have been expelled from the cell body surface by exocytosis, and the other is related to the removal of adhered and useless materials from the axopodial tip as demonstrated earlier by Kitching (1960, 1962).

IV. MEMBRANE DYNAMICS AND AXOPODIAL CONTRACTION DURING FEEDING

In *Actinophrys sol*, a rapid contraction of axopodia (more than 100 μm/sec) is observed to occur when small flagellates are captured for ingestion (Ockleford and Tucker, 1973). *Echinosphaerium nucleofilum* (Suzaki et al., 1980) and *E. akamae* (Shigenaka et al., 1982) are also reported to show similar contraction when prey protozoans attach to the axopodia. The prey organisms adhere to any region of the axopodial surface they touch. In consequence, the ability to catch prey is not considered to be a special characteristic of the axopodial tip (Suzaki et al., 1980; Ando and Shigenaka, 1989). Food capturing in *Echinosphaerium* is initiated by attachment of a prey organism to the axopodium, followed by formation of cytoplasmic protuberances and discharge of the extrusomes. The function of extrusomes is not fully understood, but it has been suggested that they contain enzymes and/or substances which assist in quieting the prey organisms or fastening it to the axopodia (Tilney and Porter, 1965). In the axopodia of other heliozoans, *Actinophrys, Heterophrys,* and *Ciliophrys*, intra-axopodial extrusomes (conicysts in *Heterophrys* and muciferous bodies in *Ciliophrys*) are thought to be indispensable for prey capturing since the number of these granules decreased conspicuously during feeding (Davidson, 1976; Hausmann and Patterson, 1982). At the ultrastructural level, the extrusomes in *Echinosphaerium* are similar to the muciferous body in *Ciliophrys*. Because the prey organisms stop their flagellar movement and appear to be paralyzed within a few seconds of becoming attached to the axopodia of *Heterophrys*, it is assumed that the conicysts and the muciferous bodies contain certain kinds of paralyzing substances (Davidson, 1976). In *Echinosphaerium*, on the other hand, prey organisms adhering to an axopodium or ingested into a food vacuole did not appear to be paralyzed but continued their active ciliary beating. Hence, the extrusomes may contain not a paralyzing but an adhesive substance.

The prey organism adhering to an axopodium will be conveyed to the proximal region of the axopodium by one of the following two ways or by a combination of them. The first one is termed axopodial contraction and is characterized by rapid contraction of the axopodium to convey the prey to the cell body surface. It is known that the axopodia of many heliozoan species such as *Echinosphaerium, Actinophrys, Heterophrys,* and *Ciliophrys* are contractile (Davidson, 1969, 1973, 1974; Ockleford and Tucker, 1973; Suzaki et al., 1980). In *Heterophrys*, axopodial contraction can be induced by physical and chemical stimuli, and also by exposure to a microflash (Davidson, 1973). In food capture, single axopodia contract when a prey organism is captured. Upon mechanical stimulation, the axopodia (about 50 μm in length) contract in about 20 msec. This can be inhibited with Ca^{2+}-free seawater but not by treatment with cytochalasin B (Davidson, 1973). It was suggested by Davidson (1974) that neither depolarization of microtubules nor alteration of surface tension produced the contractile force in *Heterophrys*, since cold or chemically induced microtubule depolymerization caused a loss of axopodial rigidity but did not result in contraction. In *Ciliophrys*, all axopodia contract simultaneously within a fraction of a second when the organism receives a mechanical shock. Axopodial contraction in this species does not occur in Ca^{2+}-free seawater (Davidson, 1969). In *Actinophrys* and *Echinosphaerium*, axopodial contraction involved in capture and ingestion of certain small flagellates takes place within a frame of the video recording (0.033 sec) and is not induced by mechanical stimulation (Kitching, 1960; Ockleford and Tucker, 1973; Suzaki et al., 1980; Shigenaka et al., 1982; Ando and Shigenaka, 1989). Such a rapid breakdown of the microtubule structure is a peculiar event to the heliozoans that has not been reported in other organisms, and the mechanism of disorganization of the microtubules is not known. As in *Heterophrys* and *Ciliophrys*, calcium ions are considered to be involved in the process of the rapid axopodial contraction in *Echinosphaerium*. Axopodial retraction is caused by the presence of Ca^{2+} and ionophore A23187 (Schliwa, 1976; Matsuoka and Shigenaka, 1984), and it is inhibited in a Ca^{2+}-free medium or in the presence of Ca^{2+}-channel blockers (Matsuoka and Shigenaka, 1984; Shigenaka et al., 1989). Electron microscopic observations revealed that a bundle of contractile filaments runs parallel to the axonemal microtubules, and that it becomes granulated and shortened after the occurrence of a rapid axopodial contraction (Shigenaka and Kaneda, 1979; Suzaki et al., 1980; Shigenaka et al., 1982; Matsuoka et al., 1985). Furthermore, the force generated by the rapid axopodial contraction of a single axopodium was recently estimated to be in the order of 10^{-9} N, at least 100 times that of the axopodial tension (Suzaki et al., 1992b). It is therefore likely that the axopodial contraction in *Echinosphaerium* is not caused by rapid disassembly of the axonemal microtubules but by shortening of the contractile filaments in a Ca^{2+}-dependent manner (Suzaki et al., 1980, 1992b; Matsuoka et al., 1985).

The second method of food conveyance is termed axopodial flow, by which the prey attached to an axopodium is transported unidirectionally toward the cell surface by means of the streaming of axopodial cytoplasm with the captured food organism (Suzaki et al., 1980). Unlike the first method, this process is observed mainly when the prey organisms are attached to the middle or nearly proximal regions of axopodia. The streaming of axopodial cytoplasm with a captured food organism may be related to the movement of intra-axopodial particles; the axopodial flow that results in conveying the prey to the cell body is possibly the result of unidirectional movement of intra-axopodial organelles.

V. MASSIVE MICROTUBULE SEVERING DURING RAPID AXOPODIAL CONTRACTION

In *Echinosphaerium* the rapid axopodial contraction takes place within a single frame of a video recording (0.033 sec), which indicates that the bundle of axonemal microtubules disorganizes almost instantaneously with a velocity higher than 2×10^{-3} µm/sec (Suzaki et al., 1980; Ando and Shigenaka, 1989). Recently, we have investigated changes in the axonemal microtubules during the process of axopodial contraction, since a rapid breakdown of the microtubular cytoskeleton is inevitably required prior to the start of active contraction. As heliozoan microtubules are known to have a strong positive birefringence (Tilney and Porter, 1965), the dynamic change of the axonemal microtubules was monitored during rapid axopodial contraction by using the microtubule birefringence as a measure to quantify microtubules *in vivo*. An attempt was made with a newly introduced theory of birefringence for concentrated microtubule suspensions (Watanabe et al., 1991) to elucidate the mechanism of massive reorganization of the heliozoan microtubules (Suzaki et al., 1992a).

The axonemal microtubules of *Echinosphaerium* have an intense positive birefringence that is clearly visualized when the organism is observed under a highly sensitive polarizing microscope equipped with rectified lenses (Inoué and Hyde, 1957; Inoué, 1986). In order to know if the axopodial contraction involves rapid disassembly of the cytoskeletal microtubules, dynamic changes of the axonemal birefringence were analyzed during the process of rapid axopodial contraction by video microscopy with an image digitizing system. The number of microtubules in an axopodium was estimated from its birefringence by applying dielectric theories for shell-covered ellipsoids in concentrated suspensions (Watanabe et al., 1991). When rapid axopodial contraction was induced by attaching a resin particle to the tip of an axopodium, the distal portion of the axonemal birefringence completely disappeared within a single frame (0.033 sec) of the video recording. Contracted axopodia usually show a size corresponding to 20–50% of the original length, and quantitative analysis showed that the fractional volume occupied by the microtubules increased at the proximal region of the axopodium concomitantly with the axopodial contraction, and it was maintained for at least 10 sec after the contraction. Electron microscopic observations showed that two bundles of axonemal microtubules crossed each other in the proximal part of a contracted axopodium. These results indicate that, at the moment of axopodial contraction, detachment of the distal part of the axoneme took place from the rest of the axoneme at its middle part, and that the distal fragment(s) of the axoneme moved rapidly into the proximal part of the axopodium without apparent disassembly of the microtubules.

Polarizing microscopy, when complemented by proper modeling, can provide otherwise inaccessible information on microtubule dynamics *in vivo* (Sato et al., 1975; Hiramoto et al., 1981a, b; Suzaki et al., 1992b). It was shown that the birefringence of microtubules can be simulated by using a realistic model that may represent the characteristic structure of the microtubules. We have introduced a new model of microtubule suspensions (3-component model) in a concentrated system for the quantitative analysis of microtubule birefringence *in vivo*. The 3-component model has potential advantages over the conventional 2-component model used by other investigators (Sato et al., 1975; Hiramoto et al., 1981a, b) since the new model fits well to the experimental data without introducing a descriptive parameter ("intrinsic birefringence" by Sato et al., 1975), and the conventional model apparently fails to cover the microtubule suspension at lower volume fractions. Furthermore, the 3-component model has been shown to give a good fit to the dielectric behavior of isolated lens tissues over a wide range of frequencies of 10^2–$5 \cdot 10^8$ Hz (Watanabe et al., 1991), which has a structural similarity to the oriented microtubule suspension in heliozoan axopodia. (The plasma membrane of the rod-like lens cell corresponds to the microtubular wall.)

From the quantitative analysis using polarizing microscopy, it was demonstrated that a rapid translocation of the cytoskeletal microtubules in heliozoan axopodia is coincident with axopodial contraction (Suzaki et al., 1992a). The distal portion of the axonemal microtubules was found to move rapidly into the proximal portion without distinct disassembly. These results, in conjunction with electron microscopical observations, showed that massive reorganization of the microtubules takes place in a contracted axopodium; this process of rapid axopodial contraction in *Echinosphaerium* is schematically shown in Figure 2. The structural alteration of the contractile filaments from their ordinary tubular form to the granular form occurs as a result of the attachment of prey organisms (Shigenaka et al., 1982), which is considered to provide the motive force for the axopodial retraction (Matsuoka et al., 1985, 1986; Shigenaka et al., 1989; Suzaki et al., 1992b). The distal part(s) of the axonemal microtubules then move into the axopodial base as a result of microtubule severing and drag force applied to the microtubules. However, it is not known how the contractile force generated by the contractile filaments is transmitted to the axonemal microtubules and how the microtubular bundle is severed in the middle part of the axoneme.

The process of the reorganization of heliozoan microtubules described here is an interesting analogy to the local severing of microtubules observed in *Chlamydomonas* (Sanders and Salisbury, 1989), in which the flagellar excision process is induced by a rapid contraction of a calcium-sensitive protein, centrin (Salisbury et al., 1984, 1986, 1987; Schulze et al., 1987). It has been reported that centrin or antigenically related proteins are present in the centrosomal regions of all eu-

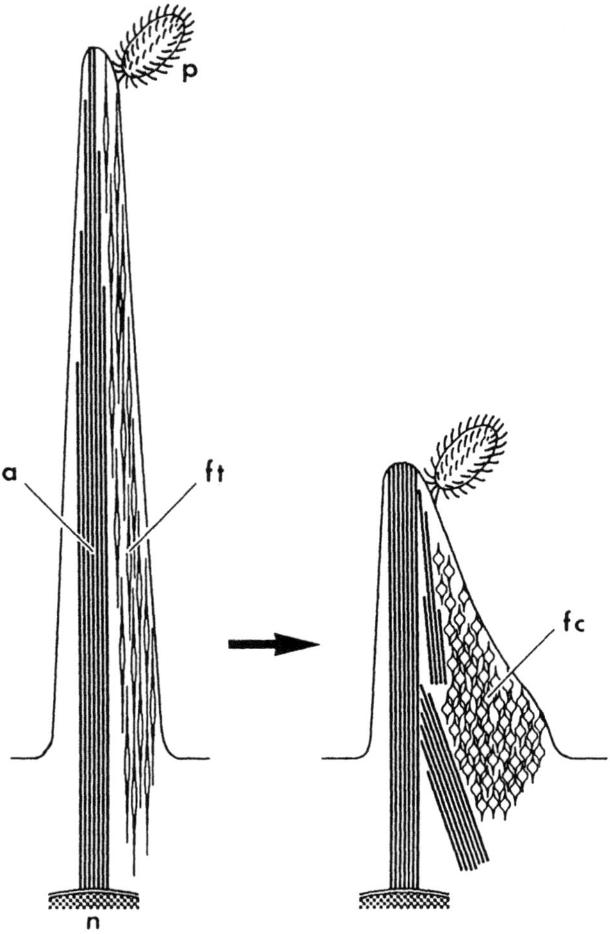

Figure 2. Diagram illustrating a possible mechanism of massive reorganization of axonemal microtubules (a) during rapid axopodial contraction. When a food protozoan (p) or an anion exchange resin particle is attached to the axopodial membrane, a motive force for contraction is generated by the structural alteration of the tubular filaments which may be associated with the undercoat of the axopodial membrane. As a result of microtubule severing, distal parts of the axoneme are then dragged into its proximal portion. n, nucleus. ft, tubular form of the contractile filaments. fc, contracted (granular) form of the contractile filaments.

karyotic cells including mammalian cells (Salisbury et al., 1986; Schulze et al., 1987; Baron and Salisbury, 1988), and it is suggested that microtubule severing is involved in the massive reorganization of cytoplasmic microtubules in mammalian cells undergoing cell movement (Kupfer et al., 1982). The rapid axopodial contraction in heliozoans is also a physiologically important cell movement because it is concerned directly with food capture and ingestion (Suzaki et al., 1980). We have previously shown that the structural alteration of the contractile filaments in the heliozoan axopodia occurs in a Ca^{2+}-dependent manner (Ando and Shigenaka, 1989; Shigenaka et al., 1989). In *Chlamydomonas*, the microtubule severing takes place within a defined region of the flagellum called the transition zone (Sanders and Salisbury, 1989). In *Echinosphaerium*, however, the site of microtubule severing is not yet defined, and the mechanism of massive reorganization of heliozoan microtubules is also an open question.

VI. CONCLUSIONS AND OUTLOOK

The crucial problem awaiting solution is the mechanism of force generation for the different kinds of membrane-associated motilities and the axopodial contraction found in heliozoan cells. There is only limited biochemical information on axopodial membranes and axonemal microtubules with intermicrotubule cross bridges. This may be because isolation procedures for the heliozoan cytoskeleton have not yet been established (Hoffman, 1977; Suzaki et al., 1980; Hoffman et al., 1983). Furthermore, the protein that constitutes the contractile filament has not yet been identified. Detachment of heliozoan axopodia was first demonstrated by Hoffman (1977) by using 10% ethanol, and quite recently we also showed that axopodia can be efficiently liberated from the cell body of *E. akamae* by treatment with 65% D_2O supplemented with Mg^{2+} and EGTA (Sugiyama et al, 1992). The isolated axopodia retained sensitivity to the calcium ions. The ultrastructure of the isolated axopodia did not deteriorate; the plasma membrane and contractile filaments remained intact, and microtubules were well preserved with their original double-helical arrays undisturbed. All these characteristics show that this preparation is very promising for further study on biochemical aspects of membrane, microtubules, and related structures in axopodia.

REFERENCES

Ando, M. and Shigenaka, Y. (1989). Structure and function of the cytoskeleton in heliozoa. I. Mechanism of rapid axopodial contraction in *Echinosphaerium. Cell Motil. Cytoskel.* 14: 288–301.

Bardele, C. F. (1976). Particle movement in heliozoan axopods associated with lateral displacement of highly ordered membrane domains. *Z. Naturforsch.* 31: 190–194.

Baron, A. T. and Salisbury, J. L. (1988). Identification and localization of a novel, cytoskeletal, centrosome-associated protein in PtK2 cells. *J. Cell Biol.* 107: 2669–2678.

Bloodgood, R. A. (1977). Motility occurring in association with the surface of the *Chlamydomonas* flagellum. *J. Cell Biol.* 75: 983–989.

Bloodgood, R. A. (1978). Unidirectional motility occurring in association with axopodial membrane of *Echinosphaerium nucleofilum. Cell Biol. Int. Rep.* 2: 171–176.

Bloodgood, R. A. (1980). Direct visualization of dynamic membrane events in cilia. *J. Exp. Zool.* 213: 293–295.

Bloodgood, R. A., Leffler, E. M., and Bojczuk, A. T. (1979). Reversible inhibition of *Chlamydomonas* flagellar surface motility. *J. Cell Biol.* 82: 664–674.

Davidson, L. (1969). The ultrastructure and motile behavior of the marine helioflagellate *Ciliophrys* sp. *J. Protozool.* 16: 14 (abstr.).

Davidson, L. (1973). Contractile axopodia of the centrohelidan heliozoan *Heterophrys marina. J. Cell Biol.* 59: 71a (abstr.).

Davidson, L. (1974). Evidence for nonsliding filament-based force generating mechanism in the microtubule-containing contractile axopodia of Actinopods (Protista). *J. Gen. Physiol.* 64: 3a (abstr.).

Davidson, L. (1976). Ultrastructure of the membrane attachment sites of the extrusomes of *Ciliophrys marina* and *Heterophrys marina* (Actinopoda). *Cell Tiss. Res.* 170: 353–367.

Edds, K. T. (1975a). Motility in *Echinosphaerium nucleofilum.* I. An analysis of particle motions in the axopodia and a direct test of the involvement of the axoneme. *J. Cell Biol.* 66: 145–155.

Edds, K. T. (1975b). Motility in *Echinosphaerium nucleofilum.* II. Cytoplasmic contractility and its molecular basis. *J. Cell Biol.* 66: 156–164.

Fitzharris, T. P., Bloodgood, R. A., and McIntosh, J. R. (1972). Particle movement in the axopodia of *Echinosphaerium*: evidence concerning the role of the axoneme. *J. Mechanochem. Cell Motil.* 1: 117–124.

Hausmann, K. and Patterson, D. J. (1982). Pseudopod formation and membrane production during prey capture by a heliozoan (Feeding by Actinophrys, II). *Cell Motil.* 2: 9–24.

Hayden, J. H. and Allen, R. D. (1985). Detection of single microtubules in living cells: particle transport can occur in both directions along the same microtubule. *J. Cell Biol.* 99: 1785–1793.

Hiramoto, Y., Hamaguchi, Y., Shoji, Y., and Shimoda, S. (1981a). Quantitative studies on the polarization optical properties of living cells. I. Microphotometric birefringence detection system. *J. Cell Biol.* 89: 115–120.

Hiramoto, Y., Hamaguchi, Y., Shoji, Y., Schroeder, T. E., Shimoda, S., and Nakamura, S. (1981b). Quantitative studies on the polarization optical properties of living cells. II. The role of microtubules in birefringence of the spindle of the sea urchin egg. *J. Cell Biol.* 89: 121–130.

Hoffman, E. J. (1977). Detachment of axopodia from the heliozoan, *Echinosphaerium nucleofilum. Exp. Cell Res.* 110: 75–78.

Hoffman, E. J., Price, D. A., and Krumpelman, J. A. (1983). Isolation of axopodial axonemes from mass-cultured *Echinosphaerium nucleofilum. Exp. Cell Res.* 110: 75–78.

Hyams, J. S. and Stebbings, H. (1979). Microtubules associated cytoplasmic transport. In: *Microtubules* (Roberts, K. and Hyams, J. S., eds.), pp. 487–530, Academic Press, London.

Inoué, S. (1986). *Video Microscopy.* Plenum Press, New York.

Inoué, S. and Hyde, W. L. (1957). Studies on polarization of light at microscope lens surfaces. II. The simultaneous realization of high resolution and high sensitivity with the polarizing microscope. *J. Biophys. Biochem. Cytol.* 3: 831–838.

Kitching, J. A. (1960). Responses of the heliozoon *Actinophrys sol* to prey, to mechanical stimulation, and to solutions of proteins and certain other substances. *J. Exp. Biol.* 37: 407–416.

Kitching, J. A. (1962). Desquamation in *Actinophrys*: induction and inhibition. *J. Exp. Biol.* 39: 359–372.

Koonce, M. P. and Schliwa, M. (1985). Bidirectional organelle transport can occur in cell processes that contain single microtubules. *J. Cell Biol.* 100: 322–326.

Kupfer, A., Luovard, D., and Singer, S. J. (1982). The polarization of the Golgi apparatus and microtubule organizing center in cultured fibroblasts at the edge of an experimental wound. *Proc. Natl. Acad. Sci. U.S.A.* 79: 2603–2607.

Matsuoka, T., Ishigame, K., Muneoka, Y., and Shigenaka, Y. (1986). External factors which induce the rapid contraction of heliozoan axopodia. *Zool. Sci.* 3: 437–443.

Matsuoka, T. and Shigenaka, Y. (1984). Localization of calcium during axopodial contraction in heliozoan, *Echinosphaerium akamae. Biomed. Res.* 5: 425–432.

Matsuoka, T., Shigenaka, Y., and Naitoh, Y. (1985). A model of contractile tubules showing how they contract in the heliozoan *Echinosphaerium. Cell Struct. Funct.* 10: 63–70.

Ockleford, C. D. and Tucker, J. B. (1973). Growth, breakdown, repair, and rapid contraction of microtubular axopodia in the heliozoan *Actinophrys sol*. *J. Ultrastruct. Res.* 44: 369–387.

Patterson, D. J. (1979). On the organization and classification of the protozoon, *Actinophrys sol* Ehrenberg, 1830. *Microbios* 26: 165–208.

Patterson, D. J. and Hausmann, K. (1981). Feeding by *Actinophrys sol* (Protista, Heliozoa): 1. Light microscopy. *Microbios* 31: 39–55.

Rebhun, L. I. (1972). Polarized intracellular particle transport: saltatory movements and cytoplasmic streaming. *Int. Rev. Cytol.* 32: 93–137.

Roth, L. E., Pihlaja, D. J., and Shigenaka, Y. (1970). Microtubules in the heliozoan axopodium. I. The gradion hypothesis of allosterism in structural proteins. *J. Ultrastruct. Res.* 30: 7–37.

Salisbury, J. L., Baron, A., Surek, B., and Melkonian, M (1984). Striated flagellar roots: isolation and partial characterization of a calcium-modulated contractile organelle. *J. Cell Biol.* 99: 962–970.

Salisbury, J. L., Baron, A., Coling, D. E., Martindale, V. E., and Sanders, M. A. (1986). Calcium-modulated contractile proteins associated with the eucaryotic centrosome. *Cell Motil. Cytoskel.* 6: 193–197.

Salisbury, J. L., Sanders, M. A., and Harpst, L. (1987). Flagellar root contraction and nuclear movement during flagellar regeneration in *Chlamydomonas reinhardtii*. *J. Cell Biol.* 105: 1799–1805.

Sanders, M. A. and Salisbury, J. L. (1989). Centrin-mediated microtubule severing during flagellar excision in *Chlamydomonas reinhardtii*. *J. Cell Biol.* 108: 1751–1760.

Sato, H., Gordon, W. E., and Inoué, S. (1975). Microtubular origin of mitotic spindle form birefringence. *J. Cell Biol.* 67: 501–517.

Schliwa, M. (1976). The role of divalent cations in the regulation of microtubule assembly: in vivo studies on microtubules of the heliozoan axopodium using the ionophore A23187. *J. Cell Biol.* 70: 527–540.

Schliwa, M. (1984). Mechanism of intracellular transport. In: *Cell and Muscle Motility* 5 (Shay, J., ed.), pp. 1–82, Plenum, New York.

Schulze, D., Robenek, H., McFadden, G. I., and Melkonian, M. (1987). Immunolocalization of a Ca^{2+}-modulated contractile protein in the flagellar apparatus of green algae: the nucleus-basal body connector. *Eur. J. Cell Biol.* 45: 51–61.

Shigenaka, Y. and Kaneda, M. (1979). Studies on the cell fusion of heliozoans. IV. An electron microscopical study on the fusion process accompanied by axopodial degradation. *Annot. Zool. Japon.* 52: 28–39.

Shigenaka, Y., Roth, L. E., and Pihlaja, D. J. (1971). Microtubules in the heliozoan axopodium. III. Degradation and reformation after dilute urea treatment. *J. Cell Sci.* 8: 127–151.

Shigenaka, Y., Watanabe, K., and Kaneda, M. (1974). Degradation and stabilization effects of Mg^{2+} ions on microtubule-containing axopodia. *Exp. Cell Res.* 85: 391–398.

Shigenaka, Y., Yano, K., and Imada, M. (1989). Axopodial contraction evoked by electrical stimulation and its ultrastructural analysis in a heliozoan *Echinosphaerium*. *Zool. Sci.* 6: 45–53.

Shigenaka, Y., Yano, K., Yogosawa, R., and Suzaki, T. (1982). Rapid contraction of the microtubule-containing axopodia in a large heliozoan *Echinosphaerium*. In: *Biological Functions of Microtubules and Related Structures* (Sakai, H., Mohri, H. and Borisy, G. G., eds.), pp. 105–114, Academic Press, Tokyo.

Sugiyama, M., Ikegawa, S., Masuyama, E., Suzaki, T., and Shigenaka, Y. (1992). Isolation and properties of the axopodial cytoskeleton of a heliozoan *Echinosphaerium akamae*. *Eur. J. Protistol.* 28: 214–219.

Suzaki, T., Ando, M., Inai, Y., and Shigenaka, Y. (1992a). Structure and function of the cytoskeleton in heliozoa. 3. Massive microtubule severing during rapid axopodial contraction in *Echinosphaerium*. *Eur. J. Protistol.* (in press).

Suzaki, T., Ando, M., Ishigame, K., Shigenaka, Y., and Sugiyama, M. (1992b). Structure and function of the cytoskeleton in Heliozoa 2. direct measurement of the force of rapid axopodial contraction in *Echinosphaerium*. *Eur. J. Protistol.* 28: 380–383.

Suzaki, T. and Shigenaka, Y. (1982). Intra-axopodial particle movement and axopodial surface motility in *Echinosphaerium akamae*. In: *Biological Functions of Microtubules and Related Structures* (Sakai, H., Mohri, H., and Borisy, G. G., eds.), pp. 91–103, Academic Press, Tokyo.

Suzaki, T., Shigenaka, Y., and Takeda, Y. (1978). Studies on the cell division of heliozoans. II. Active role of axopodia on division process and transformation of remnants of cytoplasmic connecting bridge to new axopodia. *Cell Struct. Funct.* 3: 209–218.

Suzaki, T., Shigenaka, Y., Watanabe, S., and Toyohara, A. (1980). Food capture and ingestion in the large heliozoan, *Echinosphaerium nucleofilum. J. Cell Sci.* 42: 61–79.

Tilney, L. G. (1968). Studies on the microtubules in Heliozoa. IV. The effect of colchicine on the formation and maintenance of the axopodia and the redevelopment of pattern in *Actinosphaerium nucleofilum* (Barrett). *J. Cell Sci.* 3: 549–562.

Tilney, L. G., Hiramoto, Y., and Marsland, D. (1966). Studies on the microtubules in Heliozoa. III. A pressure analysis of the role of these structures in the formation and maintenance of the axopodia of *Actinosphaerium nucleofilum* (Barrett). *J. Cell Biol.* 29: 77–95.

Tilney, L. G. and Porter, K. R. (1965). Studies on microtubules in Heliozoa. I. The fine structure of *Actinosphaerium nucleofilum* (Barrett), with particular reference to the axial rod structure. *Protoplasma* 60: 317–344.

Troyer, D. (1975). Possible involvement of the plasma membrane in saltatory particle movement in heliozoan axopods. *Nature* 254: 696–698.

Watanabe, M., Suzaki, T., and Irimajiri, A. (1991). Dielectric behavior of the frog lens in the 100 Hz to 500 Mhz range: simulation with an allocated ellipsoidal-shells model. *Biophys. J.* 59: 139–149.

Watters, C. (1968). Studies on the motility of the Heliozoa. I. The locomotion of *Actinosphaerium eichhorni* and *Actinophrys* sp. *J. Cell Sci.* 3: 231–244.

THE HOST–SYMBIONT INTERFACE IN CILIATE–ALGAE ASSOCIATIONS:
INHIBITION OF MEMBRANE FUSION

Werner Reisser

I. INTRODUCTION

Structure and biochemical features of contact areas between partners are most important for the evolutionary success of all types of organismic associations.

Advances in Cell and Molecular Biology of Membranes, Volume 2B
Membrane Traffic in Protozoa, pages 433–445

ISBN: 1-55938-628-2

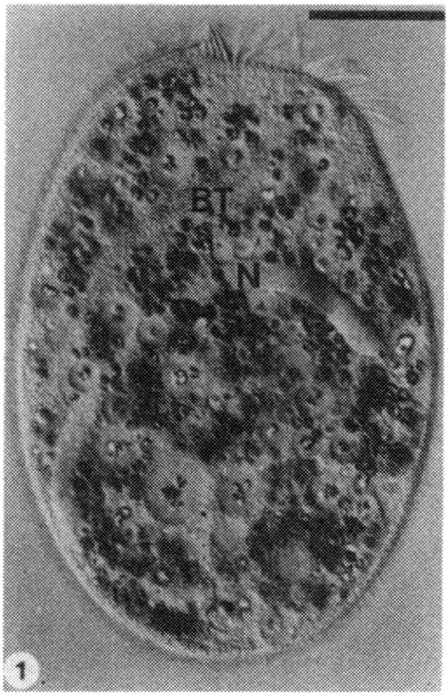

Figure 1. *Climacostomum virens* with symbiotic chlorellae. Algae in individual perialgal vacuoles are distributed throughout the host cytoplasm. BT = buccal tube, n = nucleus, bar: 50 µm. From Reisser and Kurmeier (1984) by permission of Fischer Verlag (Stuttgart).

Those interfaces have multifunctional tasks. They represent not only border lines in terms of topology but are also responsible for an effective exchange of nutrients and various kinds of signals between partners. What is more, in permanent associations, they may also have an additional stabilizing function by protecting partners against mutual attacks by digestive enzymes. This is mainly true for endocytobiotic units, that is, associations of a host cell and smaller partners which are enclosed in its cytoplasm by special membrane surrounded compartments.

Among endocytobiotic units the associations of freshwater ciliates and chlorococcalean algae, the green *Climacostomum* (Figs. 1, 2) or the green *Paramecium*, belong to the best studied systems (for an overview see Reisser, 1986). As to structure and function of the host–symbiont interface most data are available for the green *Paramecium*, that is, the association of *P. bursaria* (Ciliata) with *Chlorella* spp. (Chlorophyceae; Reisser, 1990). Accordingly, in the following I shall rely predominantly on this system by giving a short survey on ultrastructure and biochemistry of its host–symbiont interface and then discussing its origin and protective function.

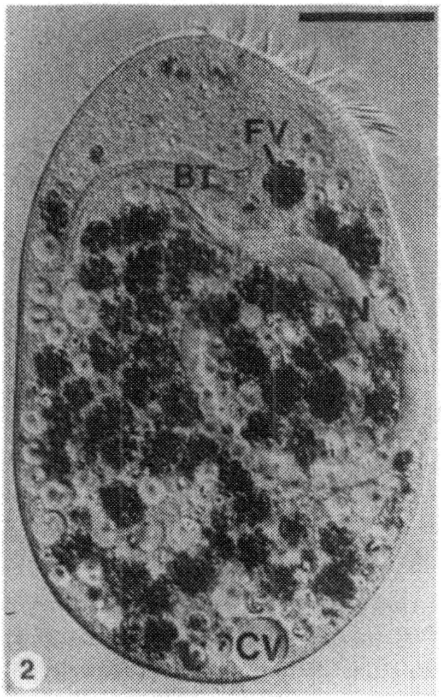

Figure 2. *Climacostomum virens* with nonsymbiotic chlorellae. Algae are crowded together in digestive vacuoles. BT = buccal tube, CV = contractile vacuole, FV = digestive (food) vacuole, bar: 50 µm. From Reisser and Kurmeier (1984) by permission of Fischer Verlag (Stuttgart).

II. THE GREEN *PARAMECIUM*

Green *Paramecium* is a hereditary endocytobiotic association of a holotriche ciliate and up to 300–400 coccoid green algae which is common in freshwater habitats (Reisser, 1986). Under natural conditions the ciliate always contains algae. Those are preferentially situated in the periphery of the host cell (Fig. 3). Each alga is individually enclosed in a special compartment which is called a perialgal vacuole. The alga divides in this vacuole into 2 or 4 daughter cells, which immediately after division are enclosed again by individual membranes. When the ciliate divides, algae are statistically distributed to its daughter cells (Fig. 4).

The symbiotic chlorellae photosynthesize by fixing the respiratory CO_2 of the host and producing oxygen, which in turn is used for host respiration. In addition they release various mono- and disaccharides such as xylose, glucose, and maltose which are metabolized by the ciliate. However, as could be shown for the green

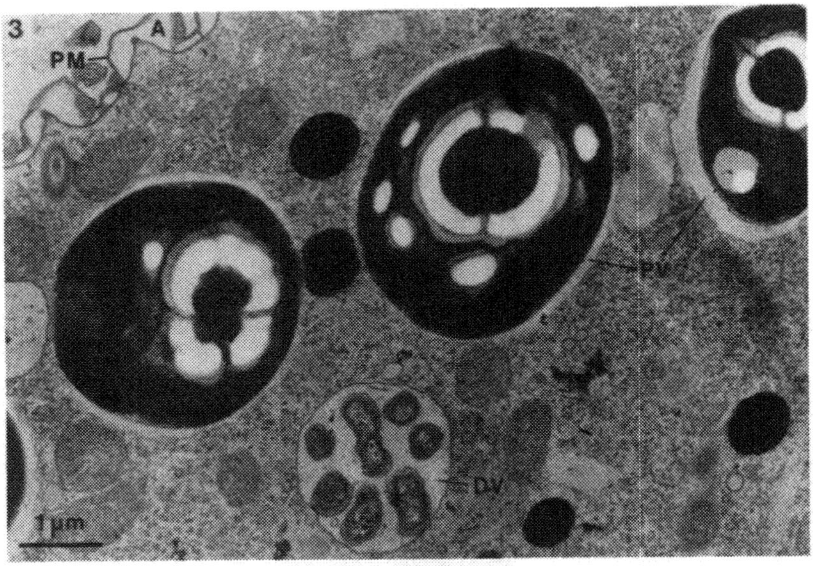

Figure 3. Perialgal (PV) and digestive (DV) vacuoles in *Paramecium bursaria*. Note peripheral location of PVs and nearby DV with bacteria. A = alveolus, PM = plasma membrane. From Reisser (1988) by permission of Springer Verlag (Heidelberg).

Paramecium (Reisser and Herbig, 1987), this internal nutrient supply cannot cover all nutritional demands of the host. Even under optimal light conditions it has to rely on an external food supply via phagocytosis of bacteria and other particulate matter (see also chapters by Allen and Fok and Fok and Allen, as well as by Hausmann and Radek in this volume).

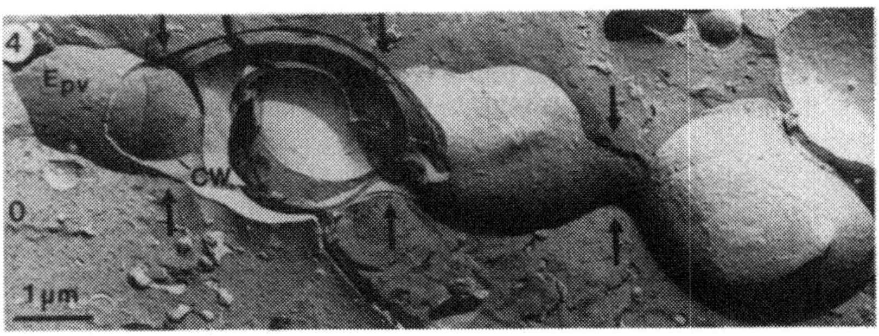

Figure 4. Division of perialgal vacuole in *Paramecium bursaria*. After division of the symbiotic *Chlorella* into four autospores the perialgal vacuole invaginates around them (arrows). Note close contact of the perialgal vacuole membrane (E_{pv} = EF-face) to the cell wall (CW) surface of autospores. From Meier et al. (1984) by permission of Springer Verlag (Heidelberg).

III. FEATURES OF THE HOST–SYMBIONT INTERFACE

Two types of vacuoles exist in green ciliates, digestive and perialgal ones, which differ in many features (Table 1; Figs. 1, 2). Whereas digestive vacuoles participate in cyclosis and accordingly vary in size and type of stored lytic enzymes, perialgal vacuoles show only small variations in size, are often located at the periphery of the host cell, and do not fuse with primary or secondary lysosomes. Thus a perialgal vacuole protects the enclosed alga from being digested by host lytic enzymes. Even under conditions which stress the whole symbiotic unit—such as transfer to an inorganic medium during permanent darkness—algae in perialgal vacuoles cannot be digested by the ciliate.

Concomitant to the division of an alga in a perialgal vacuole new perialgal vacuole membrane material is synthesized and encloses every young cell by an individual new protective membrane (Fig. 4). This compartmentation of one alga in one vacuole guarantees a maximum host–symbiont interface area which is—besides its protective task—the basis of an optimized exchange of signals and nutrients between symbiotic partners. For details on the physiology of green *Paramecium* and the role of the perialgal vacuole membrane in signal transduction during photobehavior of the unit, see Reisser, 1988.

The protective function of the perialgal vacuole membrane is a key prerequisite for the stability of endocytobiotic systems with algae. It has been studied repeatedly as, for example, in green *Paramecium* (Meier et al., 1980, 1984; Karakashian and Rudzinska, 1981), green *Climacostomum* (Reisser and Kurmeier, 1984), and green *Hydra* (Hohmann et al., 1980; O'Brien, 1982). In *Paramecium bursaria*, the ultrastructure of both digestive and perialgal vacuole membranes shows intriguing

Table 1. Features of Perialgal and Digestive Vacuoles in *Paramecium bursaria*

Features	*Vacuole*	
	Digestive	*Perialgal*
Enclosed algae	1–many	1
Vacuole diameter [μm]	2.1–12.9	2.5–4.5
Participation in cyclosis	+	–
Membrane structure:		
Variation during lifetime	+	–
Mean number of IMP[a] [μm^{-2}] on		
EF-face	320–680	650
PF-face	1430–2700	3160
Fusion with primary and secondary lysosomes	+	–
Acid phosphatase activity	+	–

Note: [a]IMP = Intramembranous particles in freeze-fractured membranes.

organizational parallels between the perialgal vacuole membrane and the membrane of a that type of digestive vacuole which is in a stage of cyclosis when fusion or pinching off of other vesicles does not occur (Meier et al., 1984; Meier and Wiessner, 1989; Table 1). Since the perialgal vacuole membrane appears to be closely associated to the algal cell wall surface (Fig. 4), it might also be speculated that there exist steric or charge interactions between algal surface carbohydrates and membrane proteins. These interactions might reduce the fluidity of the perialgal vacuole membrane. In consequence contact and fusion with other vesicles such as primary or secondary lysosomes would become rather unlikely and the enclosed alga be protected against lytic attacks.

IV. FORMATION OF THE HOST–SYMBIONT INTERFACE

A. Cell-to-Cell Recognition

The special properties of the perialgal vacuole membrane are revealed by an analysis of processes which are involved in its formation.

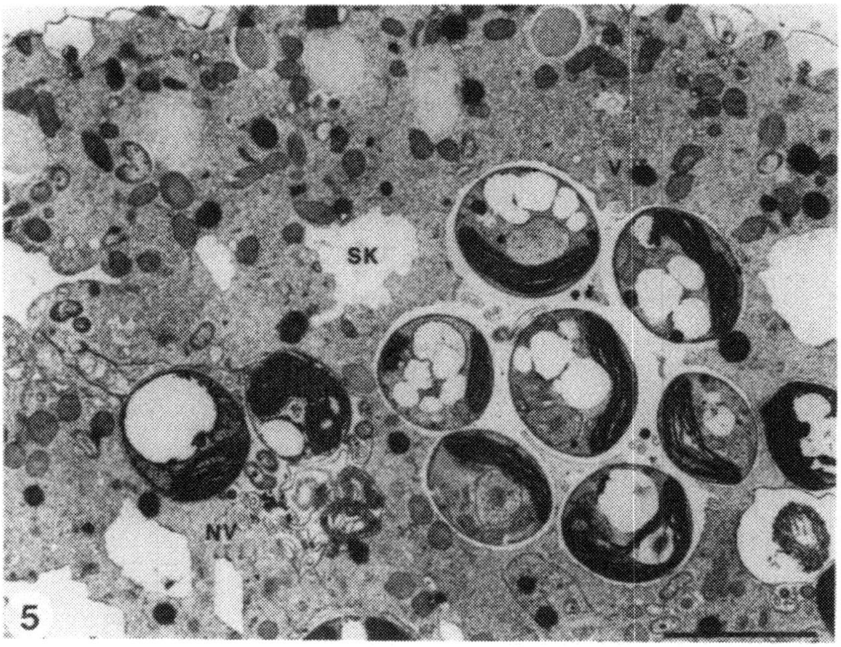

Figure 5. Shrinking vacuole (V) of *Paramecium bursaria* with ingested *Chlorella* spp. Note close attachment of the vacuolar membrane to the algal cell wall surfaces. NV = digestive vacuole, SK = Schewiakoff crystal, bar: 5 µm. From Reisser (1990) by permission of Springer Verlag (Heidelberg).

A common feature of stable endocytobiotic associations is the specificity of partners, for example, in the green *Paramecium* a stable host–symbiont interface between ciliate and algal symbionts is formed with only a limited number of algal taxa. This means that signals which are exchanged between potential partners only in few cases lead to the formation of a protective perialgal vacuole membrane around algae. This phenomenon is designated as "specificity" and communication between cells in this context is termed "recognition" (Reisser, 1988).

The current discussion on recognition phenomena usually neglects the fact that recognition works permanently in stable hereditary endosymbiotic systems, that is, it does not only work when the system is formed from formerly separated partners. In fact—at least for the *Paramecium bursaria* system—the latter case is true only under laboratory conditions, since in the natural milieu neither host nor algae can exist separately (Reisser, 1988). Recognition takes place every time an alga divides within its perialgal vacuole. Daughter cells are immediately and individually enclosed by the type of special membrane material which forms the perialgal vacuoles (Fig. 4). Recognition also works when the ciliate takes up different types of material such as bacteria or algae for food (Figs. 5, 6). When algae have been taken up, only those which send off the proper signals are recognized as suitable for symbiosis formation and are enclosed in perialgal vacuoles (Fig. 1), whereas others which are not able to trigger the proper answer are sequestered into digestive vacuoles (Fig. 2). Mechanisms of cell-to-cell recognition which decide on the formation of the special membrane type of perialgal vacuoles are far from being understood in detail. Available data allow us only to outline some key points.

B. Signaling Features of the *Chlorella*

Obviously the formation of a stable and protective perialgal vacuole membrane depends on special features of both algae and host. Probably algal cell wall surface structures play a decisive role in recognition phenomena. It has been shown that the cell wall surfaces of different *Chlorella* species and strains differ in surface charge and binding capacity to different lectins and antibodies (Table 2; Reisser, 1990). Every time the cell wall surface of an alga triggering the formation of a perialgal vacuole is changed—either by cell wall degrading enzymes or by coupling of lectins or antibodies—its signaling surface pattern is also changed and the alga is not enclosed by a perialgal vacuole type membrane but by a digestive vacuole type one. Experiments suggest that in the *Paramecium bursaria* system suitable algae are characterized by a fingerprint-like pattern of cell surface carbohydrates. This pattern has a signaling function and can change during the algal cell cycle. Only autospores have the signaling quality for triggering the formation of perialgal vacuoles (Table 3), either during infection experiments or when mother cells divide within an established system. Mother cells themselves are not endowed with a signaling surface pattern. During infection experiments they are enclosed in digestive vacuoles (Reisser et al., 1982). However, in an established system they

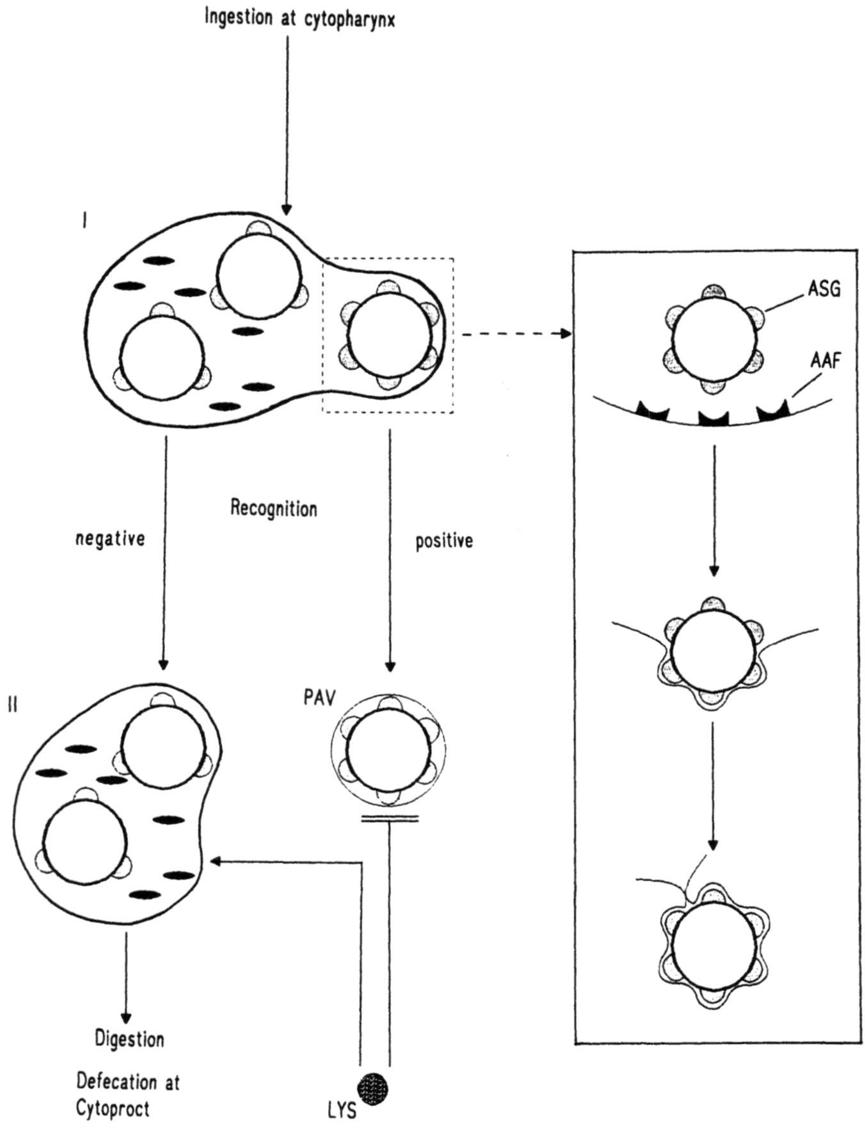

Figure 6. Cell-to-cell recognition of *Paramecium bursaria* and *Chlorella* sp. I. The shrinking digestive vacuole encloses algae (circles) suitable or unsuitable for symbiosis formation and bacteria (disks). II. Digestive vacuole of a later stage of cyclosis, AAF = algal agglutinating factor, ASG = algal surface group, PAV = perialgal vacuole, LYS = lysosome. Modified after Reisser (1990).

Table 2. Binding of Concanavalin A to the Cell Wall Surfaces of Symbiotic and Nonsymbiotic *Chlorella* spp. after Pretreatment with Cellulase and Pectinase and following Infection of Alga-Free *Paramecium bursaria*

Treatment with	*Lectin Bound[a] by Symbiotic Nonsymbiotic Chlorella[b]*		*Infection by Symbiotic Nonsymbiotic Chlorella[b]*	
Control			+[c]	–
Control	100	100	–	–
Cellulase[c]	116	161	–	–
Pectinase[c]	83	137	–	–

Notes: [a]Relative units per standard surface area, +/– 5%.

[b]Symbiotic *Chlorella* isolated from *Paramecium bursaria*, nonsymbiotic *Chlorella* = *Chlorella vulgaris* var. *vulgaris*.

[c]Incubation with enzymes for 60 min, + = establishment of symbiotic association. Modified after Reisser (1990).

are protected once the perialgal vacuole has formed. This observation again indicates a conservative structure of the perialgal vacuole membrane which obviously does not change much during its lifetime.

C. Signaling Features of the *Paramecium*

Although a lot of conclusive data indicate the signaling feature of the algal cell surface comparably little information is available on mechanisms by which those signals are received and processed by the putative host. Usually among different species of a host genus only one species is observed to form a hereditary and stable symbiotic relationship with an algal partner. Thus in the genus *Paramecium* only *P. bursaria* forms a symbiotic unit with *Chlorella* spp., in the genus *Climacostomum* only *C. virens*. Infection experiments with aposymbiotic *P. bursaria* suggest that the host's physiological condition has some influence on the success of infection but reliable quantitative data are lacking (Reisser et al., 1980; Meier and Wiessner, 1988).

Table 3. Infection of Alga-Free *Paramecium bursaria* with Autospores and Mother Cells of Symbiotic *Chlorella* sp. Isolated from Green *Paramecium bursaria* and Treated with Different Antibodies[a]

		Algae Treated with		
			Antiserum to	
Infection with Untreated	*Algae[b]*	*Control Serum*	*Autospores*	*Mother Cells*
Autospores	85	80	20	75
Mother cells	5	5	5	5

Notes; [a]Numbers give percentage of permanently infected ciliates.

[b]Untreated mother cells bind about 100% more concanavalin A per standard surface area than autospores. After Reisser (1990).

Table 4. Activity of the Algal Agglutination Factor Prepared from Green *Paramecium bursaria*

Agglutination of	*Minimum Amount of the Factor Needed for Agglutination*[a,b]
Symbiotic *Chlorella* isolated from	
Ciliates	
Paramecium bursaria[c]	1.00
Euplotes daidaleos	4.31
Lichen	
Trapelia coarctata	
(3.80)[d]	5.31
Nonsymbiotic	
Chlorella vulgaris	
(211–11b)[d]	3.56
Erythrocytes (sheep)	0.16

Notes: [a]Relative units referred to standard cell surface area.
[b]Agglutinating activity in *Paramecium bursaria* : *Paramecium tetraurelia* = 3 : 1.
[c]Enclosure in perialgal vacuole.
[d]Strain number of Algal Collection Göttingen. After Reisser (1990).

A better insight into the role which host features play during recognition events was gained when in the green *Paramecium* a factor was discovered which binds to algal cell wall surfaces and thus is called algal agglutinating factor (AAF; Reisser, 1990). This factor is present not only in the cell of *P. bursaria* but also in aposymbiotic *Paramecium* strains such as *P. tetraurelia*, although in lesser amounts. It binds to a different extent to the cell wall surfaces of both symbiotic and aposymbiotic *Chlorella* spp. and thus agglutinates them. Highest efficiency of the factor is shown with chlorellae isolated from *P. bursaria*, that is, to agglutinate those algae lesser amounts of the factor are needed than for the agglutination of, for example, free-living *Chlorella* spp. (Table 4). Obviously there are more binding sites to AAF on the cell surface of the original symbiotic chlorellae than on other algae.

AAF is very specific. Its agglutinating capacity cannot be blocked by a pretreatment of algae with bovine serum albumin or various commercially available proteins and lipids. Data suggest that AAF has proteinous characteristics and a molecular weight of about 30 kDa. It is only loosely bound to the ciliate membrane.

Since AAF also agglutinates erythrocytes, its binding to the algal surface may be a lectin-type reaction. Accordingly agglutination of algae by AAF can be inhibited by a pretreatment of algae with galactose and lactose, but not by proteolytic enzymes or glutaraldehyde.

D. The Zipper Mechanism

Data on the green *Paramecium* give for the first time conclusive indications of a lectin-type cell-to-cell recognition mechanism in an endocytobiotic association of eukaryotic partners. This mechanism has been called "zipper-mechanism" (Reisser, 1990) because for a positive recognition it postulates a step-by-step interaction of ciliate membrane-bound lectins with carbohydrates bound to the algal cell wall surface (Fig. 6). Only those algae are enclosed which have a sufficient number of suitable carbohydrate groups at their surface to start the zipper mechanism. However, when not enough contacts between algal surface groups and host membrane components can be established, the alga is not enclosed in an individual vacuole but remains in a digestive vacuole and recognition is negative by definition. This always happens when the amount of recognition groups (1) on the algal cell wall surface and/or (2) in the ciliate membrane is below a certain threshold. This model explains why *Paramecium* species other than *P. bursaria* do not form symbiotic associations with algae: Their amount of AAF is too small (Table 4). Accordingly, *P. bursaria* encloses only those algae in perialgal vacuoles which are endowed with a sufficient number of surface-bound carbohydrates reacting with AAF.

The lectin-based model of the formation of a special protective membrane type in the green *Paramecium* is supported by many further observations.

1. Algae, bacteria, and other particulate matter which is phagocytosed by the ciliate is enclosed in digestive vacuoles. These vacuoles start cyclosis by fusion with acidosomes. For details of cyclosis events in *Paramecium* sp. see Allen, 1984 and chapters by Allen and Fok and by Fok and Allen in this book. After this they shrink before fusing with lysosomes. Since it could be shown that perialgal vacuoles occur shortly after the acidification of the vacuole by acidosomes, but before lysosomes fuse and shed digestive enzymes into the vacuole (Meier and Wiessner, 1989), it is conceivable that this shrinking of the vacuole renders possible maximum surface contact between the vacuolar membrane and enclosed cells and thus recognition events. The same principle holds when algae in perialgal vacuoles divide and daughter cells are immediately surrounded by new perialgal membrane material.
2. Membranes of perialgal and digestive vacuoles in *Paramecium bursaria* differ significantly in size and distribution of intramembranous particles and stability of structure (Table 1; Meier et al., 1984). The membrane of a digestive vacuole changes considerably during cyclosis due to multiple fusion and pinching off events by other membrane-bounded compartments of the cell such as acidosomes and lysosomes. Comparative changes obviously do not occur in perialgal vacuole membranes. A possible explanation for this difference may be the close attachment of the perialgal vacuole membrane to the algal cell wall surface. Because of the lectin-type binding mechanism the fluidity of the perialgal vacuole membrane may be decreased to such an extent that any fusion or pinching off by

vesicles is not possible. Accordingly, fusion of perialgal vacuoles with primary or secondary lysosomes or any digestive activity has never been observed in perialgal vacuoles of *Paramecium bursaria* or in *Climacostomum virens* (Karakashian and Rudzinska, 1981; Reisser and Kurmeier, 1984; Reisser, 1986). Algae once enclosed in them are completely protected and thus are able to establish a stable hereditary symbiotic association.

V. CONCLUSIONS

The protective function of the perialgal vacuole membrane is a prerequisite for the stability and the evolutionary success of the symbiotic system. This holds not only for ciliate–algae associations but can be observed also in other endocytobiotic systems with algae (Reisser, 1990). Obviously this principle of compartmentation is basic in the evolution of not only symbiotic but also of parasitic associations of various kinds of organisms. In many parasitic systems recognition of a suitable partner triggers its enclosure into a special vacuole which protects it—at least for some time—against host digestive enzymes. This has been reported, for example, for macrophage-*Mycobacterium* systems (Armstrong and D'Arcy Hart, 1971; D'Arcy Hart et al., 1972) and also for macrophage–*Toxoplasma* systems (Jones and Hirsch, 1972).

The perialgal vacuole also plays a decisive role in metabolic interactions of partners and is involved in signal transduction during photobehavior of the green ciliate. Thus the green *Paramecium* is an excellent model system to study the role of membranes in cell-to-cell interaction and recognition.

REFERENCES

Allen, R. D. (1984). *Paramecium* phagosome membrane: from oral region to cytoproct and back again. *J. Protozool.* 31: 1–6.

Armstrong, J. A. and D'Arcy Hart, P. (1971). Response of cultured macrophages to *Mycobacterium tuberculosis*, with observations on fusion of lysosomes with phagosomes. *J. Exp. Med.* 134: 713–740.

D'Arcy Hart, P., Armstrong, J. A., Brown, C. A., and Draper, P. (1972). Ultrastructural study of the behavior of macrophages toward parasitic mycobacteria. *Infect. Immun.* 5: 803–807.

Hohmann, T. C., McNeil, P. L., and Muscatine, L. (1982). Phagosome-lysosome fusion inhibited by algal symbionts of green *Hydra viridis. J. Cell Biol.* 94: 56–63.

Jones, T. C. and Hirsch, J. G. (1972). The interaction between *Toxoplasma gondii* and mammalian cells. II. The absence of lysosomal fusion with phagocytotic vacuoles containing living parasites. *J. Exp. Med.* 136: 1173–1194.

Karakashian, S. J. and Rudzinska, M. A. (1981). Inhibition of lysosomal fusion with symbiont-containing vacuoles in *Paramecium bursaria. Exp. Cell Res.* 131: 387–393.

Meier, R., Lefort-Tran, M., Pouphile, M., Reisser, W., and Wiessner, W. (1984). Comparative freeze-fracture study of perialgal and digestive vacuoles in *Paramecium bursaria. J. Cell Sci.* 71: 121–140.

Meier, R., Reisser, W., Wiessner, W., and Lefort-Tran, M. (1980). Freeze-fracture evidence of differences between membranes of perialgal and digestive vacuoles in *Paramecium bursaria*. *Z. Naturforsch.* 35c: 1107–1110.

Meier, R. and Wiessner, W. (1988). Infection of alga-free *Paramecium bursaria* with symbiotic *Chlorella* sp. isolated from green paramecia. I. Effect of the incubation period. *Eur. J. Protistol.* 24: 69–74.

Meier, R. and Wiessner, W. (1989). Infection of alga-free *Paramecium bursaria* with symbiotic *Chlorella* sp. isolated from green paramecia. II. A timed study. *J. Cell Sci.* 93: 571–579

O'Brien, T. (1982). Inhibition of vacuolar membrane fusion by intracellular symbiotic algae in *Hydra viridis*. *J. Exp. Zool.* 223: 211–218.

Reisser, W. (1986). Endosymbiotic associations of freshwater protozoa and algae. In: *Progress in Protistology* (Corliss, J. O. and Patterson, D. J., eds.), Vol. 1, pp. 195–214, Biopress, Bristol.

Reisser, W. (1988). Signals in the *Paramecium bursaria–Chlorella* sp. association. *NATO ASI Ser.* H17: 271–282.

Reisser, W. (1990). Participation of algal cell wall surface structures in the formation of the host-symbiont-interface of endocytobiotic systems. In: *Experimental Phycology* (Wiessner, W., Robinson, D. G., and Starr, R. C., eds.), Vol. 1, pp. 55–68, Springer Verlag, Berlin, Heidelberg, New York.

Reisser, W. and Herbig, E. (1987). Studies on the ecophysiology of endocytobiotic associations of ciliates and algae. I. Carbohydrate budgets of green and alga-free *Paramecium bursaria* under laboratory and natural growth conditions. *Endocyt. Cell Res.* 4: 305–316.

Reisser, W. and Kurmeier, B. (1984). The endosymbiotic unit of *Climacostomum virens* and *Chlorella* sp.: symbiotic features of the association and host–symbiont regulatory mechanisms. *Protistologica* 20: 265–270.

Reisser, W., Meier, R., and Wiessner, W. (1980). Cytological studies on the endsymbiotic unit of *Paramecium bursaria* Ehrbg. and *Chlorella* spec. I. The infection of alga-free *Paramecium bursaria* with symbiotic chlorellae isolated from green paramecia as influenced by the nutritional conditions of the ciliate. *Arch. Protistenk.* 123: 326–333.

Reisser, W., Radunz, A., and Wiessner, W. (1984). The participation of algal surface structures in the cell recognition process during infection of aposymbiotic *Paramecium bursaria* with symbiotic chlorellae. *Cytobios* 33: 39–50.

LIPID COMPOSITION OF MEMBRANES INVOLVED IN MEMBRANE TRAFFIC IN *TETRAHYMENA*

Shigenobu Umeki and Yoshinori Nozawa

Advances in Cell and Molecular Biology of Membranes, Volume 2B
Membrane Traffic in Protozoa, pages 447–465

ISBN: 1-55938-628-2

I. INTRODUCTION

Tetrahymena pyriformis, a free-living ciliated protozoan, has proved to be a suitable model system for various studies in cell biology. Especially because of the ease of isolating membrane-bounded organelles and feasible manipulation of membrane lipid composition, this lower eukaryotic cell makes a great contribution toward the understanding of membrane biogenesis. In this chapter we will describe the composition, metabolism, and dynamic intracellular movement of membrane lipids in *T. pyriformis*.

II. LIPID COMPOSITION OF SUBCELLULAR MEMBRANE COMPONENTS

A. Isolation of Membrane Components

The *Tetrahymena* cell has most of the typical subcellular organelles found in higher eukaryotic cells and also contains some other specific organelles, such as cilia, an oral apparatus, and mucocysts. To examine the membrane lipid dynamics between different membrane types, including cilia, pellicles, mitochondria, and microsomes, as well as two soluble fractions (ciliary supernatant and postmicrosomal supernatant), it is important to recover as many organelles in purified form as possible. The overall procedure of isolation is presented in Figure 1 (Nozawa and Thompson, 1971a; Nozawa, 1975). The harvested *T. pyriformis* cells are suspended in cold 0.2 M phosphate buffer (pH 7.2) containing 3 mM EDTA and 0.1 M NaCl and then centrifuged at 100 *g* for 5 min. The chilled cells are resuspended at a concentration of 6–8 × 10^6 cells/ml in 12–15 ml of cold phosphate buffer and gently homogenized by hand in a loose-fitting glass homogenizer until most of the cilia have been detached from the cells. After centrifugation of the homogenate at 1,020 *g* for 5 min, the resulting supernatant is recentrifuged at 14,600 *g* for 5 min, producing a pellet of cilia and ciliary supernatant. The deciliated cells are resuspended in 4–5 ml of the buffer and vigorously homogenized by hand in a tightly fitting glass homogenizer. Then the homogenate is fractionated on a discontinuous buffered sucrose gradient (10 ml of 0.34 M, 15 ml of 1.0 M, and 15 ml of 1.72 M sucrose) by centrifugation at 4,080 *g* for 5 min. Three major zones are separated: zone A, a top band down through the 0.34 M sucrose layer; zone B, a discrete band of pellicles at the interface between 1.0 and 1.72 M; and zone C, a small pellet of unbroken cells at the bottom. The pellicle (zone B) and particulate (zone A) fractions are carefully collected with a syringe. The pellicles containing the plasma membrane and outer and inner alveolar membranes are obtained by centrifugation at 4,080 *g* for 5 min.

The particulate fraction (zone A), which is composed of mitochondria and microsomes, is diluted by the buffer and spun down at 19,600 *g* for 20 min to sediment mitochondria for extraction. The resulting supernatant is further centri-

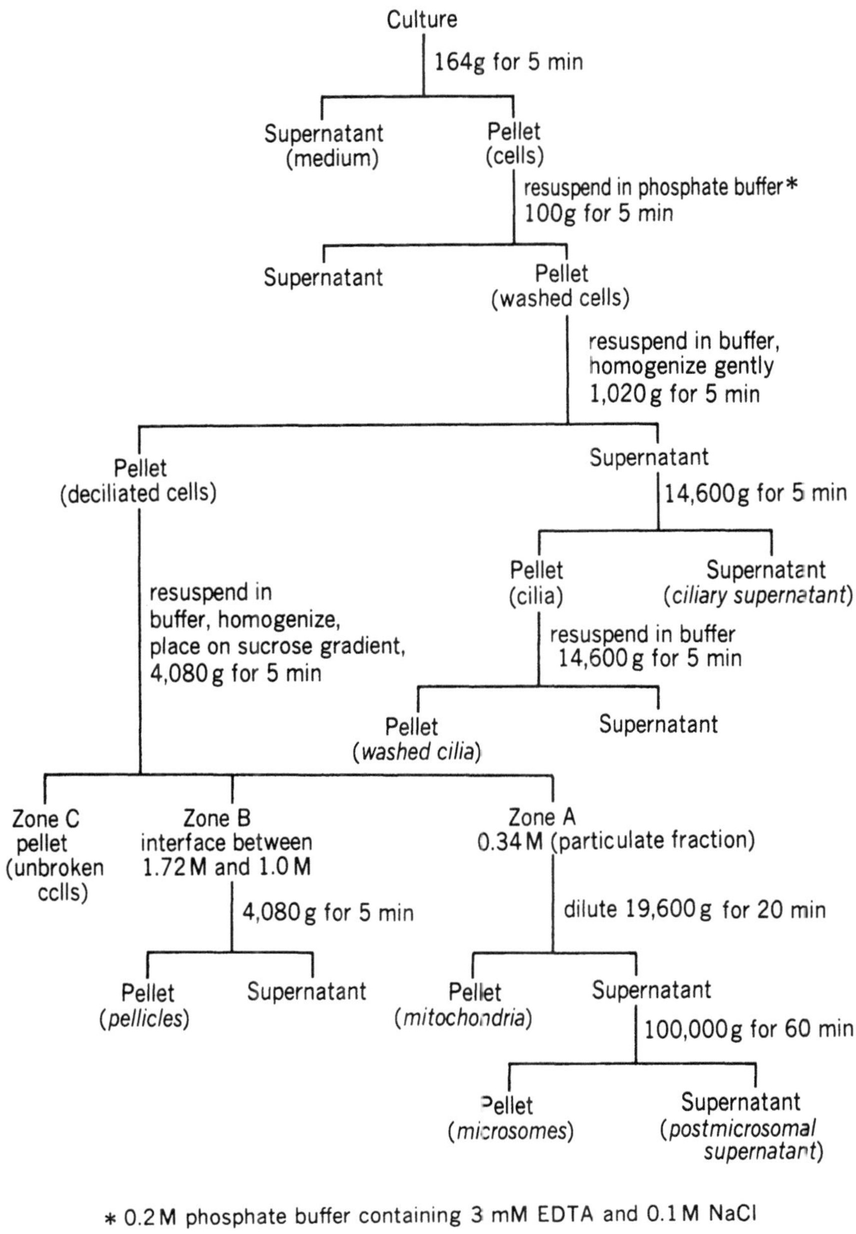

Figure 1. A procedure for the systematic isolation of various organelles from *T. pyriformis*. (From Nozawa and Thompson, 1971a).

fuged at 100,000 *g* for 60 min, yielding a pellet comprised of microsomes and a postmicrosomal supernatant fraction.

B. Phospholipid Composition of Various Membrane Fractions

In addition to containing major membrane phospholipids (phosphatidylethanolamine, PE; phosphatidylcholine, PC), *Tetrahymena* also contains some characteristic lipids, including phosphonolipids (Nozawa and Thompson, 1971a; Hori and Nozawa, 1982), the specific sterol analog tetrahymanol (Mallory et al., 1963) and taurolipids (Kaya, 1992). Representative examples of specific lipids are shown in Figure 2. Two types of diacyl glycerophosphonolipid (GPnL) (Fig. 2A) and ether-GPnL (Fig. 2B) have been found to exist, in which the phosphorylethanolamine is replaced by 2-aminoethylphonate (ciliatine, AEP). The occurrence of sphingophosphonolipids in nature is well documented (Hori and Nozawa, 1982); two major lipids are ceramide 2-aminoethylphosphonate (CAEP) and ceramide *N*-methyl-2-aminoethylphosphonate (CMAEP) (Fig. 2C) (Thompson, 1967).

Several lines of evidence indicate that the different subcellular organelles of a living cell have their own characteristic lipid composition in order to maintain individual specific functions. In this respect, *Tetrahymena* is not different from most other eukaryotic cells. Details of the phospholipid composition patterns in various subcellular components of *T. pyriformis* (strain WH-14) are presented in Table 1. Although there are sizable quantitative differences from one organelle to

Table 1. Phospholipid Composition of Various Membrane Fractions from *T. pyriformis* WH-14 cells

Membrane Fractions	*Phosphonate (% of Total Lipid Phosphorus)*	*Mol % of Total Phospholipids*						*Tetrahymanol (Mol/Mol Lipid Phosphorus)*	*Glyceryl Ether (Mol/100 Mol Lipid Phosphorus)*
		GPnL	*PE*	*PC*	*Lyso-PC*	*Lyso-GPnL and Lyso-PE*	*CL*		
Whole cells	29	23	37	33	2	0	5	0.057	29.7
Cilia	67	47	11	28	1	9	1	0.30	52.6
Ciliary supernatant	44	35	16	19	8	13	1	0.16	23.1
Pellicles	42	30	34	25	5	3	2	0.084	32.8
Mitochondria	26	18	35	35	2	0	10	0.048	24.7
Nuclear envelopes	—	23	26	31	6	6	3	0.036	—
Microsomes	33	23	34	35	1	3	1	0.041	18.3
Postmicrosomal supernatant	26	22	30	34	5	4	2	0.016	27.4

Notes: GPnL, glycerophosphonolipids; PE, phosphatidylethanolamine; PC, phosphatidylcholine; CL, cardiolipin (from Nozawa and Thompson, 1971a)

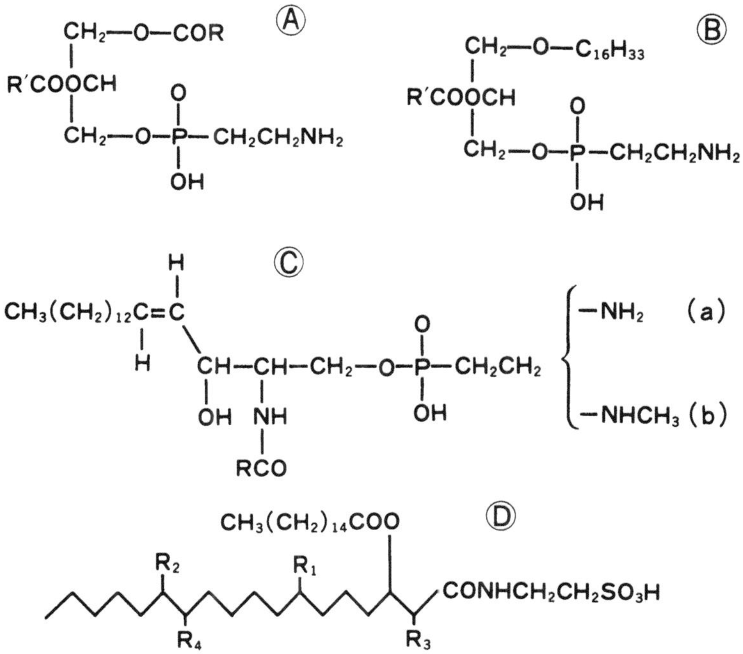

Taurolipid	R_1	R_2	R_3	R_4
Taurolipid A	OH	OH	H	H
γ-Acyltaurolipid A	$CH_3(CH_2)_{14}COO$	OH	H	H
Taurolipid B	OH	OH	OH	H
Taurolipid C	OH	OH	OH	OH

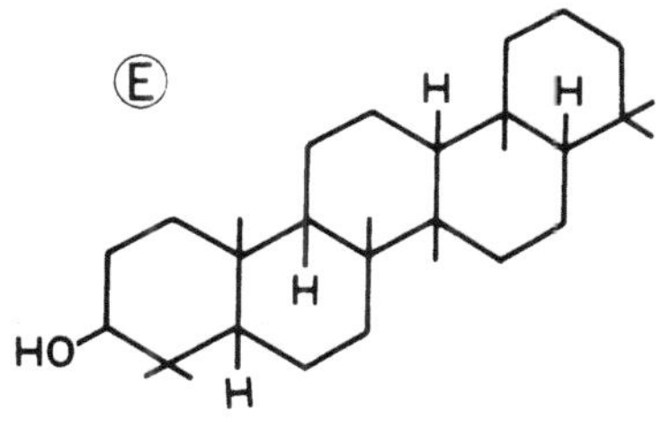

Figure 2. Chemical structures of glycerophosphonolipids (GPnL) (A and B), sphingophosphonolipids (C), taurolipids (D), and tetrahymanol (E).

another, all parts of the cell contained the major lipid components, PE and PC. *Tetrahymena* is unique in having a prominent level of phosphonolipids. These account for nearly 50 mol% of the total phospholipids in ciliary membrane lipids, more than twice the level found in whole cells (Nozawa and Thompson, 1971a). These findings are supported by several other studies of various organelles (Smith et al., 1970; Jonah and Erwin, 1971). The next highest proportions of GPnLs were found in the ciliary supernatant and pellicles, respectively, while the internal membranes, including mitochondria, microsomes, and nuclear envelopes, contain lower amounts. It is of great interest that the GPnLs specifically associate with a carbon–phosphorus bond and glyceryl ethers are localized in the surface membranes. The GPnLs have one more specificity. The acyl chain compositions of the GPnLs are clearly distinguished from those of two other major phospholipids, PE and PC, by their high content of γ-linolenic acid ($C18:3^{\Delta 6,9,12}$) and an unusual fatty acid, cilienic acid ($C18:2^{\Delta 6,11}$). The glyceryl ethers are located at the C-1 of GPnL whose C-2 is largely occupied by γ-linolenic acid.

Like GPnLs, taurolipids also are unusual lipids characteristic of *Tetrahymena*. As shown in Fig. 2D, these lipids consist of taurine, nonhydroxy fatty acids, and tri-, tetra-, or pentahydroxystearic acid and are named taurolipid A [2-(3-acyloxy-7,13-dihydroxyoctadecanoylamino)ethanesulfonic acid], 7-acyltaurolipid A [2-(3,7-diacyloxy-13-trihydroxyoctadecanoylamino)ethanesulfonic acid], taurolipid B [2-(3-acyloxy-2,7,13-hydroxyoctadecanoylamino)ethanesulfonic acid], and taurolipid C [2-(3-acyloxy-2,7,12,13-tetrahydroxyoctadecanoylamino)ethanesulfonic acid] (Kaya, 1992). In the taurolipids A, B, and C, the nonhydroxy fatty acid moiety is esterified with the hydroxy group at C-1 of tri-, tetra-, and pentahydroxystearic acid. Taurolipids B and C differ from taurolipid A by one extra hydroxy group, which is found at C-1 in the taurolipid B and at C-2 and C-12 in taurolipid C. In 7-acyltaurolipid A, in addition, the nonhydroxy fatty acids were esterified with the hydroxy group at C-3 and C-7 of trihydroxystearic acid. More than 60% of the total taurolipids of the cell are observed in the mitochondria-lysosome fractions, followed by the nuclei (15%) and microsomes (9%) (Kaya, 1992).

The *Tetrahymena* cell contains no typical sterols but does possess equivalent amounts of a sterol-like neutral lipid, tetrahymanol (Fig. 2E). The tetrahymanol principally existing in the growing cell is a saturated pentacyclic triterpenoid alcohol and carries out the function fulfilled by cholesterol in mammalian cells. The lipids extracted from cilia, ciliary supernatant, and pellicles, all of which are associated with the surface structures of the cell, contain tetrahymanol in amounts exceeding those measured for intracellular membranes, such as mitochondria, microsomes, and nuclear envelopes, as shown in Table 1. The molar ratio of tetrahymanol to phospholipids in cilia was more than 7-fold higher than that in microsomes. The finding that the subcellular distributional pattern of tetrahymanol in *Tetrahymena* is to some extent similar to that of cholesterol in mammalian cells, indicates that tetrahymanol has functions similar to cholesterol in mammalian cells.

C. Modifications of Membrane Lipid Composition in Response to Environmental Changes (Temperature, Starvation/Refeeding)

Numerous studies have shown that *Tetrahymena*, like more complex organisms and mammalian cells, can accomplish homeoadaptive manipulations of membrane lipid composition, which is considered necessary for maintaining normal organelle functions (homeostasis) under various environmental conditions, such as temperature and nutrients.

Temperature

Tetrahymena can grow at varying temperatures from 15° to 39.5°C and quickly acclimates itself when transferred from a high to a low temperature and vice versa. When *T. pyriformis* (strain NT-1) is isothermally grown at different temperatures, not only the molecular species of the fatty acyl chains but also the composition of the phospholipid head group are altered by thermoacclimation to a low temperature (Table 2). There is an increase in 2-aminoethylphosphonolipid (AEPL) concomitant with a decreased proportion of PE as the growth temperature decreases, whereas the level of PC remains fairly consistent. Cardiolipin, on the other hand, is exceptional and shows no change in the proportions during thermal acclimation (Fukushima et al., 1976). The physical property (fluidity) of mitochondrial membranes richest in cardiolipin is unaffected by an increase in fatty acid desaturation after a growth temperature down-shift (Yamauchi et al., 1981), indicating that cardiolipin would exert a rigidifying effect on the mitochondrial membranes of acclimated cells.

The prominent effects of a lower growth temperature on fatty acid compositions in the subcellular fractions have been shown to be an increase in cilienic, linoleic

Table 2. Lipid Composition of Various Membranes from *T. pyriformis* NT-1 Cells Grown at Different Temperatures

	Cilia			*Pellicles*			*Microsomes*		
	15°C	*24°C*	*39.5°C*	*15°C*	*24°C*	*39.5°C*	*15°C*	*24°C*	*39.5°C*
Phospholipid composition									
2-Aminoethylphosphonolipids	40.0	37.0	36.4	30.8	32.9	19.1	22.9	20.8	14.7
Phosphatidylethanolamines	20.8	21.4	20.0	33.2	36.2	49.2	34.1	35.5	43.9
Phosphatidylcholines	15.8	18.1	30.0	18.9	21.4	23.5	30.0	26.7	32.4
Tetrahymenol/lipid phosphorus (molar ratio)	0.137	0.260	0.178	0.064	0.090	0.098	0.053	0.055	0.078
Fatty acid composition of phospholipids									
C14:0	6.0	3.9	6.7	8.4	9.6	8.9	7.2	7.4	7.3
C16:0	16.8	19.4	16.8	12.7	15.8	16.4	8.9	10.4	13.2
$C16:1^{\Delta 9}$	14.4	7.7	8.3	9.9	7.4	7.1	8.8	9.8	8.7
C18:0	2.5	3.8	6.1	0.7	2.0	3.0	0.5	1.4	2.2
$C18:1^{\Delta 9}$	6.9	8.0	11.7	7.4	8.5	12.6	8.1	7.1	13.9
$C18:2^{\Delta 6,11}$	11.7	7.3	3.9	8.0	5.8	3.2	6.6	4.9	3.3
$C18:2^{\Delta 9,12}$	9.0	9.7	8.1	16.2	13.8	10.5	19.5	17.7	13.0
$C18:3^{\Delta 6,9,12}$	27.2	28.2	20.7	28.0	27.4	18.5	31.8	31.2	21.4

Source: Fukushima et al., 1976.

(C18:2$^{\Delta9,12}$), and γ-linolenic acids, accompanied by a decrease in palmitic acid (C16:0). Similar fluctuations have been found to occur in the compositions of phospholipids as well as fatty acyl chains during down-shift of the growth temperature from high to low degree (Nozawa and Kasai, 1978). Although there were few significant changes in phospholipid class composition of the whole cell within a temperature down-shift of 10 hr duration, in which no cell division was observed, there was a marked decrease in the PE level closely associated with an increased level of AEPL. On the other hand, a quick response of the fatty acid composition to the downward shift was observed. The proportion of palmitoleic acid started to increase immediately after the shift and then reached its maximum peak at 4 hr, being concomitant with a concurrent inverse change in the proportion of palmitic acid. After the shift polyunsaturated fatty acids, such as γ-linolenic and linoleic acids, tended to rise gradually.

Such rearrangement of phospholipid fatty acid compositions during the thermoadaptation seen in the *Tetrahymena* whole cell is conspicuous in the internal membranes (mitochondria and microsomes), with the rise in γ-linolenic acid being well counterbalanced by a reduction in palmitic acid (Maruyama et al., 1982). A slow rearrangement of the surface membrane (pellicle) fatty acid pattern is terminated at nearly the same time, when cold acclimation has been achieved and cell division was regained. This delay in the manifestation of the fatty acid changes in the surface membrane may result from the slow transfer of phospholipids containing fatty acids, which may be rearranged by the microsomal membranes to the cell surface. These findings suggest that the cell possesses a variety of highly selective mechanisms, allowing those organelles which accept phospholipids of microsomal origin to retain a surprising degree of independence in responding to environmental changes.

When *Tetrahymena* was pulse-labeled with ^{14}C-palmitic acid at intervals after a temperature down-shift (39.5° to 15°C), the ^{14}C-palmitoleic acid (C16:1) content immediately rose after the shift and 2 hr later reached the maximum level, being comparable to palmitoyl-CoA desaturase activity (Nozawa and Kasai, 1978). Umeki et al. (1982) demonstrated that, after a temperature shift from 39.5° to 15°C, there was an immediate increase in the activities of palmitoyl-CoA and stearoyl-CoA desaturases in *Tetrahymena* microsomes, resulting from induction of the terminal component of the desaturase system. Changes in overall desaturase activity after the shift are thought to be principally due to induction of the terminal component of the desaturase system (Umeki and Nozawa, 1983; Nozawa and Umeki, 1988).

Starvation/Refeeding

Tetrahymena has been proved to have the ability to survive by decreasing the cell size for about 1 day even in nutrient-deficient medium (Thompson et al., 1972). The characteristic profile in lipid composition during starvation is a drastic increase in the molar ratio of tetrahymanol to phospholipids in cilia and microsomes. The

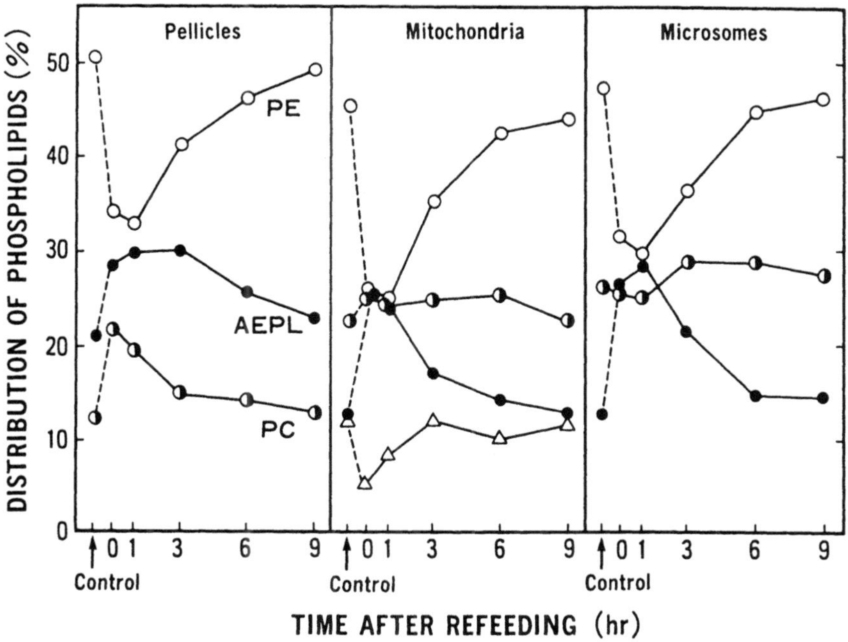

Figure 3. Refeeding-induced modification of phospholipid composition of various membranes of *T. pyriformis* (strain NT-1). The 24-hr starved cells were transferred into proteose-peptone medium enriched with glucose and yeast extract. ●, AEPL; ○, PE; , ◑ PC; △, cardiolipin. (From Nozawa et al., 1980).

proportion of AEPL largely increases, with a corresponding decrease in PE, which results from the hypersusceptibility of diacyl-PE to lipolytic attack by phospholipases as compared with alkyl forms of PC and AEPL (Fig. 3) (Nozawa et al., 1980). These findings suggest that AEPL, together with tetrahymanol, subserves as an essential lipid component for survival of the starved cell. With regard to phospholipid fatty acid composition, starvation produces significant changes, such as increases in oleic (C18:$1^{\Delta 9}$), linoleic and γ-linolenic acids and a concurrent decrease in palmitoleic acid. However, following transfer of starved cells into an enriched growth medium, this altered lipid composition rapidly returns to the initial profile of the control cells prior to starvation. In particular, a linear increase in the formation of palmitoleic acid within 6 hr after refeeding parallels that in microsomal palmitoyl-CoA desaturase activity. All the phospholipid and fatty acid constituents individually play a crucial role in starvation stress against the cell. Nozawa et al. (1980) suggested that PC and AEPL participate in the acclimation to starvation/refeeding by modifying the γ-linolenic acid content and that PE and PC participate by regulation of palmitoleic acid content by palmitoyl-CoA desaturase.

III. MEMBRANE LIPID METABOLISM

A. Regulation of Membrane Phospholipid Synthesis

The major phospholipids found in *Tetrahymena* membranes are PE, PC, and AEPL (Nozawa and Thompson, 1971a). Mitochondrial membranes also contain cardiolipin. This cell is capable of synthesizing PE and PC by the major pathway of de novo phospholipid synthesis closely associated with the acyltransferases, such as glycerophosphate and monoacylglycerophosphate acyltransferases and lysophospholipid acyltransferase (Okuyama et al., 1977). PC is formed from diacylglycerol and CDP-derivatives in similar way to that reported for mammalian cells (Liang and Rosenberg, 1966; Smith and Law, 1970). Furthermore, it has been observed that there are, in *Tetrahymena*, two major enzymes, CDP-choline:1,2-diacylglycerol cholinephosphotransferase and CDP-ethanolamine:1,2-diacylglycerol ethanolaminephosphotransferase, catalyzing the final step of de novo synthesis of PC and PE, respectively (Smith and Law, 1970; Kameyama et al., 1980). PE, once made, undergoes a stepwise methylation to PC, with *S*-adenosylmethionine functioning as the methyl group donor. However, no methylation of AEP has been detected. AEP formation has been shown to result from an intramolecular rearrangement of phosphonoenolpyruvate via decarboxylation and subsequent transamination (Horiguchi, 1972). Liang and Rosenberg (1966) provided evidence suggesting that AEPL is synthesized from diacylglycerol and cytidine monophosphate-AEP (CMP-AEP) in a cell-free preparation of *Tetrahymena*, possibly being catalyzed by CMP-AEP:diacylglycerol AEP-transferase.

It is obvious that the phospholipid metabolism is modified by a variety of environmental factors. As described in the preceding section, fatty acyl-CoA desaturase, which is involved in the formation of unsaturated fatty acids, is a key enzyme for quick thermoadaptation, the level in *Tetrahymena* being temperature controlled. Indeed, palmitoyl-CoA desaturase, which catalyzes the conversion of palmitic to palmitoleic acid, requires palmitoyl-CoA but never uses palmitoyl phospholipid as a substrate. In an experiment using microsomes from cells labeled with [1-^{14}C]stearic acid (C18:0), Nagao et al. (1978) indicated the possibility of a direct desaturation of phospholipid from oleic to linoleic acid and from linoleic to γ-linolenic acid. In the case of *Tetrahymena*, Kameyama et al. (1980) showed the appearance of a microsomal direct phospholipid desaturation system from oleoyl-PC to linoleoyl-PC and emphasized its important role in controlling membrane lipid adaptation to a lower growth temperature.

In an experiment regarding the incorporation of [^{3}H]- or [^{14}C]palmitic acid, [^{14}C]acetate, and [^{32}P]P_i into *Tetrahymena* membrane phospholipids, Kameyama et al. (1984) investigated the involvement of deacylation-reacylation in the phospholipid fatty acyl metabolism during temperature down-shift. Their results suggest a thermoadaptive mechanism by which preexisting fatty acids of membrane phospholipids seem to be deacylated and then, after the modification to adequate unsaturated fatty acids by desaturation and/or elongation, be reacylated into lysophospholipids.

The first step of adaptive modification of the fatty acyl chain profile is considered to be deacylation by phospholipase A_1 at the C-1 of phospholipids. It has been observed that the activity of phospholipase A_1, rather than A_2, is rich in the plasma membrane of *Tetrahymena* (Arai et al., 1985). The increased unsaturated fatty acyl-CoAs following a down-shift are reacylated in phospholipids, leading us to consider the significance of the reacylation during thermoadaptation. Yoshioka et al. (1984) measured acyltransferase activities for the synthesis of PC and PE from 2-acyl-GPC and -GPE with acyl-CoAs in *Tetrahymena* and suggested that the changes in acyl-CoA specificities in the reacylation enzyme activities during a temperature down-shift contribute little to the increase in unsaturated fatty acids in phospholipids, although reacylating enzymes from cells grown isothermally at a lower temperature partially participate in those increases.

In the light of these studies, we conclude that fatty acyl-CoA desaturase plays a crucial role in the thermal acclimation of membrane phospholipid acyl chain composition; the enhanced activity during thermoadaptation supplies amounts of unsaturated fatty acids and subsequently these are inserted through the deacylation-reacylation process.

B. Metabolism of Tetrahymanol and Specific Lysosomal Taurolipids

Tetrahymanol

Tetrahymena does not synthesize sterols but does synthesize tetrahymanol in large amounts, presumably to replace sterols as structural components of membranes. Tetrahymanol is formed from the nonoxidative cyclization of squalene, which is synthesized from mevalonate. When the organism is grown in the presence of cholesterol or ergosterol, synthesis of tetrahymanol is repressed (Ferguson et al., 1971) and these sterols are substituted for tetrahymanol. The cell metabolizes cholesterol to cholesta- 5,7,22-trien-3β-ol but not to ergosterol. These results suggest that cholesterol and ergosterol presumably satisfy any structural requirements normally met by tetrahymanol.

Exposure of *Tetrahymena* to a nutrient-deficient medium produces a gradual decrease in the cell size with a concomitant reduction to about 20% of the normal cell volume and surface area (Thompson et al., 1972). This shrinkage was reflected by an equivalent decrease in phospholipids per cell but, surprisingly, a 3-fold increase in the molar ratio of tetrahymanol to phospholipids was observed. The latter data indicate that the selective accumulation of a particular neutral lipid, tetrahymanol, has some survival value for the cell living in nature. Evidence supporting this suggestion comes from the report of Ramesha and Thompson (1982) referring to an increase in the relative proportion of tetrahymanol during a temperature down-shift.

Taurolipids

Tetrahymena has the ability to synthesize not only phospholipids and tetrahymanol but also specific taurolipids, taurine-containing lipids (Fig. 2D) (Kaya, 1992). Localization of the activity of acid phosphatase as a marker enzyme of lysosomes agrees well with that of the taurolipids, indicating that taurolipids may be specific lysosomal lipids.

Kaya et al. (1985) investigated the dynamics of the radioactivity of [1-^{14}C]stearic acid added to the growth medium of *Tetrahymena*. They found incorporation of radioactivity into both nonhydroxy fatty acid (55%) and trihydroxystearic acid (45%) fragments, suggesting that stearic acid is a possible precursor of trihydroxystearic acid. When *T. thermophila*, with a very small amount of taurolipid A, was incubated with taurolipid A isolated from *T. pyriformis*, the taurolipid A was rapidly deacylated and then the taurolipid B content in the cell became elevated. Using mass and nuclear magnetic resonance spectrometry, Kaya and Sato (1991) defined the total biosynthesis pathway of taurolipids in *Tetrahymena*. First, one molecule of stearic acid was combined with one molecule of taurine to the structure of lipotaurine, the last precursor of the taurolipids, forming the initial product, 2-(octadecanoylamino)ethanesulfonic acid. After subsequent reactions, including the stepwise hydroxylation of the C-7 and C-13 positions and desaturation between the C-2 and C-3 positions, lipotaurine [2-(7,13-dihydroxy-2-transoctadecanoylamino)ethanesulfonic acid] was made. The resulting lipotaurine in turn is converted to C-3 hydroxy (lysotaurolipid A) $\rightarrow$ C-2,3 hydroxy(lysotaurolipid B) $\rightarrow$ C-2,3,12 hydroxy(lysotaurolipid C) derivatives. Taurolipid A (B or C) was formed by acylation of the C-3 position of lysotaurolipid A (B or C). In addition, the C-7 acylation of taurolipid A resulted in 7-acyltaurolipid A.

Like the fatty acid composition of phospholipids, that of taurolipids in *Tetrahymena* also varies in response to changes in the growth temperature. In *T. pyriformis* (strain NT-1) grown at 39°C, the proportion of palmitic acid, which is a major fatty acid of taurolipid A, is 57% of the total fatty acids, while that in 15°C grown cells is 19%, leading us to conclude that this is a specific function of taurolipid A in lysosomes during thermoadaptation (Kaya, 1992).

C. Phospholipases Involving Membrane Lipid Modification

Phospholipid turnover is vital for the maintenance of normal cellular functions. Not only has a rapid labeling of the phospholipids been observed after a pulse with radioactively labeled phosphate, but the incorporated radioactivity was seen to disappear rapidly again from the phospholipid fraction. These observations have led to the concept of phospholipid turnover, in which a constant supply of newly synthesized phospholipids is balanced by the removal of existing molecules. Like mammalian cells, *Tetrahymena* possesses a very potent assortment of degrading enzymes, phospholipases, which become highly activated upon disruption of the cell (Thompson, 1967). Phospholipase activity occurs widely throughout the cell but is most concentrated in the mitochondria-lysosome fractions and in surface membranes (Arai et al., 1985). There are three major phospholipases: Ca^{2+}-dependent alkaline phospholipase A_1 (located mainly in the surface membrane) and Ca^{2+}-independent acidic phospholipase A_1 and C (located predominantly in mitochondria-lysosome fractions). Alkaline phospholipase A_1 can hydrolyze PE but has hardly any effect on PC, which is a good substrate for acidic phospholipase A_1 and C. From the results of studies using specifically labeled PC and PE it was shown that the hydrolysis of phospholipids in *Tetrahymena* lysosomes was initiated either by deacylation at the C-1 position with phospholipase A_1 (major pathway) or by cleavage of phosphodiester with phospholipase C (minor pathway). 2-Acyl lysophospholipid, the first product formed by phospholipase A_1, was attacked by (lyso)phospholipase C to form 2-monoacylglycerol. Exogenous radioactive phospholipids were converted into triacylglycerol in *Tetrahymena*, suggesting that 2-monoacylglycerol, one of the end products formed by the phospholipase system, could be directly converted into triacylglycerol through a process similar to that in mammalian intestinal cells, by which 2-monoacylglycerol can be converted to diacylglycerol and triacylglycerol by direct esterification (monoacylglycerol pathway). On the other hand, free fatty acids derived *in vivo* from the C-1 of phospholipids by phospholipase A_1 are apparently subject to degradation to acetyl-CoA by either α- or β-oxidation.

No detectable amount of phosphatidic acid is made during hydrolysis of labeled PC and PE by *Tetrahymena* homogenates, indicating that phospholipase D is not active in the cell. Furthermore, *Tetrahymena* lysosomes lack phospholipase A_2, lysophospholipase, and monoacylglycerol lipase activities based on the following results: (1) there was no appreciable formation of radioactive fatty acids from the substrates labeled at the C-2 position, and (2) neither radioactive lysophospholipids nor monoacylglycerol was formed when substrates labeled at the C-1 position were employed (Arai et al., 1986).

It is generally known that the phosphonolipids, which possess the carbon–phosphorus bond, are extremely resistant to chemical and enzymatic hydrolysis. At present, there is no evidence to indicate the existence of enzymes degrading the carbon–phosphorus bond itself in organisms, such as *Tetrahymena* and the sea anemone, which contain phosphonolipids in their membranes (Hori and Nozawa, 1982). LaNauze et al. (1970) found a so-called phosphonatase, an enzyme capable of cleaving the carbon–phosphorus bond of phosphonolipids. On the other hand, the degradation of the lipid moiety in phosphonolipids has been demonstrated in several cell lines. *Tetrahymena* can break down GPnL to release the AEP base, but at a slower rate than phospholipid degradation (Hori and Nozawa, 1982). CAEP is hydrolyzed by phospholipase C from *Clostridium perfringens*, yielding AEP and ceramide (Hori et al., 1968).

A significant change in phospholipase activity takes place in response to environmental factors. The phospholipase A_1 activities hydrolyzing PC and PE in *Tetrahymena* grown at 15°C for 7 hr after the growth temperature down-shift from 39° to 15°C were 2-fold higher than those in cells grown isothermally at 39°C.

Tetrahymena predominantly modifies the fatty acyl chain composition of the phospholipid C-1 position during thermal acclimation (see Section II.C). These findings suggest that phospholipase A_1, like fatty acyl-CoA desaturases, plays a pivotal role in thermoadaptation by increased phospholipid deacylation, which is closely associated with a rise in the enzyme content.

IV. MEMBRANE LIPID TRAFFIC

A. Membrane Formation

Membrane biogenesis, including biosynthesis, transfer, and sorting of synthesized molecules, is an essential feature of cellular development and growth. In addition to proteins, lipids are important cellular constituents, occurring in the biomembranes of a variety of cells. The intracellular locations of the major pathways of phospholipid biosynthesis in these cells have been determined by subcellular fractionation. Using this method, the endoplasmic reticulum has been shown to be the key site for biosynthesis and the initial assembly of membrane phospholipids in mammalian cells. This has been supported by several investigations of intracellular phospholipid assembly in eukaryotic cells. In an experiment concerning the dynamics of radiolabeled precursors in *Acanthamoeba palestinensis*, Chlapowski and Band (1971) showed that the following transport of label took place in specific activity of the labeled phospholipids: nuclei and endoplasmic reticulum → Golgi apparatus → collapsed vesicles → plasma membranes. At the same time, Nozawa and Thompson (1971b) reported that incorporation of ^{14}C-palmitic acid into the microsomal phospholipids in *T. pyriformis* (strain WH-14) is rapid (within 1 min), whereas appearance in the surface membrane (cilia) took 6 hr (Fig. 4). The incorporation pattern at 6 hr after incubation with a labeled precursor suggests the involvement of intracellular phospholipid transfer among different membranes.

B. Transfer (Exchange) of Phospholipids between Membranes

In a eukaryotic cell, there are significant changes in lipid composition and lipid asymmetry in the various organelle membranes, indicating that specific inter- and intramembrane sorting and transfer processes that ensure the correct ultimate location of these lipid molecules within the cell must exist. Possible mechanisms that could be involved in the intermembrane (or intramembrane) transfer processes include protein-mediated transfer and vesicular transport (or a flip-flop process). In this section, the intermembrane transferring systems of phospholipids are discussed. In addition to the unidirectional transfer system, an exchange mechanism involving bidirectional transfer of phospholipids between two different membranes seems to occur in mammalian and non-mammalian cells (Wirtz, 1974). In growing *Tetrahymena* and even in non-growing ones in which no lipid biosyn-

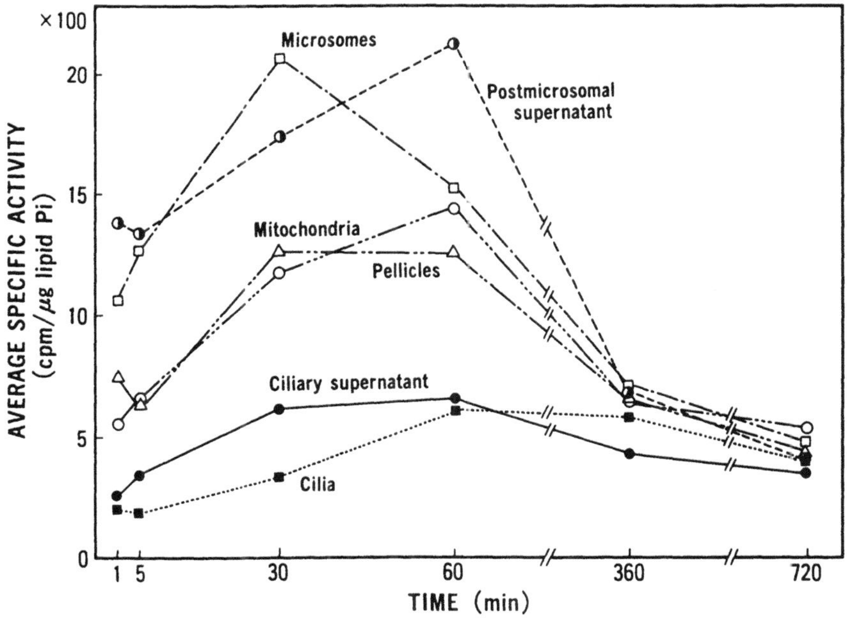

Figure 4. Incorporation of [^{14}C]palmitic acid into phospholipids of various membranes in *T. pyriformis* (strain WH-14). (From Nozawa and Thompson, 1971b).

thesis has occurred, a similar transfer of phospholipid molecules among various membranes has been observed (Nozawa and Thompson, 1979). A considerable delay in the appearance of newly synthesized phospholipids in the surface membranes (cilia), as indicated by previous results, seems to be attributable to their relative inaccessibility.

In an experiment using *T. pyriformis* (strain NT-1), Iida et al. (1978) investigated the traffic profile of spin-labeled PC between microsomes containing spin-labeled PC and other acceptor membranes. As shown in Fig. 5, the transfer rates decreased in the following order: microsomes > pellicles > cilia. Furthermore, they found that the spin-labeled PC transfer to more fluid subcellular membranes (prepared from 15°C grown cells) occurred much more rapidly than that to less fluid subcellular membranes (from 39°C grown cells). These results suggest not only that phospholipid transfer may be enhanced as acceptor membrane fluidity increases, but also that it may be carried out via an exchange process rather than via membrane fusion. No stimulation of lipid transfer was observed when suboptimal pH medium was used, suggesting the possible occurrence of phospholipid exchange proteins in *Tetrahymena* (Iida et al., 1978). However, lipid transfer by a vesicular route was not analyzed in this cell.

Living organisms have in common a great variety of subcellular membrane structures which can be distinguished both morphologically and biochemically.

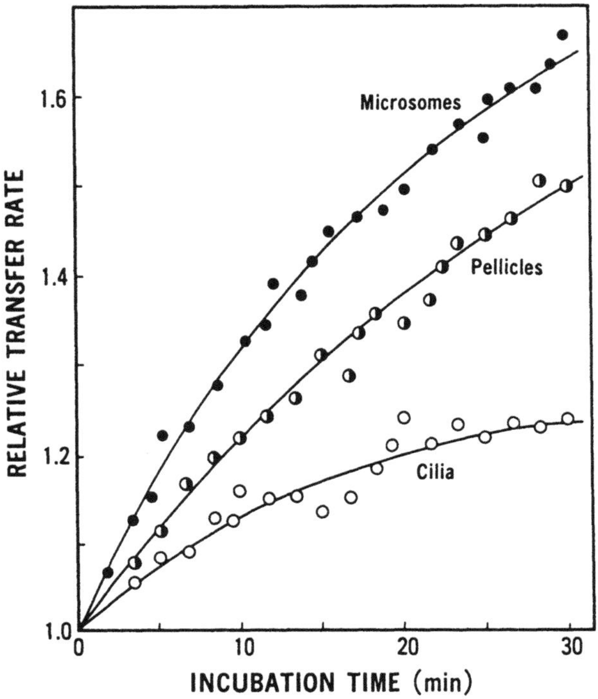

Figure 5. Transfer of spin-labeled phosphatidylcholine between different membranes. (From Iida et al. 1978).

Each of them has specific transfer systems composed of specific lipids to finally produce some characteristic membrane properties. Thompson et al. (1971) investigated the distribution pattern of radiolabeled precursors in various organelle membranes of *Tetrahymena* and found that the accumulation of radioactive lipids in the surface membranes (remote from their origin) may be due to the specificity in the selection at the acceptor membranes rather than to the specificity in the transferring system. All the subcellular membranes may receive an optimal assortment of lipids from the biosynthesis sites or from other donating membranes and rearrange the accepted lipids to maintain their membrane homeostasis. To establish the exact rationale for membrane lipid traffic in close association with diverse dynamic lipid composition, as observed in various organelle membranes, however, requires much additional work.

V. CONCLUSIONS AND OUTLOOK

The lower eukaryote *Tetrahymena*, in addition to usual lipids (such as phosphatidylcholine and phosphatidylethanolamine, as present also in mammalian cells), membrane lipids also contain phosphonolipids and sterol-like tetrahymanol, both of which are resistant to enzyme attack, and taurolipids. The former two lipids are abundant in the surface membrane (pellicle) and the latter lipid is principally localized in lysosomes. Therefore, these unique lipids exhibiting specific localization may be useful as markers to examine the membrane traffic in *Tetrahymena*. Although the food vacuole exerts the most active membrane traffic, a detailed analysis of its membrane lipid composition and metabolism has to await the establishment of a procedure to obtain highly purified preparations.

ABBREVIATIONS

AEP: 2-aminoethylphosphonate
AEPL: 2-aminoethylphosphonolipid
CAEP: ceramide 2-aminoethylphosphonate
CDP: cytidine diphosphate
CMAEP: ceramide *N*-methyl-2-aminoethylphosphonate
CMP: cytidine monophosphate
CoA: coenzyme A
GPC: *sn*-glycerol-3-phosphorylcholine
GPE: *sn*-glycerol-3-phosphorylethanolamine
GPnL: glycerophosphonolipid
PC: phosphatidylcholine
PE: phosphatidylethanolamine

REFERENCES

Arai, H., Inoue, K., Natori, Y., Banno, Y., Nozawa, Y., and Nojima, S. (1985). Intracellular phospholipase activities of *Tetrahymena pyriformis*. *J. Biochem.* 97: 1525–1532.

Arai, H., Inoue, K., Nishikawa, K., Banno, Y., Nozawa, Y., and Nojima, S. (1986). Properties of acid phospholipases in lysosome and extracellular medium of *Tetrahymena pyriformis*. *J. Biochem.* 99: 125–133.

Chlapowski, F. J. and Band, R. H. (1971). Assembly of lipids into membranes in *Acanthamoeba palestinensis*. II. The origin and fate of glycerol-^{3}H-labeled phospholipids of cellular membranes. *J. Cell Biol.* 50: 534–651.

Ferguson, K. A., Conner, R. L., and Mallory, F. B. (1971). The effect of ergosterol on the fatty acid composition of *Tetrahymena pyriformis*. *Arch. Biochem. Biophys.* 144: 448–450.

Fukushima, H., Martin, C. E., Iida, H., Kitajima, Y., Thompson, G. A., and Nozawa, Y. (1976). Changes in membrane lipid composition during temperature adaptation by a thermotolerant strain of *Tetrahymena pyriformis*. *Biochim. Biophys. Acta* 431: 165–179.

Hori, T., Arakawa, I., Sugita, M., and Itasaka, O. (1968). Biochemistry of shellfish lipids. IX. Enzymatic hydrolysis of ceramide 2-aminoethylphosphonate and sphingoethanolamine. *J. Biochem.* 64: 533–536.

Hori, T. and Nozawa, Y. (1982). Phosphonolipids. In: *New Comprehensive Biochemistry* (Hawthorne, J. N. and Ansell, G. B., eds.), Vol. 4, pp. 95–128, Elsevier, Amsterdam.

Horiguchi, M. (1972). Biosynthesis of 2-aminoethylphosphonic acid in cell-free preparations from *Tetrahymena*. *Biochim. Biophys. Acta* 261: 102–113.

Iida, H., Maeda, T., Ohki, K., Nozawa, Y., and Ohnishi, S. (1978). Transfer of phosphatidylcholine between different membranes in *Tetrahymena* as studied by spin labeling. *Biochim. Biophys. Acta* 508: 55–64.

Jonah, M. and Erwin, J.A. (1971). The lipids of membranous cell organelles isolated from the ciliate, *Tetrahymena pyriformis. Biochim. Biophys. Acta* 231: 80–92.

Kameyama, Y., Yoshioka, S., and Nozawa, Y. (1980). The occurrence of direct desaturation of phospholipid acyl chain in *Tetrahymena pyriformis:* thermal adaptation of membrane phospholipid. *Biochim. Biophys. Acta* 618: 214–222.

Kameyama, Y., Yoshioka, S., and Nozawa, Y. (1984). Mechanism for adaptive modification during cold acclimation of phospholipid acyl chain composition in *Tetrahymena.* I. Principal involvement of deacylation-reacylation. *Biochim. Biophys. Acta* 793: 28–33.

Kaya, K. (1992). Chemistry and biochemistry of taurolipids. *Progr. Lipid Res.* (in press).

Kaya, K., Uchida, K., an Kusumi, T. (1985). Identification of taurine-containing lipid in *Tetrahymena pyriformis* NT-1. *Biochim. Biophys. Acta* 835: 77–82.

Kaya, K. and Sato, T. (1991). Definition of total biosynthesis pathway of taurolipids in *Tetrahymena* cells. *Biochim. Biophys. Acta* 1074: 101–104.

LaNauze, J. M., Rosenberg, H., and Shaw, D.C. (1970). The enzymic cleavage of the carbon–phosphorus bond: purification and properties of phosphonatase. *Biochim. Biophys. Acta* 212: 332–350.

Liang, C. R. and Rosenberg, H. (1966). The biosynthesis of the phosphonic analogue of cephalin in *Tetrahymena. Biochim. Biophys. Acta* 125: 548–562.

Mallory, F. B., Gordon, J. T., and Conner, R. L. (1963). The isolation of a pentacyclic triterpenoid alcohol from a protozoan. *J. Am. Chem. Soc.* 85: 1362–1363.

Maruyama, H., Banno, Y., Watanabe, T., and Nozawa, Y. (1982). Studies on thermal adaptation in *Tetrahymena* membrane lipids: modification of positional distribution of phospholipid acyl chains in plasma membranes, mitochondria and microsomes. *Biochim. Biophys. Acta* 711: 229–244.

Nagao, S., Fukushima, H., and Nozawa, Y. (1978). Studies on *Tetrahymena* membranes: substrates for desaturation of fatty acyl chains in *Tetrahymena pyriformis* microsomes. *Biochim. Biophys. Acta* 530: 165–174.

Nozawa, Y. (1975) Isolation of subcellular membrane components from *Tetrahymena. Meth. Cell Biol.* 10: 105–133.

Nozawa, Y. and Kasai, R. (1978). Mechanism of thermal adaptation of membrane in *Tetrahymena pyriformis* NT-1: possible evidence for temperature-mediated induction of palmitoyl-CoA desaturase. *Biochim. Biophys. Acta* 529: 54–66.

Nozawa, Y., Kasai, R., and Sekiya, T. (1980). Modifications of membrane lipid composition following the nutritional shift-up of starved cells: a comparison with membrane biogenesis in *Tetrahymena. Biochim. Biophys. Acta* 603: 347–365.

Nozawa, Y. and Thompson, G. A. (1971a). Studies of membrane formation in *Tetrahymena pyriformis.* II. Isolation and lipid analysis of cell fractions. *J. Cell Biol.* 49: 712–721.

Nozawa, Y. and Thompson, G. A. (1971b). Studies of membrane formation in *Tetrahymena pyriformis.* III. Lipid incorporation into various cellular membranes of logarithmic phase cultures. *J. Cell Biol.* 49: 722–730.

Nozawa, Y. and Thompson, G. A. (1979). Lipids and membrane organization in *Tetrahymena.* In: *Biochemistry and Physiology of Protozoa* (Levandowsky, M. and Hutner, S. H., eds.), Vol. 2, pp. 275–338, Academic Press, New York.

Nozawa, Y. and Umeki, S. (1988). Regulation of membrane fluidity in unicellular organisms. In: *Physiological Regulation of Membrane Fluidity* (Aloia, R. C., Curtain, C. C., and Gordon, L. M., eds.), pp. 239–257, Alan R. Liss, New York.

Okuyama, H., Yamada, K., Kameyama, Y., Ikezawa, H., Fukushima, H., and Nozawa, Y. (1977). Studies on lipid metabolism in *Tetrahymena pyriformis*: properties of acyltransferase systems. *Arch. Biochem. Biophys.* 178: 319–326.

Ramesha, C. S. and Thompson, G. A. (1982). Changes in the lipid composition and physical properties of *Tetrahymena* ciliary membranes following low-temperature acclimation. *Biochemistry* 21: 3612–3617.

Smith, J. D. and Law, J. H. (1970). Phosphatidylcholine biosynthesis in *Tetrahymena pyriformis*. *Biochim. Biophys. Acta* 202: 141–152.

Smith, J. D., Snyder, W. R., and Law, J. H. (1970). Phospholipids in *Tetrahymena* cilia. *Biochem. Biophys. Res. Commun.* 39: 1163–1169.

Thompson, G. A. (1967). Studies of membrane formation in *Tetrahymena pyriformis*. I. Rates of phospholipid biosynthesis. *Biochemistry* 6: 2015–2022.

Thompson, G. A., Bambery, R. J., and Nozawa, Y. (1971). Further studies of the lipid composition and biochemical properties of *Tetrahymena pyriformis* membrane systems. *Biochemistry* 10: 4441–4447.

Thompson, G. A., Bambery, R. J., and Nozawa, Y. (1972). Environmentally produced alterations of the tetrahymanol: phospholipid ratio in *Tetrahymena pyriformis* membranes. *Biochim. Biophys. Acta* 260: 630–638.

Umeki, S., Fukushima, H., Watanabe, T., and Nozawa, Y. (1982). Temperature acclimation mechanisms in *Tetrahymena pyriformis*: effects of decreased temperature on microsomal electron transport. *Biochem. Int.* 4: 101–107.

Umeki, S. and Nozawa, Y. (1983). Variations in the activity of fatty acyl-CoA desaturases and electron-transport components in *Tetrahymena* microsomes with temperature and age of culture. *J. Therm. Biol.* 8: 353–360.

Wirtz, K. W. A. (1974). Transfer of phospholipids between membranes. *Biochim. Biophys. Acta* 344: 95–117.

Yamauchi, T., Ohki, K., Maruyama, H., and Nozawa, Y. (1981). Thermal adaptation of *Tetrahymena* membranes with special reference to mitochondria: role of cardiolipin in fluidity of mitochondrial membranes. *Biochim. Biophys. Acta* 649: 385–392.

Yoshioka, S., Kameyama, Y., and Nozawa, Y. (1984). Mechanism for adaptive modification during cold acclimation of phospholipid acyl chain composition in *Tetrahymena*. II. Activities of 2-acyl-*sn*-glycerol-3-phosphorylcholine and 2-acyl-*sn*-glycerol-3-phosphorylethanolamine acyltransferase involving the reacylation. *Biochim. Biophys. Acta* 793: 34–41.

INDEX

Note: Pages 1-226 (in roman type) apply to Volume 2A.
Pages *227-464* in italic type) apply to Volume 2B.

www.ingramcontent.com/pod-product-compliance
Ingram Content Group UK Ltd.
Pitfield, Milton Keynes, MK11 3LW, UK
UKHW020250290926
15470UKWH00002B/852

9 781559 386289